Gene Regulation and Therapeutics for Cancer

Editors

Surinder K. Batra, Ph.D

and

Moorthy P. Ponnusamy, Ph.D

Department of Biochemistry and Molecular Biology
University of Nebraska Medical Center, Omaha, NE, USA

CRC Press is an imprint of the
Taylor & Francis Group, an **informa** business

A SCIENCE PUBLISHERS BOOK

Cover illustration reproduced by kind courtesy of the editors.

CRC Press
Taylor & Francis Group
6000 Broken Sound Parkway NW, Suite 300
Boca Raton, FL 33487-2742

© 2021 by Taylor & Francis Group, LLC
CRC Press is an imprint of Taylor & Francis Group, an Informa business

No claim to original U.S. Government works

Version Date: 20200723

International Standard Book Number-13: 978-1-138-71242-3 (Hardback)

This book contains information obtained from authentic and highly regarded sources. Reasonable efforts have been made to publish reliable data and information, but the author and publisher cannot assume responsibility for the validity of all materials or the consequences of their use. The authors and publishers have attempted to trace the copyright holders of all material reproduced in this publication and apologize to copyright holders if permission to publish in this form has not been obtained. If any copyright material has not been acknowledged please write and let us know so we may rectify in any future reprint.

Except as permitted under U.S. Copyright Law, no part of this book may be reprinted, reproduced, transmitted, or utilized in any form by any electronic, mechanical, or other means, now known or hereafter invented, including photocopying, microfilming, and recording, or in any information storage or retrieval system, without written permission from the publishers.

For permission to photocopy or use material electronically from this work, please access www.copyright.com (http://www.copyright.com/) or contact the Copyright Clearance Center, Inc. (CCC), 222 Rosewood Drive, Danvers, MA 01923, 978-750-8400. CCC is a not-for-profit organization that provides licenses and registration for a variety of users. For organizations that have been granted a photocopy license by the CCC, a separate system of payment has been arranged.

Trademark Notice: Product or corporate names may be trademarks or registered trademarks, and are used only for identification and explanation without intent to infringe.

Library of Congress Cataloging-in-Publication Data
Names: Batra, Surinder K., editor.	Ponnusamy, Moorthy P., editor.
Title: Gene regulation and therapeutics for cancer / editors, Surinder K. Batra and Moorthy P. Ponnusamy.
Description: First edition.	Boca Raton, FL : CRC Press, [2020]	Includes bibliographical references.
Identifiers: LCCN 2020013566	ISBN 9781138712423 (hardcover)
Subjects: MESH: Neoplasms--therapy	Molecular Targeted Therapy	Neoplasms--genetics	Gene Expression Regulation, Neoplastic--drug effects
Classification: LCC RC270.8	NLM QZ 266	DDC 616.99/406--dc23
LC record available at https://lccn.loc.gov/2020013566

Visit the Taylor & Francis Web site at
http://www.taylorandfrancis.com

and the CRC Press Web site at
http://www.routledge.com

Preface

Physicians and research scientists have been working "hand in hand" to find every possible way to prevent and treat patients from the deadly disease "cancer" since the early 1900. The fact that "cancer is a complex disease involving several drivers and passenger mutations in multiple "oncogene and tumor suppressor" genes makes it more challenging to identify an appropriate drug to manage numerous signalling". One among five thousand pre-clinically investigated drugs will take a journey time of around 10-12 years to travel from lab bench to pharmacy shelf with 2.5 bn investment. Hence, it is necessary to understand the root cause of its molecular mechanism controlled by the genes and their product proteins to better manage and to design targeted therapies.

In the first edition of this book, we focused on discussing the genes that regulate several cancers, and available and optimized targeted therapies along with first-line treatment options to treat such cancers. These chapters comprehensively cover 13 different topics reviewing different cancers such as head and neck, lung, pancreatic, prostate, breast cancers, and paediatric Medulloblastoma. Several research scientists and physicians who are experts in pre-clinical drug testing in mouse models and translating its benefits to cancer centres discussed these chapters. We extensively discuss the role of transcriptional dysregulations in cancer. The authors provide a strong rationale to target such a transcriptional effector and the re-programming process driving the cancer. Most targeted therapeutics developed to date have been focused on targeting proteins encoding cell surface or cytoplasmic kinases that function in intracellular signalling cascades while leaving a sub-group of cell-surface proteins such as mucins. One of the chapters addresses the regulation of mucins and targeted therapy in pancreatic cancer.

Furthermore, the authors also discuss the role and biomarker potential of few secretory factors such as MUC5AC and exosomes in cancers. This chapter deals with the recent development and early detection strategies using liquid biopsies in pancreatic cancer. Because early detection is one of the significant challenges in pancreatic cancer, it can also be used to predict treatment response. In particular, summarized studies on the development and contributions of PD-1-mediated immunomodulatory effects in pancreatic cancer. One of the chapters summarizes the targeting approach of subgroup-specific cancer epitopes for effective treatment of paediatric Medulloblastoma.

Furthermore, two individual chapters focus on prostate cancer biology, and contemporary updates on clinically developed therapeutics for castration-resistant prostate cancer were also discussed. In the final section of this book, the applications of cancer stem cells as a promising target in cancer therapy to abrogate aggressiveness, drug-resistance, and recurrence nature, as observed in most solid tumors are highlighted.

We strongly believe this edition, with all the recent information, will be very knowledgeable for students, scientists, and physicians who are working in the field of gene regulation and targeted therapy for different cancers. We sincerely hope that the discussed inhibitors, chemotherapy, or targeted therapy agents will serve as a platform to design future clinical trials and aid in curing cancer patients.

Surinder K. Batra
Moorthy P. Ponnusamy

Contents

Preface   iii

1. Programmed Death 1 Receptor (PD-1)-mediated Immunomodulatory
   Effects in Pancreatic Cancer   1
   Ashu Shah, Catherine Orzechowski and Maneesh Jain

2. Nuclear Factor Kappa-B: Bridging Inflammation and Cancer   23
   *Mohammad Aslam Khan, Girijesh Kumar Patel, Haseeb Zubair,
   Nikhil Tyagi, Shafquat Azim, Seema Singh, Aamir Ahmad and
   Ajay Pratap Singh*

3. The S100A7/8/9 Proteins: Novel Biomarker and Therapeutic
   Targets for Solid Tumor Stroma   50
   *Sanjay Mishra, Dinesh Ahirwar, Mohd W. Nasser and
   Ramesh K. Ganju*

4. Nuclear Receptor Coactivators: Mechanism and Therapeutic
   Targeting in Cancer   73
   Andrew Cannon, Christopher Thomson, Rakesh Bhatia and Sushil Kumar

5. Liquid Biopsies for Pancreatic Cancer: A Step Towards Early
   Detection   108
   *Joseph Carmicheal, Rahat Jahan, Koelina Ganguly, Ashu Shah and
   Sukhwinder Kaur*

6. Targeting Subgroup-specific Cancer Epitopes for Effective
   Treatment of Pediatric Medulloblastoma   132
   Sidharth Mahapatra and Naveen Kumar Perumal

7. Aberrant Methylation of UC Promoters in Human Pancreatic
   Ductal Carcinomas   151
   Michiyo Higashi and Seiya Yokoyama

8. Receptor Tyrosine Kinase Signaling Pathways as a Goldmine
   for Targeted Therapy in Head and Neck Cancers   163
   *Muzafar A. Macha, Satyanarayana Rachagani, Sanjib Chaudhary,
   Zafar Sayed, Dwight T. Jones and Surinder K. Batra*

9. Molecular Drivers in Lung Adenocarcinoma: Therapeutic Implications — 185
   Imayavaramban Lakshmanan and Apar Kishor Ganti

10. Molecular Mediator of Prostate Cancer Progression and Its Implication in Therapy — 207
    Samikshan Dutta, Navatha Shree Sharma, Ridwan Islam and Kaustubh Datta

11. Therapeutic Options for Prostate Cancer: A Contemporary Update — 234
    Sakthivel Muniyan, Jawed A. Siddiqui and Surinder K. Batra

12. Regulation and Targeting of MUCINS in Pancreatic Cancer — 264
    Shailendra K. Gautam, Abhijit Aithal, Grish C. Varshney and Parthasarathy Seshacharyulu

13. Targeted Therapy for Cancer Stem Cells — 291
    Rama Krishna Nimmakayala, Saswati Karmakar, Garima Kaushik, Sanchita Rauth, Srikanth Barkeer, Saravanakumar Marimuthu and Moorthy P. Ponnusamy

Index — 321

CHAPTER 1

Programmed Death 1 Receptor (PD-1)-mediated Immunomodulatory Effects in Pancreatic Cancer

Ashu Shah[1], Catherine Orzechowski[1] and Maneesh Jain[1,2]*

[1] Department of Biochemistry and Molecular Biology, University of Nebraska Medical Center, Omaha, NE, USA
[2] Fred and Pamela Buffett Cancer Center, University of Nebraska Medical Center, Omaha, NE, USA

Introduction

Pancreatic ductal adenocarcinoma (PDAC) is one of the most lethal, treatment refractory malignancies that has emerged as the third leading cause of cancer related deaths in the United States. The overall five-year survival rate for PDAC patients is dismally low at 8% with an estimated 55,440 new cases and 44,330 deaths in the year 2018 [1]. Existing therapies for PDAC include chemotherapy, radiotherapy, and radical surgery. However, the failure to diagnose PDAC at an early stage makes the treatment options ineffective in almost 80% of the patients. Even after surgery, recurrence occurs in 80% of the patients. Till now, only five FDA approved drugs and one combination therapy exists for PDAC patients [2]. Chemotherapy combined with radiation has not shown much success in the patients. The reasons attributed to inadequacy of treatment options is complex molecular landscape of pancreatic tumors [2].

Pancreatic cancer is driven by an accumulation of several activating mutations in the oncogene KRAS and inactivating mutations in tumor suppressor genes TP53, CDKN2A (p16), and SMAD4 in the normal pancreatic duct epithelium. Activating KRAS mutations are present in approximately 95% of PDAC cases and is responsible for activation of PI3K-Akt, notch pathway, hedgehog signaling, and STAT3 pathways which are potent drivers of tumor initiation and maintenance. In addition, shortened telomerase, genomic instability and epigenetic alterations play significant roles in the

---

*Corresponding author: mjain@unmc.edu

progression of pancreatic cancer [3]. The disease is believed to originate from a spectrum of precursor lesions including pancreatic intraepithelial neoplasms (PanINs), intraductal papillary neoplasm (IPMN), and mucinous cystic neoplasm (MCN) which over several years develop into aggressive PDAC that invades surrounding tissues and metastasize to different organs.

PDAC is characterized by dense stroma (desmoplasia) comprising pancreatic stellate cells, fibroblasts, vascular, glial, smooth muscle cells, fat cells, epithelial cells, and immune cells along with extracellular matrix (ECM) and extracellular molecules surrounding epithelial cells. PDAC progression is driven by a complex interplay between tumor cells and the surrounding cells of stroma [4].

Immune System in Pancreatic Cancer

PDAC is characterized by infiltration of both innate and adaptive immune cells including monocytes, macrophages (M1, M2), dendritic cells, B cells (B1, B2, B_{reg}), T cell subtypes (T_{eff}, T_{reg}), NK cells and myeloid derived suppressor cells (MDSC). Although the proportion of these immune cells may vary but their number, location, and stage of maturation in the tumor microenvironment, and their ultimate functional differentiation may impact tumor growth and progression [5, 6]. Intricate cross-talk of infiltrating immune cells, with tumor cells and other stromal cells, result in the establishment of a microenvironment rich in immunosuppressive myeloid and lymphoid subtypes [7, 8]. PDAC is enriched in M2 macrophages, neutrophils, $CD4^+$ T cells, T_{reg} and relatively less number of $CD8^+$ T cells [9, 10]. Therefore, PDAC has generally been considered as a poorly immunogenic cancer [11]. However, studies in KPC and other pancreatic cancer mouse models have highlighted the sensitivity of pancreatic cancer to $CD8^+$ T cell mediated cytotoxicity [12, 13]. Also, recent studies have clustered PDAC patients into three subtypes on the basis of genetic and transcriptional signatures and these subtypes display differences in their immune cytolytic activity suggesting the importance of immune activity in PDAC [14]. Conversely, multiple studies indicate that PDAC is characterized by lesser immune cell infiltration compared to other cancers such as melanoma, NSCLC, and HNSC where immunotherapy has been successful [15-17].

Checkpoint Blockade Receptor Programmed Death-1 Receptor (PD-1) in Pancreatic Cancer

Tumor cells exploit intrinsic mechanisms of immune evasion which include reduced antigen presentation, increased expression of immunosuppressive molecules such as PD-L1 and accumulation of antigen specific T_{regs} in the tumor microenvironment resulting in immunosuppression [18]. The immune system is characterized by its ability to distinguish between normal cells in the body and tumor cells through the expression of costimulatory and coinhibitory molecules on immune cells called immune checkpoints. These immune checkpoint molecules play a key role in immunoregulation and immune homeostasis through on-off switch mechanisms and protect the host against autoimmunity. However, tumor cells use these checkpoint molecules to protect themselves from an attack by the immune system. One such checkpoint protein, PD-1, is a coinhibitory receptor which is present on conventional T cells in conjunction with other costimulatory and co-inhibitory molecules and is an important regulator of T cell activation. PD-1 operates as off switch to limit T cell activation by interacting with its ligand, PD-L1, present on normal cells. Under inflammatory conditions like cancer, receptor PD-1 overexpression on T cells is congruent with upregulation of its ligand PD-L1 on antigen presenting cells (APC) and tumor cells [19]. The interaction between PD-1 receptor and its ligand PD-L1 leads to suppressed T cell activation and proliferation which promotes an immunosuppressed condition in the inflammatory microenvironment.

Extensive research over the last two decades on dense desmoplastic stroma in PDAC has elucidated the role of the inflammatory milieu in preventing specific and significant immune response within the tumor. Many immunosuppressive mechanisms operate simultaneously in the PDAC microenvironment, thus making PDAC inaccessible to immunotherapy. Nomi et al. investigated, for the first time, the expression of PD-L1 on human pancreatic cancer cells [20] and showed an association of PD-L1 with poor survival outcomes in PDAC patients. This was associated with lesser number of CD8$^+$ T cells in tumor infiltrating lymphocytes of patients with increased PD-L1 expression. Higher PD-L1 levels in 51 human PDAC samples have been correlated with tumor growth rather than progression in contrast to other cancers where PD-L1 expression has been shown to be associated with advanced stage of disease [21, 22]. However, these findings need to be validated in studies with a larger cohort of PDAC patients. The importance of PD-1/PD-L1 axis in pancreatic cancer was further evaluated by testing anti-PD-L1 antibody in the Panc02 subcutaneous mouse model. Treatment with anti-PD-L1 antibody after two weeks of tumor development resulted in a significant reduction of tumor growth along with an increase in tumor reactive CD8$^+$ T cells infiltration and IFNγ, Granazyme B and perforin secretion in the tumor. Furthermore, combination of anti-PD-L1 antibody with the FDA approved chemotherapeutic drug Gemcitabine resulted in a substantial decrease in tumor growth as compared to gemcitabine treatment alone [23]. Similar results were obtained with anti-PD-L1 and anti-PD-L2 antibodies in a mouse model, with orthotopically implanted Panc02 cells where antibody treatment resulted in significant tumor suppression along with enhanced intratumoral infiltration of IFNγ-producing CD8$^+$ T cells and reduced FoxP3$^+$ T$_{reg}$ cells [24]. However, these preclinical observations were not recapitulated in subsequent clinical studies where PDAC patients were found to be refractory to anti-PD-1 therapy [25]. A possible explanation for this can be the difference in heterogeneity of murine and human tumors;

while murine Panc02 tumors are hypermutated, human pancreatic tumors have low mutational load [26]. Another study evaluated the efficacy of anti-PD-1 antibodies in KPC mouse model of PDAC which closely recapitulates the immune environment of human disease. In both KPC mouse and human PDAC, expression of PD-L1 is independent of the presence of tumor-specific CD8+ T cells [23]. In spite of very high expression of immune checkpoint molecules PD-1, CTLA4 in the tumor microenvironment and upregulation of PD-L1 in tumor associated dendritic cells and macrophages, anti-PD-1 antibodies did not show any effect on tumor growth. The contrasting results with anti-PD-1 antibody in two different models of PDAC represent the difference in tumor biology and malignancy. Interestingly, anti-PD-1 antibodies in combination with anti-CD40 agonist showed significant tumor cell killing by activated CD8+ T cells and enhanced survival of tumor-bearing KPC mice by 80%. Nevertheless, these studies demonstrated the relevance of immune checkpoint receptor, PD-1 in PDAC and the induction of tumor immune response combined with checkpoint blockade might provide a curative treatment for PDAC.

PD-1 Regulation in Pancreatic Cancer

PD-1 and its ligand PD-L1 are expressed on infiltrating immune cells and tumor cells, respectively, and are regulated through intrinsic immune resistance and adaptive resistance mechanisms. Oncogenes activating Akt and STAT3 pathways have been shown to regulate constitutive PD-L1 expression in triple negative breast cancer and NSCLC [27, 28]. This phenomenon is referred to as intrinsic immune resistance. In contrast, IFN- secreted by CD8+ T cells stimulates the upregulation of PD-L1 on tumor cells, as shown in melanoma, HNSCC, and other solid tumors [19, 29, 30], resulting in adaptive immune resistance. These resistance mechanisms have been acquired by tumors for immune evasion. As a result, anti-PD-1 immunotherapy has shown an effect in PD-L1 induced immunosuppressed patients. Surprisingly, in PDAC no correlation was observed between IFNγ and PD-L1 expression on tumor cells, although intratumoral dendritic cells and macrophages exhibited slightly elevated PD-L1 in tumors with high IFNγ secretion [23]. Possibly, this difference in expression levels of PD-L1 on tumor cells could be another determinant for unresponsiveness to anti-PD-1/PD-L1 antibody efficacy in PDAC unlike other cancers. In contrast, one recent study showed uniform and higher expression of PD-L1 on mouse tumor cells Panc02 and UN-KC-6141, both *in vitro* and *in vivo* orthotopic tumors, human pancreatic tumor cell lines, and 60-90% of human PDAC tumor samples coupled with a corresponding increase in PD-1 on tumor infiltrating CD8+ T cells [31-33]. This discrepancy in levels of PD-L1 in pancreatic tumors might be attributed to different anti-PD-L1 antibodies being used for detection. Furthermore, the role of epigenetic mechanisms in regulating PD-L1 expression in PDAC is just beginning to be unravelled. In contrast to the normal pancreas, PD-L1 (CD274) promoter region in both human and mouse pancreatic tumor cells exhibited enhanced H3K4 methylation by histone methylase MLL1 suggesting its transcriptional activation in PDAC. This observation also opens a new avenue of epigenetic targeting for improved therapeutic efficacy of immune checkpoint inhibitors in nonresponding PDAC patients. In fact, inhibition of MLL1 activity resulted in enhanced efficacy of anti-PD-1/PD-L1 immunotherapy in pancreatic cancer but, the combination of epigenetic and immune checkpoint inhibitor did not show a synergistic effect which may need further optimization in terms of dose and timing of treatments [30]. Very recently, the role of microbiome in mediating PDAC progression through modulating tumor microenvironment and induction of immune suppression has been deciphered in the mouse model [32]. The same study showed upregulation of PD-1 in intratumoral $CD4^+$ and $CD8^+$ T cells after microbiome ablation which became more evident when the reduction in tumor growth was observed by combination therapy of oral antibiotic (microbiome ablation) and anti-PD-1 antibody.

Immune Cells Mediated PD-1 Regulation in Pancreatic Cancer

PDAC immune suppressive tumor microenvironment has been proposed as a major hallmark of cancer pathogenesis [34, 35]. How the multifaceted interactions between tumor and tumor infiltrating myeloid and lymphoid cells help cancer cells to evade immune surveillance, are not fully understood which might explain the reason for not much success gained with current immunotherapeutic strategies. Here, we discuss in detail how these interactions induce immunosuppression through PD-1/PD-L1 axis.

Myeloid Cells in Mediating PD-1/PD-L1 Activation in PDAC Microenvironment

The initial observation that myeloid derived cells suppress T cell activation was made two decades ago and has been well established in various cancers. Immature myeloid cells have the potential to differentiate between various subsets such as macrophages, dendritic cells, endothelial cells, neutrophils, and myeloid derived suppressor cells (MDSC) in tumors. Of importance, these different myeloid subsets function through multiple mechanisms such as nitric oxide production, STAT5 signaling inhibition, T cell function suppression by nitrosylation of essential amino acids, switching $CD4^+$ T cell differentiation to T_{reg} and induction of immunosuppression in tumor microenvironment [4, 36-38]. High numbers of tumor infiltrating myeloid cells have been shown to be associated with relapse and poor survival of PDAC patients [9, 39, 40]. Different signaling pathways such as JAK-STAT, CSF1-CSF1R, and MAPK form a crosslink between the immune system and non-immune cells and inhibit the immune cell activation along with promoting the factors for immunosuppression (Fig. 1) [31, 41]. Several attempts have been

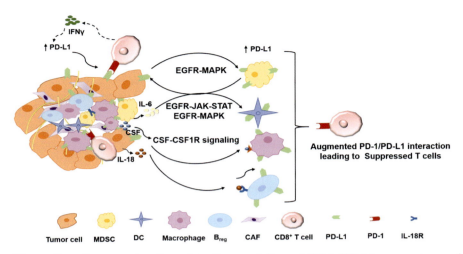

Fig. 1: Signaling axes implicated in the regulation of PD-1/PD-L1 expression and downstream events in pancreatic tumor microenvironment. IFNγ secreted from activated CD8+ T cells increases the expression of PD-L1 on tumor cells which in turn binds to PD-1 on T cells and result in their functional exhaustion. EGFR-MAPK, JAK-STAT and CSF1-CSF1R signaling axes augment PD-L1 expression on myeloid derived suppressor cells (MDSC), dendritic cells and macrophages respectively, leading to PD-1/PD-L1 mediated exhaustion of CD8+ T cells. IL-18 Secreted from tumor cells induce PD-L1 expression on B$_{reg}$ cells by binding to IL-18R.

made to target these signaling pathways using pharmacological inhibitors in haematological malignancies [42]. Role of these signaling axes in regulating PD-1/PD-L1 expression on myeloid cells have been dissected and blocking these pathways have increased the efficacy of anti-PD-1/PD-L1 antibodies in tumor regression *in vitro* and *in vivo* mouse models [31, 42]. For example, inhibition of the CSF1-CSF1R signaling resulted in elevated expression of PD-L1 and CTLA-4 in the tumors, and the treatment increased the efficacy of the anti-PD-1 antibody in the mouse model of PDAC [43]. This was attributed to the selective depletion of the CD206^{Hi} population of tumor infiltrating macrophages and the reprograming of remaining TAMs for reduced immunosuppression and decreased PD-1 expression with CSF1R inhibition. Recently, selective inhibition of IFNγ-STAT1 signaling in JAK-STAT pathway resulted in an increase in CTL infiltration and induction of T cell response in an orthotropic mouse model of PDAC. This was possibly linked to the reduction of IFNγ induced PD-L1 expression on pancreatic tumors and thus diminished PD-1-PD-L1 interaction in the tumor microenvironment [31]. STAT3, another arm of JAK-STAT pathway and a key molecule for Kras driven PDAC progression, upon activation augments immunosuppressive conditions in the tumor microenvironment and inhibits T cell activation thus enhancing PD-1/PD-L1 mediated immune suppression in the pancreatic tumor microenvironment [44]. Reg3g through HMOX1 (heme-oxygenase), dendritic cell maturation inhibiting enzyme, activation and induction of

JAK2/STAT3 signaling pathway, promoted the upregulation of PD-1 and PD-L1 on DCs and T cells respectively in orthotopic pancreatic tumor mice and thereby abolished cross priming and cytotoxic activities of T cells [45]. Previous studies have demonstrated the role of Reg3g activated HMOX1 in regulating immune suppression by causing dendritic cell impairment [46]. The role of EGFR-MAPK signaling in the upregulation of PD-L1 expression on various myeloid cell populations including macrophages, MDSCs, stromal cells and to some extent on pancreatic tumor cells has also been described [47]. Confirmation of the role of EGFR ligand in PD-L1 regulation has been deciphered using MEK inhibitors which resulted in the reduction of PD-L1 expression. Interestingly, MEK inhibitors increased the efficacy of anti-PD-1 antibody, reducing the tumor burden in a mouse model.

Taken together, these findings suggest the importance of combining signaling inhibitors with checkpoint inhibitor for the regression of well-established PDAC tumors.

Regulatory B- and T-cells Mediated PD-1 Regulation in Pancreatic Cancer

Regulatory B (B_{reg}) and T (T_{reg}) cells play an important role in maintaining immune tolerance by suppressing the proinflammatory response of B and T cells. The role of B cells became more evident when their transfer into mammary tumor-bearing B cell deficient mice (BCDM) led to T_{reg} proliferation and depletion of $CD8^+$ T cells and NK cells in the tumor microenvironment and thereby enhanced tumor growth [48]. Similarly, in another study, adoptive transfer of WT B cells in oxaliplatin treated BCDM mice interfered with oxaliplatin-mediated tumor reduction while PD-L1 or IL10 deficient B cells enhanced the efficacy of oxalplatin in tumor regression, thus suggesting an independent role of IL-10 and PD-L1 in B_{reg} mediated immunosuppression [49, 50]. Several studies have shown B_{reg} mediated conversion of $CD4^+$ T cells to T_{regs} which subsequently attenuates the immune response in various cancers [51, 52]. Accumulating evidence now suggests that infiltration of B_{regs} in the tumor microenvironment of several solid cancers leads to an immunosuppressive phenotype. B_{regs} exhibit immunosuppression through PD-L1, lymphotoxin, IL-10, TGFβ secretion and interaction with other immune cells. In PDAC patients, a significantly higher number of B_{regs} in peripheral blood than healthy controls has been correlated with tumor node metastasis (TNM) stage [53]. In addition, $IL10^+$ B_{regs} displayed higher IL-18R expression which on binding to IL-18 secreted from pancreatic tumor cells, induced PD-L1 expression on B_{regs}. T_{regs} participate in the negative regulation of immune response by suppressing effector T cell proliferation and activation. Presence of T_{regs} has been associated with disease progression in PDAC and correlates with poor prognosis [54]. Indeed, the level of tumor infiltrating T_{regs} was higher in PD-L1 positive pancreatic tumors than PD-L1 negative ones [55], suggesting a plausible role of PD-L1 in immune downregulation

Table 1: Active clinical trials investigating PD-1-PD-L1 axis targeting alone or in combination immunotherapy approaches for pancreatic cancer

Type of therapy	Name	Target population	Clinical trial number	Phase (status)
Single agent checkpoint inhibitor PD-1/PD-L1	Pembrolizumab	Resectable or Borderline resectable pancreatic cancer	NCT02305186	PI
		Advanced or metastatic cancers (1 Pancreatic cancer patient)	NCT01295827	PI
	MDX-11-5 (anti-PD-L1)	Pancreatic cancer	NCT00729664	PI completed, no objective responses in patients with colorectal or pancreatic cancer, efficacy in 50% of pancreatic cancer patients observed (7/14)
	Durvalumab (anti-PD-L1)	Resected PDA patients	NCT03038477	PI, suspended
PD-1 combination with other checkpoint molecules/small molecule drugs	Pembrolizumab+ CXCR4 antagonist (BL-8040)	Metastatic PDAC	NCT02826486	PII
	Nivolumab + Cabiralizumab	Advanced pancreatic cancer	NCT02526017	PI
	Durvalumab + Galunisertib	Metastatic pancreatic cancer	NCT02734160	PI
	Durvalumab + Tremelimumab	Metastatic PDAC	NCT02558894	PII

	Pembrolizumab+ mFOLFOX	GI cancers (including pancreatic cancers)	NCT02268825	PI
	Ulocuplumab + Nivolumab	Pancreatic cancer	NCT02472977	PI, lack of efficacy so terminated
	Durvalumab + TGF beta receptor inhibitor Galunisertib (LY2157299)	Metastatic pancreatic cancer	NCT02734160	PI
	Nivolumab + Chemotherapy (NivoPlus)	Advanced pancreatic cancer	NCT02423954	PII, Terminated, Investigator no longer at site to enroll patients or write up data
	Gemcitabine and CT-011	Pancreatic Neoplasms	NCT01313416	PII, suspended, drug clinical issue
	Nivolumab With Nab-Paclitaxel +/- Gemcitabine	Pancreatic Neoplasms	NCT02309177	PI
Combination with chemotherapy	Pembrolizumab+ Paricalcitol	Stage IV pancreatic cancer patients	NCT03331562	PII
	ACP-196 in combination with Pembrolizumab	Advanced or Metastatic pancreatic cancer patients	NCT02362048	PII
	BL-8040 + Pembrolizumab	Metastatic pancreatic cancer patients	NCT02826486	PII
	IDO-1 Inhibitor Epacadostat in combination with Pembrolizumab	Advanced pancreatic cancer with chromosomal Instability/Homologous Recombination Repair Deficiency (HRD)	NCT03432676	PII, withdrawn

(Contd.)

Table 1: (Contd.)

Type of therapy	Name	Target population	Clinical trial number	Phase (status)
Combination with radiotherapy	Radiation with Nivolumab with or without Ipilimumab	Metastatic PDAC	NCT02866383	PII
	Anti-PD-1 in combination with radiotherapy	Pancreatic cancer	NCT03374293	PII
	Pemrolizumab+ neoadjuvant chemoradiation	Resectable or Borderline resectable pancreatic cancer	NCT02305186	PI
Combination with vaccines	Durvalumab + "Booster" RT	Metastatic Adenocarcinoma pancreas cancer	NCT02885727	PII withdrawn (funding issue)
	CT-011+p53 genetic vaccine	Advanced pancreatic cancer	NCT01386502	withdrawn
Multiple combinations	Cyclophosphamide+ Pembrolizumab+ GVAX+SBRT	Pancreatic cancer	NCT02648282	PII
	Cyclophosphamide+ Nivolumab+GVAX+ SBRT	Borderline resectable pancreatic cancer	NCT03161379	PII
	Losartan and Nivolumab in combination with Folfirinox and SBRT	Localized pancreatic cancer	NCT03563248	PII

Cabiralizumab in combination with nivolumab, with or without chemotherapy	Advanced pancreatic cancer	NCT03336216	PII
MEDI4736 in combination with nab-paclitaxel and gemcitabine	Metastatic PDAC	NCT02583477	P1b/II
Cyclophosphamide (CY)+Nivolumab + GVAX	Previously treated Metastatic Adenocarcinoma of the pancreas	NCT02243371	PII

through T_{regs}. Despite extensive $T_{effector}$ cell infiltration in the PDAC microenvironment, there remains a bias towards immune suppression due to an increase in the numbers of tumor infiltrating CD4$^+$ T_{regs} in comparison to those present in peripheral blood and also, a concomitant increase in PD-1 overexpressing CD4$^+$ and CD8$^+$ T cells suggests PD-1 mediated regulation and T cell exhaustion [56]. However, another study showed that blockade of PD-L2, another ligand for PD-1, enhanced anti-tumor response by reducing T_{reg} infiltration in tumors and decreased IL-10 mRNA levels as compared to anti-PD-L1 antibody which predominantly affected IFNγ expression. These observations suggest that PD-1-PD-L2 and PD-1-PD-L1 axes exert T_{reg} mediated immunosuppression in PDAC by distinct mechanisms [24]. Of interest, the depletion of T_{regs} in animal model of Panc02 tumor has shown enhanced tumor regression [57].

Pathobiological Significance of PD-1 Expression in Pancreatic Cancer

Expression of PD-1 and PD-L1 has been extensively examined in PDAC tissues. PD-1, present on the T cells, has been identified as the receptor for ligand PD-L1 (B7-H1) which is upregulated under inflammatory conditions on antigen presenting cells such as dendritic cells, macrophages, and cancer cells [20, 58] . PD-1/PD-L1 interaction leads to cancer antigen-specific T cell exhaustion and attenuation of CD4$^+$ and CD8$^+$ T cell activation thus leading to immunosuppression in the tumor microenvironment and immune evasion of the tumor. Higher expression of PD-L1 in poorly differentiated human pancreatic tumors than well differentiated tumors has been correlated with tumor progression through increased IL-10 secretion and the inhibition of Th1 cytokines inducing immunosuppression [59]. Many other mechanisms for PD-L1-mediated tumor progression in PDAC have been suggested. For example, PD-L1 was reported to directly promote pancreatic tumor cell proliferation via enhanced JNK signaling and cyclin D1 upregulation [60, 61]. High levels of PD-L1 expression on tumors were correlated with enhanced infiltration of T_{regs} and significantly shorter postoperative survival in PDAC patients [55]. Surgically resected PDAC patients with high PD-L1 were associated with a higher pathological stage and lower two-year survival rate (32%) as compared to PD-L1 negative patients (65%). Several studies revealed the correlation of PD-L1 expression with the pathological stage of PDAC and implicate its role in PDAC prognosis and survival analysis [20, 62, 63]. However, PD-L1 in combination with B7-1 served as better postoperative prognosis marker than PD-L1 alone. One study examined the levels of CD8$^+$ TILs, FoxP3$^+$$T_{regs}$, PD-1, and PD-L1 expression in conjunction with stromal density and activity and examined their correlation with four clinical endpoints [overall survival-OS; progression free survival PFS; local progression-free survival-LPFS; distant metastases free-survival-DMFS], in a cohort of 145 patients following surgical resection. In univariate analysis, patients with high infiltration of CD8⁺ T cells or PD-1⁺ TILs had better prognosis (for all four clinical endpoints) compared to patients with low intratumoral CD8⁺ T cells or PD-1⁺ TILs while the expression of FoxP3 and PD-L1 did not correlate with clinical outcome [64]. However, prognostic value of PD-1⁺ TILs or CD8⁺ T cells varied with respect to their location in intratumoral compartments. High stromal infiltrating PD-1⁺CD8⁺TILs served as better prognostic marker than intraepithelial and peripheral compartment TILs which had high PD-1 expression but lacked CD8⁺ T cells. Furthermore, patients with higher PD-1⁺CD8⁺ TILs infiltration in lymphoid aggregates were associated with better OS, LPFS and DMFS in comparison to patients with low PD-1 and CD8 expression. In line with this, another study using a cohort of 94 Chinese PDAC patients demonstrated correlation between high PD-L1 expression and lymph node metastasis, while low PD-L1 patients with high number of PD-1⁺ T cells exhibited an improved overall survival [65]. In addition, the patients with high PD-1 and dense stroma had much more favorable prognosis (median OS > 24 months) as compared to those with low PD-1 and moderate stroma (median OS = 7 months). The presence of a very low number of PD-L1⁺ cells with tumoral CD8⁺ T cells overexpressing other inhibitory receptors in PDAC may account for the failure of PD-1/PD-L1 blockade therapies. Nonetheless, disparities exist in results obtained with PD-L1 expression levels predicting response to anti-PD-1 therapy in PDAC [66]. Qualitative analysis of PDAC tumors suggested the presence of PD-1⁺ CD8 T cells with minimal clonality in a subset of patients. In addition, they observed that tertiary lymphoid structures (TLS) of tumor stroma were enriched in PD-1⁺/PD-L1⁺ T cells which are usually associated with positive prognosis. Lesser clonality of PDAC tumors might explain their resistance to anti-PD-1 therapy in comparison to other high clonal tumors like colorectal cancer and melanoma which respond to anti-PD-1 therapy.

The observed variability in the association between the expressions of PD-1/PD-L1 with various clinicopathological characteristics across these studies can be attributed to several factors including small size of patient population or analysis being restricted to tissue microarray or small sections which may not represent the heterogeneity present in complex desmoplastic tumor microenvironment. This warrants further studies on larger cohorts and accounting for heterogeneity in the tumor microenvironment. Overall, the correlation of PD-1 and PD-L1 with clinical outcomes has been variable across studies. Further clarity on their role in prognosis and survival will help dictate the effectiveness of anti-PD-1/PD-L1 antibodies in immune therapy in PDAC.

Resistance to Anti-PD-1 Based Immunotherapy in Pancreatic Cancer

It is becoming increasingly evident that PDAC is a poorly immunogenic malignancy which stands out as the dominant reason for the failure of checkpoint inhibitor immunotherapies. Given that a majority of checkpoint blockade trials have targeted the PD-1/PD-L1 axis, it is important to understand the underlying mechanism of the single agent anti-PD-1/PD-L1 immunotherapy failure in PDAC. The response to checkpoint inhibitor-based immunotherapy is dependent on immunological characteristic of the tumor, based on which, there exist two categories: immune active and immune quiescent tumors [67]. Immune active tumors are characterized by rich immune CD8$^+$ T cell infiltration and are, therefore, sensitive to checkpoint inhibitors while immune quiescent ones have poor effector T cell infiltration and thereby do not respond to immune checkpoint blockade therapies. PDAC being a relatively quiescent tumor might provide limited targets for anti-PD-1/PD-L1 or other checkpoint blockade antibodies. Several hypotheses have been proposed for the limited success with anti-PD-1/PD-L1 therapies in PDAC. According to the immunoediting theory by Pardoll, tumor antigens, also called neoantigens, are acquired by a series of genetic mutations and other events. Thus, tumor antigens become very different from normal cellular antigens and are recognized by host immune system for elimination [68]. However, T cell exhaustion by multiple co-inhibitory receptors and ligands such as PD-1, CTLA-4, PD-L1, IDO, IL-10, TIM3, and TGFβ lead to an ineffective anti-tumor immune response. Therefore, antibodies against these molecules would allow activation of a patient's own immune system and restore anti-cancer immune response. Nevertheless, despite the presence of multiple inhibitory and co-inhibitory receptors on exhausted T cells, single agent checkpoint blockade agents showed limited success in PDAC [25, 69]. Rather, combining immune activation strategies with anti-PD-1 antibodies may provide better efficacy against pancreatic cancer patients and show promise in the clinic. Loss of tumor antigens due to mutations or epigenetic silencing can also contribute to the failure of T-cells in immunotherapy [70]. Additionally, PDAC, being a relatively quiescent tumor, is characterized by poor T cell infiltration and therefore might provide limited target inaccessibility for anti-PD-1/PD-L1 or other checkpoint blockade antibodies [67]. PDAC frequently metastasizes to liver which is relatively more tolerogenic and may present an altogether different immune milieu from the pancreas, thereby a varying degree of PD-1 expression might influence its efficacy in those patients [71]. Several efforts have been made to understand PDAC in the context of immune microenvironment. Of note, genomic and transcriptional stratification of 134 PDAC patients was done on the basis of cytolytic T cell activity and delineated a new link between genomic alterations and immune activation. Also, in contrast to other tumors, no direct correlation was found between neoepitope load and T cell cytolytic activity which might explain the lack of clinical response to anti-PD-1 therapies in PDAC [14]. Feng et al. have thoroughly discussed the relation of immune profile with PD-1/PD-L1 blockade therapies according to which PDAC patients with immune desert and immune excluded phenotype do not respond to anti-PD-1 therapy due to low number of CD8$^+$T and that too functionally exhausted [72]. Sharma et al. have described in detail how cancer cells exploit intrinsic and extrinsic mechanism of resistance, which include lack of antigenic proteins or their presentation, genetic T cells exclusion or their inaccessibility, inhibitory immune checkpoints and immunosuppressive cells, and limit the success of anti-PD-1 or PD-L1 antibody therapies [73]. A very recent study stratified a cohort of 110 PDAC patients, using next generation sequencing and tissue microarray staining, in to three subtypes as immune escape, immune rich and immune exhausted, for prognosis and indicated that immune exhausted subtype displayed high PD-L1 expression along with higher frequency of PIK3CA mutations and higher rate of PTEN loss in one subpopulation. This stratification might be useful in designing novel approaches for the use of immunotherapy and overcome immunosuppression in PDAC microenvironment [74]. Besides genetic and immunological factors for determining resistance to anti-PD-1 therapies, transcriptional mechanisms regulating PD-1 expression have also been implicated in resistance to checkpoint blockade therapies in several other metastatic tumors. However, transcriptional profiling also poses serious technical challenges in terms of instability and fragmentation of RNA. PD-1/PD-L1 based therapeutic strategies seem to be highly challenging for immune-quiescent pancreatic cancer and require further understanding of immune resistance to overcome obstacles associated with checkpoint blockade.

Current Progress in Anti-PD-1 Therapy in Pancreatic Cancer

Pembrolizumab (Keytruda) and Nivolumab (Opdivo) were the first two anti-PD-1 antibodies approved by FDA for the treatment of metastatic melanoma patients in the year 2014, announced as a breakthrough therapy and have been tested in several clinical trials of solid cancers, including PDAC to evaluate therapeutic potential [25, 75, 76]. Pembrolizumab eventually stood out with a significantly higher response rate and prolonged patient survival than Ipilimumab, the anti-CTLA4 (inhibitory receptor on T cells) mAb. The results of clinical trials evaluating anti-PD-1/PD-L1 therapy in 50 MMR deficient cases (NCT01876511 and NCT02465060) including PDAC patients are awaited [77]. Anti-PD-1/PD-L1 immunotherapy reduces immunosuppression but combining it with other checkpoint blockade antibodies or immune activation mechanism might offer an additional advantage to PDAC patients in terms of improving clinical responses. We searched ongoing clinical trials in pancreatic cancer for anti-PD-1 antibodies alone and/or in combination with chemotherapy, radiotherapy, vaccine or multiple combinations and those are enlisted in Table 1.

Induction of endogenous anti-tumor T cell response by the use of therapeutic vaccines and/or neoadjuvant treatment has resulted in specific treatment with minimal toxicity in PDAC patients and genetically engineered mouse model of pancreatic cancer [56, 78]. GVAX (granulocyte-macrophage colony-stimulating factor (GM-CSF) secreting, allogeneic PDAC vaccine), a vaccine-based immunotherapy, when administered to PDAC patients demonstrated for the first time that immune therapies can convert immune quiescent tumors to immunogenic tumors, amenable to checkpoint blockade therapies [11, 79]. This was the first study to show that vaccination induced the formation of intratumoral lymphoid aggregates and gene signature analysis of the isolated aggregates was indicative of diminished T_{regs} activity, augmented Th17 pathway, and upregulation of PD-1 and PD-L1 axis. In another preclinical study, GVAX treatment showed upregulation of PD-L1 on pancreatic tumor cells and treatment with anti-PD-1 antibodies resulted in IFNγ secretion from infiltrated effector T cells [80] which is in line with observations made by Lutz et al. in human subjects. An alternative approach for developing neo-antigen based vaccine, based on neoepitopes, accumulated as a result of genetic mutations, for PDAC could also lead to optimal anti-tumor response by combination with immune checkpoint blockade molecules [67, 81].

Conclusions

While considerable research efforts have suggested that PD-1/PD-L1 axis is a potential target for combating immunosuppression in PDAC, this optimism has been tempered by the limited success of clinical studies involving anti-PD-1/PD-L1 antibodies. The inadequate clinical efficacy of anti-PD-1 antibodies in PDAC patients can be due to immune quiescent nature of tumor and inappropriate patient selection criteria in clinical trials. PDAC is immunologically cold due to the limited repertoire of neoantigens, immunosuppressive driver mutations, and stromal obstruction to immune cell infiltration. However, recent studies demonstrating the utility of anti-stromal agents in enhancing the immune cell trafficking have raised the possibility of combination therapy involving such agents with anti-PD-1/PD-L1 therapies [82]. Further, the recent findings also suggest that the focus for immunotherapy development should be more on neoantigen quality rather than quantity [83]. Another challenge that remains to be addressed is the variability observed in various studies examining correlation between PD-1/PD-L1 expression and clinicopathological outcomes in PDAC patients. While this can be attributed to the differences in immunostaining methods for the detection of PD-1/PD-L1 expression, some of the variability can be due to the inherent heterogeneity of tumors in terms of tumor grade stage and histology. Furthermore, there is no uniformity across studies in defining the PD-1/PD-L1 expression in specific cellular compartment: while some studies have defined PD-L1 expression on tumor cells, other studies have examined PD-L1 expression of fibroblasts and other stromal cells like macrophages. There is also variability in the sensitivity of detection of expression across various studies due to differences in staining methods and reagents. One recent study reported enhanced detection of PD-L1 expression using a sensitive nanoparticle based immunodetection method but these studies need validation in larger sample set [84]. Recent advancements in understanding intrinsic and extrinsic mechanisms of immune resistance along with PDAC patient stratification into different immune subtypes suggest upregulation of other checkpoint molecules [73, 74]. Given the non-redundant nature of immune checkpoints, it will be of merit to examine anti-PD-1/PD-L1 immunotherapies in combination with other immune checkpoint blockade agents. Much work also needs to be done to define how various therapeutic modalities like chemotherapy and radiation therapy impact PD-1/PD-L1 expression. Tissue samples from the completed and ongoing immunotherapy trials can help address these questions and may lead to more effective combination immunotherapies against PDAC.

Acknowledgement

These authors are supported, in part, by the following grants from the National Institute of Health: U01 CA213862, R01 CA195586 and U01 CA200466.

References

1. Siegel, R.L., K.D. Miller, A. Jemal. Cancer statistics. CA Cancer J Clin, 2018. 68(1): 7-30.
2. Adamska, A., O. Elaskalani, A. Emmanouilidi, M. Kim, N.B. Abdol Razak, P. Metharom, et al. Molecular and cellular mechanisms of chemoresistance in pancreatic cancer. Adv Biol Regul, 2018. 68: 77-87.
3. Jamieson, N.B., D.K. Chang, S.M. Grimmond, A.V. Biankin. Can we move towards personalised pancreatic cancer therapy? Expert Rev Gastroenterol Hepatol, 2014. 8(4): 335-8.
4. Kerkar, S.P., N.P. Restifo. Cellular constituents of immune escape within the tumor microenvironment. Cancer Res, 2012. 72(13): 3125-30.
5. Galon, J., W.H. Fridman, F. Pages. The adaptive immunologic microenvironment in colorectal cancer: a novel perspective. Cancer Res, 2007. 67(5): 1883-6.
6. Galon, J., F. Pages, F.M. Marincola, M. Thurin, G. Trinchieri, B.A. Fox, et al. The immune score as a new possible approach for the classification of cancer. J Transl Med, 2012. 10: 1.
7. Basso, D., D. Bozzato, A. Padoan, S. Moz, C.F. Zambon, P. Fogar, et al. Inflammation and pancreatic cancer: molecular and functional interactions between S100A8, S100A9, NT-S100A8 and TGFbeta1. Cell Commun Signal, 2014. 12: 20.
8. Wormann, S.M., K.N. Diakopoulos, M. Lesina, H. Algul. The immune network in pancreatic cancer development and progression. Oncogene, 2014. 33(23): 2956-67.
9. Ino, Y., R. Yamazaki-Itoh, K. Shimada, M. Iwasaki, T. Kosuge, Y. Kanai, et al. Immune cell infiltration as an indicator of the immune microenvironment of pancreatic cancer. Br J Cancer, 2013. 108(4): 914-23.

10. Bailey, P., D.K. Chang, K. Nones, A.L. Johns, A.M. Patch, M.C. Gingras, et al. Genomic analyses identify molecular subtypes of pancreatic cancer. Nature, 2016. 531(7592): 47-52.
11. Martinez-Bosch, N., J. Vinaixa, P. Navarro. Immune evasion in pancreatic cancer: from mechanisms to therapy. Cancers (Basel), 2018. 10(1): 1-16.
12. Clark, C.E., S.R. Hingorani, R. Mick, C. Combs, D.A. Tuveson, R.H. Vonderheide. Dynamics of the immune reaction to pancreatic cancer from inception to invasion. Cancer Res, 2007. 67(19): 9518-27.
13. Ene-Obong, A., A.J. Clear, J. Watt, J. Wang, R. Fatah, J.C. Riches, et al. Activated pancreatic stellate cells sequester $CD8^+$ T cells to reduce their infiltration of the juxtatumoral compartment of pancreatic ductal adenocarcinoma. Gastroenterology, 2013. 145(5): 1121-32.
14. Balli, D., A.J. Rech, B.Z. Stanger, R.H. Vonderheide. Immune cytolytic activity stratifies molecular subsets of human pancreatic cancer. Clin Cancer Res, 2017. 23(12): 3129-3138.
15. Lanitis, E., D. Dangaj, M. Irving, G. Coukos. Mechanisms regulating T-cell infiltration and activity in solid tumors. Ann Oncol, 2017. 28(suppl_12): xii18-xii32.
16. Fridman, W.H., F. Pages, C. Sautes-Fridman, J. Galon. The immune contexture in human tumours: impact on clinical outcome. Nat Rev Cancer, 2012. 12(4): 298-306.
17. Protti, M.P., L. De Monte. Immune infiltrates as predictive markers of survival in pancreatic cancer patients. Front Physiol, 2013. 4: 210.
18. Ghirelli, C., T. Hagemann. Targeting immunosuppression for cancer therapy. J Clin Invest, 2013. 123(6): 2355-7.
19. Dong, H., S.E. Strome, D.R. Salomao, H. Tamura, F. Hirano, D.B. Flies, et al. Tumor-associated B7-H1 promotes T-cell apoptosis: a potential mechanism of immune evasion. Nat Med, 2002. 8(8): 793-800.
20. Nomi, T., M. Sho, T. Akahori, K. Hamada, A. Kubo, H. Kanehiro, et al. Clinical significance and therapeutic potential of the programmed death-1 ligand/programmed death-1 pathway in human pancreatic cancer. Clin Cancer Res, 2007. 13(7): 2151-7.
21. Chang, Y.L., C.Y. Yang, Y.L. Huang, C.T. Wu, P.C. Yang. High PD-L1 expression is associated with stage IV disease and poorer overall survival in 186 cases of small cell lung cancers. Oncotarget, 2017. 8(11): 18021-30.
22. Rosenblatt, J., D. Avigan. Targeting the PD-1/PD-L1 axis in multiple myeloma: a dream or a reality? Blood, 2017. 129(3): 275-279.
23. Winograd, R., K.T. Byrne, R.A. Evans, P.M. Odorizzi, A.R. Meyer, D.L. Bajor, et al. Induction of T-cell immunity overcomes complete resistance to PD-1 and CTLA-4 blockade and improves survival in pancreatic carcinoma. Cancer Immunol Res, 2015. 3(4): 399-411.
24. Okudaira, K., R. Hokari, Y. Tsuzuki, Y. Okada, S. Komoto, C. Watanabe, et al. Blockade of B7-H1 or B7-DC induces an anti-tumor effect in a mouse pancreatic cancer model. Int J Oncol, 2009. 35(4): 741-9.
25. Brahmer, J.R., S.S. Tykodi, L.Q. Chow, W.J. Hwu, S.L. Topalian, P. Hwu, et al. Safety and activity of anti-PD-L1 antibody in patients with advanced cancer. N Engl J Med, 2012. 366(26): 2455-65.
26. Jones, S., X. Zhang, D.W. Parsons, J.C. Lin, R.J. Leary, P. Angenendt, et al. Core signaling pathways in human pancreatic cancers revealed by global genomic analyses. Science, 2008. 321(5897): 1801-6.

27. Chen, M., B. Pockaj, M. Andreozzi, M.T. Barrett, S. Krishna, S. Eaton, et al. JAK2 and PD-L1 amplification enhance the dynamic expression of PD-L1 in triple-negative breast cancer. Clin Breast Cancer, 2018. e1205-1215.
28. Abdelhamed, S., K. Ogura, S. Yokoyama, I. Saiki, Y. Hayakawa. AKT-STAT3 pathway as a downstream target of EGFR signaling to regulate PD-L1 expression on NSCLC cells. J Cancer, 2016. 7(12): 1579-1586.
29. Lyford-Pike, S., S. Peng, G.D. Young, J.M. Taube, W.H. Westra, B. Akpeng, et al. Evidence for a role of the PD-1:PD-L1 pathway in immune resistance of HPV-associated head and neck squamous cell carcinoma. Cancer Res, 2013. 73(6): 1733-41.
30. Spranger, S., R.M. Spaapen, Y. Zha, J. Williams, Y. Meng, T.T. Ha, et al. Up-regulation of PD-L1, IDO, and T(regs) in the melanoma tumor microenvironment is driven by CD8(+) T cells. Sci Transl Med, 2013. 5(200): 200ra116.
31. Lu, C., A. Talukder, N.M. Savage, N. Singh, K. Liu. JAK-STAT-mediated chronic inflammation impairs cytotoxic T lymphocyte activation to decrease anti-PD-1 immunotherapy efficacy in pancreatic cancer. Oncoimmunology, 2017. 6(3): e1291106.
32. Pushalkar, S., M. Hundeyin, D. Daley, C.P. Zambirinis, E. Kurz, A. Mishra, et al. The pancreatic cancer microbiome promotes oncogenesis by induction of innate and adaptive immune suppression. Cancer Discov, 2018. 8(4): 403-416.
33. Lu, C., A.V. Paschall, H. Shi, N. Savage, J.L. Waller, M.E. Sabbatini, et al. The MLL1-H3K4me3 axis-mediated PD-L1 expression and pancreatic cancer immune evasion. J Natl Cancer Inst, 2017. 109(6): 1-12.
34. Makohon-Moore, A., C.A. Iacobuzio-Donahue. Pancreatic cancer biology and genetics from an evolutionary perspective. Nat Rev Cancer, 2016. 16(9): 553-65.
35. Vonderheide, R.H., L.J. Bayne. Inflammatory networks and immune surveillance of pancreatic carcinoma. Curr Opin Immunol, 2013. 25(2): 200-5.
36. Bronte, V. Myeloid-derived suppressor cells in inflammation: uncovering cell subsets with enhanced immunosuppressive functions. Eur J Immunol, 2009. 39(10): 2670-2.
37. Gabrilovich, D.I., S. Nagaraj. Myeloid-derived suppressor cells as regulators of the immune system. Nat Rev Immunol, 2009. 9(3): 162-74.
38. Huang, B., P.Y. Pan, Q. Li, A.I. Sato, D.E. Levy, J. Bromberg, et al. Gr-1+CD115+ immature myeloid suppressor cells mediate the development of tumor-induced T regulatory cells and T-cell energy in tumor-bearing host. Cancer Res, 2006. 66(2): 1123-31.
39. Balaz, P., H. Friess, Y. Kondo, Z. Zhu, A. Zimmermann, M.W. Buchler. Human macrophage metalloelastase worsens the prognosis of pancreatic cancer. Ann Surg, 2002. 235(4): 519-27.
40. Kurahara, H., H. Shinchi, Y. Mataki, K. Maemura, H. Noma, F. Kubo, et al. Significance of M2-polarized tumor-associated macrophage in pancreatic cancer. J Surg Res, 2011. 167(2): e211-9.
41. Yu, H., M. Kortylewski, D. Pardoll. Crosstalk between cancer and immune cells: role of STAT3 in the tumour microenvironment. Nat Rev Immunol, 2007. 7(1): 41-51.
42. Maude, S.L., S. Dolai, C. Delgado-Martin, T. Vincent, A. Robbins, A. Selvanathan, et al. Efficacy of JAK/STAT pathway inhibition in murine xenograft models of early T-cell precursor (ETP) acute lymphoblastic leukemia. Blood, 2015. 125(11): 1759-67.
43. Zhu, Y., B.L. Knolhoff, M.A. Meyer, T.M. Nywening, B.L. West, J. Luo, et al. CSF1/CSF1R blockade reprograms tumor-infiltrating macrophages and

improves response to T-cell checkpoint immunotherapy in pancreatic cancer models. Cancer Res, 2014. 74(18): 5057-69.
44. Liu, X., J. Wang, H. Wang, G. Yin, Y. Liu, X. Lei, et al. REG3A accelerates pancreatic cancer cell growth under IL-6-associated inflammatory condition: involvement of a REG3A-JAK2/STAT3 positive feedback loop. Cancer Lett, 2015. 362(1): 45-60.
45. Liu, X., Z. Zhou, Q. Cheng, H. Wang, H. Cao, Q. Xu, et al. Acceleration of pancreatic tumorigenesis under immunosuppressive microenvironment induced by Reg3g overexpression. Cell Death Dis, 2017. 8(9): e3033.
46. Chauveau, C., S. Remy, P.J. Royer, M. Hill, S. Tanguy-Royer, F.X. Hubert, et al. Heme oxygenase-1 expression inhibits dendritic cell maturation and proinflammatory function but conserves IL-10 expression. Blood, 2005. 106(5): 1694-702.
47. Zhang, Y., A. Velez-Delgado, E. Mathew, D. Li, F.M. Mendez, K. Flannagan, et al. Myeloid cells are required for PD-1/PD-L1 checkpoint activation and the establishment of an immunosuppressive environment in pancreatic cancer. Gut, 2017. 66(1): 124-36.
48. Zhang, Y., N. Gallastegui, J.D. Rosenblatt. Regulatory B cells in anti-tumor immunity. Int Immunol, 2015. 27(10): 521-30.
49. Schwartz, M., Y. Zhang, J.D. Rosenblatt. B cell regulation of the anti-tumor response and role in carcinogenesis. J Immunother Cancer, 2016. 4: 40.
50. Shalapour, S., J. Font-Burgada, G. Di Caro, Z. Zhong, E. Sanchez-Lopez, D. Dhar, et al. Immunosuppressive plasma cells impede T-cell-dependent immunogenic chemotherapy. Nature, 2015. 521(7550): 94-8.
51. Qian, L., W. Chen, H. Qin, C. Rui, X. Jia, Y. Fu, et al. Immune complex negatively regulates Toll-like receptor 9-mediated immune responses in B cells through the inhibitory Fc-gamma receptor IIb. Microbiol Immunol, 2015. 59(3): 142-51.
52. Wei, X., Y. Jin, Y. Tian, H. Zhang, J. Wu, W. Lu, et al. Regulatory B cells contribute to the impaired antitumor immunity in ovarian cancer patients. Tumour Biol, 2016. 37(5): 6581-8.
53. Zhao, Y., M. Shen, Y. Feng, R. He, X. Xu, Y. Xie, et al. Regulatory B cells induced by pancreatic cancer cell-derived interleukin-18 promote immune tolerance via the PD-1/PD-L1 pathway. Oncotarget, 2018. 9(19): 14803-14.
54. Hiraoka, N., K. Onozato, T. Kosuge, S. Hirohashi. Prevalence of FOXP3+ regulatory T cells increases during the progression of pancreatic ductal adenocarcinoma and its premalignant lesions. Clin Cancer Res, 2006. 12(18): 5423-34.
55. Loos, M., N.A. Giese, J. Kleeff, T. Giese, M.M. Gaida, F. Bergmann, et al. Clinical significance and regulation of the costimulatory molecule B7-H1 in pancreatic cancer. Cancer Lett, 2008. 268(1): 98-109.
56. Shibuya, K.C., V.K. Goel, W. Xiong, J.G. Sham, S.M. Pollack, A.M. Leahy, et al. Pancreatic ductal adenocarcinoma contains an effector and regulatory immune cell infiltrate that is altered by multimodal neoadjuvant treatment. PLoS One, 2014. 9(5): e96565.
57. Linehan, D.C., P.S. Goedegebuure. CD25+ CD4+ regulatory T-cells in cancer. Immunol Res, 2005. 32(1-3): 155-68.
58. Pardoll, D.M. The blockade of immune checkpoints in cancer immunotherapy. Nat Rev Cancer, 2012. 12(4): 252-64.
59. Geng, L., D. Huang, J. Liu, Y. Qian, J. Deng, D. Li, et al. B7-H1 up-regulated expression in human pancreatic carcinoma tissue associates with tumor progression. J Cancer Res Clin Oncol, 2008. 134(9): 1021-7.

60. Song, X., J. Liu, Y. Lu, H. Jin, D. Huang. Overexpression of B7-H1 correlates with malignant cell proliferation in pancreatic cancer. Oncol Rep, 2014. 31(3): 1191-8.
61. Takahashi, R., Y. Hirata, K. Sakitani, W. Nakata, H. Kinoshita, Y. Hayakawa, et al. Therapeutic effect of c-Jun N-terminal kinase inhibition on pancreatic cancer. Cancer Sci, 2013. 104(3): 337-44.
62. Hutcheson, J., U. Balaji, M.R. Porembka, M.B. Wachsmann, P.A. McCue, E.S. Knudsen, et al. Immunologic and metabolic features of pancreatic ductal adenocarcinoma define prognostic subtypes of disease. Clin Cancer Res, 2016. 22(14): 3606-17.
63. Wang, L., Q. Ma, X. Chen, K. Guo, J. Li, M. Zhang. Clinical significance of B7-H1 and B7-1 expressions in pancreatic carcinoma. World J Surg, 2010. 34(5): 1059-65.
64. Diana, A., L.M. Wang, Z. D'Costa, P. Allen, A. Azad, M.A. Silva, et al. Prognostic value, localization and correlation of PD-1/PD-L1, CD8 and FOXP3 with the desmoplastic stroma in pancreatic ductal adenocarcinoma. Oncotarget, 2016. 7(27): 40992-41004.
65. Wang, Y., J. Lin, J. Cui, T. Han, F. Jiao, Z. Meng, et al. Prognostic value and clinicopathological features of PD-1/PD-L1 expression with mismatch repair status and desmoplastic stroma in Chinese patients with pancreatic cancer. Oncotarget, 2017. 8(6): 9354-65.
66. Stromnes, I.M., A. Hulbert, R.H. Pierce, P.D. Greenberg, S.R. Hingorani. T-cell localization, activation, and clonal expansion in human pancreatic ductal adenocarcinoma. Cancer Immunol Res, 2017. 5(11): 978-91.
67. Foley, K., V. Kim, E. Jaffee, L. Zheng. Current progress in immunotherapy for pancreatic cancer. Cancer Lett, 2016. 381(1): 244-51.
68. Pardoll, D. Cancer and the immune system: basic concepts and targets for intervention. Semin Oncol, 2015. 42(4): 523-38.
69. Royal, R.E., C. Levy, K. Turner, A. Mathur, M. Hughes, U.S. Kammula, et al. Phase 2 trial of single agent Ipilimumab (anti-CTLA-4) for locally advanced or metastatic pancreatic adenocarcinoma. J Immunother, 2010. 33(8): 828-33.
70. Gubin, M.M., X. Zhang, H. Schuster, E. Caron, J.P. Ward, T. Noguchi, et al. Checkpoint blockade cancer immunotherapy targets tumour-specific mutant antigens. Nature, 2014. 515(7528): 577-81.
71. Houg, D.S., M.F. Bijlsma. The hepatic pre-metastatic niche in pancreatic ductal adenocarcinoma. Mol Cancer, 2018. 17(1): 95.
72. Feng, L., H. Qian, X. Yu, K. Liu, T. Xiao, C. Zhang, et al. Heterogeneity of tumor-infiltrating lymphocytes ascribed to local immune status rather than neoantigens by multi-omics analysis of glioblastoma multiforme. Sci Rep, 2017. 7(1): 6968.
73. Sharma, P., S. Hu-Lieskovan, J.A. Wargo, A. Ribas. Primary, adaptive, and acquired resistance to cancer immunotherapy. Cell, 2017. 168(4): 707-23.
74. Wartenberg, M., S. Cibin, I. Zlobec, E. Vassella, S. Eppenberger-Castori, L. Terracciano, et al. Integrated genomic and immunophenotypic classification of pancreatic cancer reveals three distinct subtypes with prognostic/predictive significance. Clin Cancer Res, 2018. 24(18): 4444-54.
75. Herbst, R.S., J.C. Soria, M. Kowanetz, G.D. Fine, O. Hamid, M.S. Gordon, et al. Predictive correlates of response to the anti-PD-L1 antibody MPDL3280A in cancer patients. Nature, 2014. 515(7528): 563-7.

76. Patnaik, A., S.P. Kang, D. Rasco, K.P. Papadopoulos, J. Elassaiss-Schaap, M. Beeram, et al. Phase I study of pembrolizumab (MK-3475; Anti-PD-1 monoclonal antibody) in patients with advanced solid tumors. Clin Cancer Res, 2015. 21(19): 4286-93.
77. Xu, J.W., L. Wang, Y.G. Cheng, G.Y. Zhang, S.Y. Hu, B. Zhou, et al. Immunotherapy for pancreatic cancer: a long and hopeful journey. Cancer Lett, 2018. 425: 143-51.
78. Lutz, E., C.J. Yeo, K.D. Lillemoe, B. Biedrzycki, B. Kobrin, J. Herman, et al. A lethally irradiated allogeneic granulocyte-macrophage colony stimulating factor-secreting tumor vaccine for pancreatic adenocarcinoma. A Phase II trial of safety, efficacy, and immune activation. Ann Surg, 2011. 253(2): 328-35.
79. Lutz, E.R., A.A. Wu, E. Bigelow, R. Sharma, G. Mo, K. Soares, et al. Immunotherapy converts nonimmunogenic pancreatic tumors into immunogenic foci of immune regulation. Cancer Immunol Res, 2014. 2(7): 616-31.
80. Soares, K.C., Rucki A.A. Rucki, A.A. Wu, K. Olino, Q Xiao, Y. Chai, et al. PD-1/PD-L1 blockade together with vaccine therapy facilitates effector T-cell infiltration into pancreatic tumors. J Immunother, 2015. 38(1): 1-11.
81. Rahma, O.E., J.M. Hamilton, M. Wojtowicz, O. Dakheel, S. Bernstein, D.J. Liewehr, et al. The immunological and clinical effects of mutated ras peptide vaccine in combination with IL-2, GM-CSF, or both in patients with solid tumors. J Transl Med, 2014. 12(55): 1-12.
82. Neesse, A., K.K. Frese, T.E. Bapiro, T. Nakagawa, M.D. Sternlicht, T.W. Seeley, et al. CTGF antagonism with mAb FG-3019 enhances chemotherapy response without increasing drug delivery in murine ductal pancreas cancer. Proc Natl Acad Sci USA, 2013. 110(30): 12325-30.
83. Balachandran, V.P., M. Luksza, J.N. Zhao, V. Makarov, J.A. Moral, R. Remark, et al. Identification of unique neoantigen qualities in long-term survivors of pancreatic cancer. Nature, 2017. 551(7681): 512-6.
84. Yamaki, S., H. Yanagimoto, K. Tsuta, H. Ryota, M. Kon. PD-L1 expression in pancreatic ductal adenocarcinoma is a poor prognostic factor in patients with high CD8(+) tumor-infiltrating lymphocytes: highly sensitive detection using phosphor-integrated dot staining. Int J Clin Oncol, 2017. 22(4): 726-33.

CHAPTER 2

Nuclear Factor Kappa-B: Bridging Inflammation and Cancer

Mohammad Aslam Khan[1], Girijesh Kumar Patel[1], Haseeb Zubair[1], Nikhil Tyagi[1], Shafquat Azim[1], Seema Singh[1,2], Aamir Ahmad[1] and Ajay Pratap Singh[1,2*]

[1] Department of Oncologic Sciences, Mitchell Cancer Institute, University of South Alabama, Mobile, Alabama, USA 36604
[2] Department of Biochemistry and Molecular Biology, College of Medicine, University of South Alabama, Mobile, Alabama, USA 36688

Introduction

It has long been realized that inflammation is functionally related to cancer [1, 2]. The inflammatory response to infections and tissue damage is characterized by tissue repair and regeneration, which involves cell proliferation. Not only does inflammation predispose to cancer, it also influences cancer progression, mainly through the inflammatory cells that make up the tumor microenvironment. Inflammatory cells secrete chemokines, growth factors, cytokines etc. that drive tumor growth and metastasis [3]. NF-κB is an important pro-inflammatory factor that is activated by pro-inflammatory cytokines such as interleukins. Also, NF-κB itself induces the expression of multiple genes including cytokines, chemokines and adhesion molecules. In addition to its role in inflammation [4], NF-κB plays a key role in cancer initiation and progression [5], leading to its reputation as a liaison that connects inflammation with cancer.

NF-κB was discovered by David Baltimore as a transcription factor in B cells which binds to the enhancer elements, and controls the expression of immunoglobulin kappa (κ) light chain [6]. Initially it was thought that NF-κB has a role only in B cell activation and development, but, subsequently, it has been found that NF-κB controls the expression of many other genes that are involved in many physiological processes [7]. Inside the cells, NF-κB gets activated when environmental or biological insults impinge on the cells, and this activation induces downstream signaling. At the time of infection or wound, NF-κB becomes active and upregulates secretion of pro-

*Corresponding author: asingh@health.southalabama.edu

inflammatory molecules, which attract cells of innate immune system at the location of infection/wound to combat the infection or insult. Once the issue is resolved, NF-κB signaling is downregulated to normal physiological levels [8]. Constituted activation of NF-κB, or the failure to bring it down to normal levels, has been linked to many pathological conditions such as arthritis, diabetes, asthma, inflammatory bowel disease, heart diseases, infections and several malignancies [8].

The NF-κB family consists of five members: RelA (p65), RelB, c-Rel, NF-κB1 (p105) and NF-κB2 (p100). The p105 and p100 subunits further get processed upon activation into p50 and p52, respectively. Two distinct modes of NF-κB activation have been reported; a classical/canonical activation and an alternate/non-canonical activation based on extracellular stimuli, adaptor molecules and dimers of NF-κB [9]. Unlike canonical pathway, non-canonical way of NF-κB signaling is slow and persistent. Once NF-κB is activated, subunits of NF-κB dimerize (either p50/p65 or p52/RelB), followed by translocation of hetero-dimers to the nucleus where they bind to κB sites in the promoters and enhancer regions of genes and regulate their expression [10, 11]. Deregulated NF-κB affects several hallmarks of cancers transcriptionally (Fig. 1) with profound effects on cancer development, progression and therapy resistance [12]. Based on its diverse and widespread role in cancer, targeting NF-κB pathway has always remained a desirable approach to control cancer development and progression.

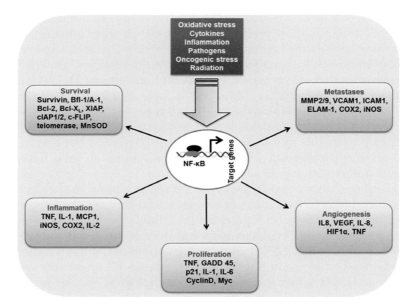

Fig. 1: Role of nuclear factor-κB (NF-κB) in cancer development and progression. Diverse array of inducers can activate NF-κB which affects tumorigenesis by regulating expression of several genes involved in different physiological processes, such as, cell proliferation, angiogenesis, metastases, and inflammation.

Inflammation and Cancer

In the year 1863, Rudolf Virchow hypothesized that cancer starts at the site of chronic inflammation [13]. Inflammation can occur during infections, exposure to the irritants or tissue damages, and is considered a protective response of body against these insults. There are two types of inflammations, innate (shorter duration) and chronic or persistent. Chronic inflammation is considered as one of the risk factors for the development of cancer [14]. Moreover, there are many additional inflammatory insults or conditions [15-25] that are associated with the development of cancer (Table 1). The two classical examples of inflammation associated-cancers are the infection of *Helicobacter pylori*, associated with gastric cancer, and the inflammatory bowel disease, associated with the development of colorectal cancer [26, 27]. Multiple roles of chronic inflammation in tumorigenesis have been suggested; these include induction of genomic instability and altered epigenetic events leading to deregulated gene expression, uncontrolled proliferation of cancer cells, aggressiveness and resistance to apoptosis [14]. The molecular mechanism(s) by which inflammatory cells promote tumorigenesis is still not very clear.

Table 1: Role of inflammation or infection in the development of human malignancies

Stimuli	Cancer types	References
Chronic pancreatitis	Pancreatic cancer	[15]
Lung inflammation and pulmonary fibrosis	Lung cancer	[16, 17]
Schistosomiasis	Bladder cancer	[18]
Exposure to asbestos	Malignant mesothelioma	[19]
Thyroiditis	Papillary thyroid carcinoma	[20]
Infection of *Helicobacter pylori*	Gastric cancer	[22]
Inflammatory bowel disease	Colorectal cancer	[23]
Infection of Kaposi's sarcoma herpes virus	Kaposi's sarcoma	[24]
Chronic prostatitis	Prostate cancer	[25]
Escherichia coli infection of prostate	Atypical hyperplasia and dysplasia of prostate	[21]

Genomic Instability

Genomic instability is thought to be one of the key drivers for neoplastic transformation [28]. Cells from immune system, primarily macrophages and neutrophils, generate reactive nitrogen species (RNS) and reactive oxygen species (ROS) which cause DNA breaks and mutations. Also, cytokines (TGF-β, tumor necrosis factor; TNF, IL-4, IL-15, IL-1β, etc.) produced by immune cells upregulate the expression of cytosine deaminase, an enzyme that has the ability to induce mutations in human genome and has been suggested to induce mutations in tumor-suppressor and/or tumor promoting genes, p53 and MYC, respectively, leading to tumorigenesis [29-31]. Inflammatory cytokines and ROS, through genetic and/or epigenetic alterations, inhibit DNA repair pathways which eventually accelerate mutational frequency in the pre-cancer cells as well as in transformed cells [31, 32].

Deregulated Proliferative Signaling

Immune cell-derived cytokines/growth factors act on the cancer cells and activate their growth by engaging effector downstream molecular signaling. Majority of inflammatory pathways can activate NF-κB and/or signal transducer and activator of transcription 3 (STAT3) pathway and these pathways induce cell proliferation by upregulating the expression of Bcl-2 family proteins as well as cell cycle associated proteins. IL-6-(derived from myeloid cells)-dependent STAT3 activation has been shown to play an important role in the early tumorigenesis of colitis-associated colorectal cancer [33]. Inflammation activates NF-κB signaling which converges into the upregulation of several cell survival related genes, and production of pro-inflammatory cytokines. Preclinical studies have suggested that TNF produced by immune and endothelial cells activates NF-κB signaling in hepatocytes which is essential for the progression of hepatocellular carcinoma [34]. Further, inhibition of NF-κB lowers the incidence of colitis-associated colorectal cancer, and deletion of *IKKβ* in myeloid cells causes reduction in the tumor size [35].

Enhanced Invasiveness

Genomic instability and uncontrolled proliferative signaling is more relevant to the initiation of tumorigenesis but a majority of cancer-associated deaths are due to metastases of tumor cells [31]. Tumor cells undergo cellular transition *i.e.* epithelial to mesenchymal transition (EMT), and this process is considered critical for the tumor cells to migrate from primary sites to other (metastatic) sites in the body. Inflammatory cytokines such as IL-6, TNF and IL-1 induce EMT by activating NF-κB and STAT3 signaling [36-38]. IL-4-activated tumor associated macrophages (TAMs) as well as CCR1[+] immature myeloid cells secrete matrix degrading enzymes which help in tumor cell invasion [39, 40]. Moreover, CCL-2 mediated recruitment of Gr1[+] and CCR2[+] inflammatory monocytes helps in the metastases of breast cancer cells and inhibition of CCL2-CCR2 signaling axis increases survival of tumor bearing mice [41]. Thus, prolonged inflammation helps in the cellular transformation and early events of tumorigenesis, whereas inflammatory tumor microenvironment induces tumor aggressiveness in the later events.

Canonical and Non-canonical Pathways of NF-κB Activation

NF-κB transcriptionally regulates expression of genes which are involved in cell survival, apoptosis, differentiation and immune responses [42, 43]. As discussed above, NF-κB family consists of five members. Both the precursors (NF-κB1/p105 and NF-κB2/p100) contain IκB-homologous regions at the C-terminal that function as NF-κB inhibitors. Proteolytic processing removes inhibitory domains, allowing the processed proteins (NF-κB1/p50 and NF-κB2/p52) to enter into the nucleus. Inside the nucleus, p50 and p52 form homodimers or heterodimers with Rel family proteins [42]. All these NF-κB associated proteins have Rel homology domain (RHD) at the N-terminal region. Some of the NF-κB proteins (p65, c-Rel and RelB) contain transcriptional activation domain (TAD) at C-terminal, and these TAD containing proteins positively regulate transcriptional activity. On the other hand, two TAD lacking proteins, p52 and p100, negatively regulate gene expression when bound as homodimer, but can activate gene expression when recruited with other TAD containing proteins like p65, RelB and cRel [44]. Activation of NF-κB pathway is either by classical/canonical or the alternate/non-canonical signaling pathways.

Canonical Pathway

Canonical pathway of NF-κB has been extensively studied in different human pathological disorders such as obesity, autoimmune diseases and cancer. Canonical pathway of NF-κB activation is quick and transient, and can be activated by diverse stimuli, such as, TNF, IL-1, toll-like receptors (TLRs), lipopolysaccharide (LPS), etc. LPS and IL-1 stimulate NF-κB activation by recruiting adaptor molecules, TNFR associated factors6 (TRAF6), myeloid differentiation primary response gene 88 (MyD88) and kinase protein, interleukin-1 receptor-associated kinase (IRAK). On the other hand, binding of TNF to tumor necrosis factor receptor (TNFR) results in the recruitment of TRAF2, TNFR1-associated death domain (TRADD), cIAP1, cIAP2, along with kinase protein, receptor interacting protein1 (RIP1), to the receptor [45]. After binding to the receptor, RIP1 undergoes polyubiquitination *via* non-degradative linkage at Lys63 (K63) which results in the recruitment of IKK complex to the receptor in the close proximity of transforming growth factor beta activated kinase 1 (TAK1) and mitogen activated protein kinase 3 (MEKK3) for its activation [45, 46]. IKK complex is composed of two catalytic subunits (IKK1/IKKα, IKK2/IKKβ) and one regulatory subunit (IKKγ, also known as NEMO). In unstimulated conditions, family of IκB proteins, IκBα, IκBβ, and IκBε, sequesters NF-κB proteins in the cytoplasm [47]. IκBα binds to the RHD of NF-κB complex and masks nuclear localization signal (NLS) which results in the cytoplasmic retention of NF-κB proteins [48]. Once cells receive stimuli, IKK complex phosphorylates IκBα proteins at

Ser-32 and Ser-36, and phosphorylated IκBα is targeted for ubiquitination and 26S proteasome degradation, ultimately rendering NF-κB (p50-p65 complex) free, which translocates to the nucleus and binds to the promoter regions of target genes [49, 50] (Fig. 2). Canonical pathway of NF-κB is very

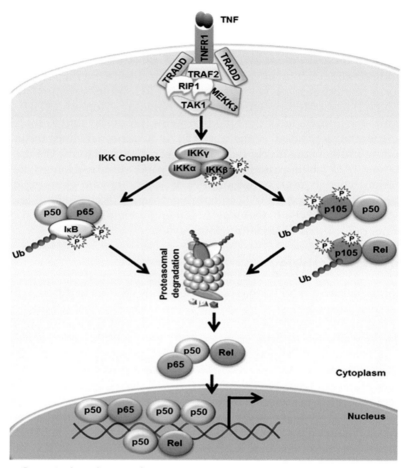

Fig. 2: Canonical pathway of NF-κB activation. Activation of NF-κB pathway is triggered by a variety of stimuli/ligands. Ligand-receptor interactions initiate the recruitment of adaptor molecules TRADD, TRAF2, cIAP1/2, and activate kinase molecules, TAK1 and MEEK, which further activate IκB kinase (IKK) complex by phosphorylating IKKβ. Once IKK complex is activated, it phosphorylates members of inhibitor of κB (IκB) family, IκBα and IκB-like molecule, p105 which sequester NF-κB proteins in the cytoplasm. In unstimulated condition, IκBα binds with p50-p65 dimers and p105 binds with either p50 or REL (RELA or c-REL). Phosphorylated IκBα and p105 are targeted for ubiquitin (Ub)-mediated proteasomal degradation, abolishing inhibitory effect of IκBα or p105. The released NF-κB dimers (p50-p65, p50-Rel and p50-p50) are free to translocate to the nucleus where they can bind at specific sites with the promoter region of target genes.

tightly regulated and can be suppressed by deubiquitinase (DUB) enzyme, A20, which removes polyubiquitin chain from the adaptor molecules and destabilizes IKK complex. Expression of A20 is under the control of NF-κB. Thus, A20 serves as negative feedback loop for NF-κB activation [51]. It has been suggested that tumor suppressive role of A20 can be compromised due to genetic mutations or its proteolytic cleavage in B cell lymphoma [52, 53]. There are some other DUBs, such as, Cezanne and cylindromatosis (CYLD), which can also negatively regulate the activation of NF-κB activation [54, 55].

Non-canonical Pathway

Primarily, non-canonical or alternate pathway of NF-κB plays important role in immune response, B cell maturation, lymphoid organogenesis bone cell differentiation, etc. Dysregulation of non-canonical pathway of NF-κB has been linked to many immune related diseases and cancers [56]. Non-canonical NF-κB pathway differs from classical pathway in terms of stimuli, cytoplasmic adaptors and subunits of NF-κB. Various stimuli (Table 2) have been reported for the activation of non-canonical NF-κB pathway [57-70].

Table 2: The list of stimuli, which can activate non-canonical NF-κB pathway

Ligands/pathogens	Receptors/molecules	References
Cluster of differentiation 40 ligand (CD40L)	CD40	[57]
B cell activating factor (BAFF)	BAFF receptor	[58]
CD70	CD27	[59]
CD30L	CD30	[60]
OX40L (CD134)	OX40 receptor	[61]
Receptor activator of nuclear factor kappa-B ligand (RANKL)	RANK	[62]
Lymphotoxin α1β2 (LTα1β2)	LTβ receptor	[63]
TNF-like weak inducer of apoptosis (TWEAK)	Fibroblast growth factor-inducible factor 14 (FN14)	[64]
Macrophage colony-stimulating factor (MCSF)	MCSFR	[65]
Influenza virus	RNA	[66]
Kaposi sarcoma-associated herpes virus	v-FLIP protein	[67]
Human T cell lymphotropic virus 1	Tax protein	[68]
Legionella pneumophila	Legionella kinase 1	[69]
Epstein-Barr virus	LMP1 protein	[70]

Unlike canonical pathway of NF-κB, this signaling pathway is comparatively slow in response to the stimulating signals [56, 71]. Here, NF-κB inducing kinase (NIK) serves as central regulator and, during unstimulated conditions, the levels of NIK protein are kept under check by TNF receptor associated factor 3 (TRAF3) which serves as an adaptor molecule for the ubiquitination and proteasomal degradation of NIK. *De novo* synthesis of NIK and its degradation reduces NIK inside the cells. Therefore, non-canonical NF-κB remains inactive [56] and once ligand binds to the receptor, it induces ubiquitination and degradation of TRAF3 and stabilizes NIK levels which eventually further activate non-canonical pathway of NF-κB [72, 73]. NIK activates IKKα homodimer complex which, in turn, phosphorylates IkB domain of p100, and processing of p100 occurs. Heterodimer of p52/RelB translocates to the nucleus to transcribe the target genes [72] (Fig. 3).

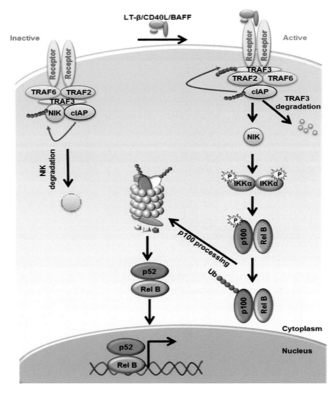

Fig. 3: Non-canonical pathway of NF-κB activation. In the resting cells, NIK undergoes proteasomal degradation and its levels remain low. Binding of stimuli such as LT-β, CD40L or BAFF to their receptors leads to induction of TRAF3 degradation and activation of NIK. NIK phosphorylates and activates IKKα which induces p100 phosphorylation and processing. Finally, p52 is generated and p52-RelB dimers translocate to the nucleus for transcription of target genes.

Several regulatory mechanisms and crosstalks have been proposed for the non-canonical pathway of NF-κB activation. During stimulation, IKKα exhibits negative feedback mechanism and destabilizes NIK levels while another kinase, TANK-binding kinase1 (TBK1), also phosphorylates and controls stability of NIK in a signal-dependent manner [74, 75]. De-ubiquitinase, OTU deubiquitinase 7B (OTUD7b), acts as negative regulator for the activation of non-canonical NF-κB pathway. After binding to TRAF3, OUTD7b deubiquitinates TRAF3 and indirectly controls the levels of NIK in a signal-dependent manner [76].

A crosstalk between canonical and non-canonical pathway also exists as NEMO/IKKγ, an important factor in canonical NF-κB pathway, has been found to be a negative regulator of NIK. NEMO-deficient cells display high expression of NIK with basal activation of the non-canonical pathway relative to control cells [77].

NF-κB and Inflammation

Inflammation is a protective immune response which helps to combat harmful entities like pathogens, irritants or dead cells. Inflammation is a highly regulated process and once it becomes deregulated, it may lead to many pathological conditions [78]. Innate immune response can be elicited by pathogen derived molecules, pathogen associated molecular patterns (PAMPs), or by danger-associated molecular patterns (DAMPs) generated through endogenous stress [78]. Inflammasome is a multiprotein complex, based on physiological and pathogenic stresses, responsible for the induction of inflammatory response. Inflammasome contains nucleotide-binding domain-like receptors (NLRs), absent in melanoma (AIM) and a zymogen, procaspase-1. Assembly of inflammasome leads to the activation of caspase-1, which enzymatically cleaves precursors of IL-1β and IL-18 to orchestrate the downstream inflammatory signals [78, 79]. Inflammasome-mediated immune response is highly regulated and it has been suggested that activation of NF-κB signaling pathway is required for the activation of NLRP3 inflammasome [80]. Recently, it has been shown that NF-κB signaling pathway can control excessive inflammatory response by restraining NLRP3 mediated inflammasome activation by inducing p62 dependent autophagy [81].

Transcriptional activation of NF-κB promotes gene expression associated with pro-inflammatory molecules such as chemokines, cytokines, matrix metalloproteinases (MMPs), adhesion molecules, cyclooxygenase-2 (cox-2), inducible nitric oxide (iNOS), etc. This inflammatory milieu activates and differentiates inflammatory T cells [4, 82]. These locally produced cytokines can activate macrophage and other monocyte lineage to execute and resolve inflammation. Furthermore, plasticity of immune cells gives rise to different subsets of polarized cells in response to the environmental signals, and underlying molecular mechanisms of macrophage polarization suggest the role of NF-κB signaling pathway. Moreover, NF-κB regulates the expression of pro-inflammatory molecules, such as IL-6, TNF-α, IL-1β, and Cox-2 in M1 macrophages [83]. Porta and colleagues reported that p50-NF-κB induces IL-10 expression (M2 marker), and mice deficient in p50-NF-κB exhibit M1-mediated inflammation [84]. *In vivo* study suggested that mice devoid of NF-κB signaling in T cells, exhibited delayed hypersensitivity responses and suppressed IFN-γ production [85]. NF-κB acts as a bridge between innate and the adaptive immune response by engaging components of adaptive arm of immunity. Furthermore, NF-κB-mediated cytokines (IL-12 and IL-23) trigger CD4$^+$ T cell differentiation into T helper 1 (Th1) and Th17 [86, 87]. NF-κB has been implicated in many inflammation-associated diseases like type 2 diabetes, rheumatoid arthritis, atherosclerosis, multiple sclerosis, asthma, etc. [87, 88]

NF-κB in Cancer

NF-κB is perhaps the most studied transcription factor associated with human malignancies. NF-κB signaling pathway governs diverse cellular processes in cancer, such as, cell transformation, proliferation, angiogenesis, metastases and therapy resistance. Majority of cancers have constitutively active NF-κB, and inhibition of NF-κB signaling results in reduced growth of cancer cells, suggesting cancer cells' addiction to NF-κB [89].

Role in Tumor Cell Proliferation

NF-κB has been shown to control proliferation of normal as well as tumor cells by binding to the promoter of genes responsible for cell proliferation. Study published by Brantley and colleagues stated that NF-κB induces proliferation of mammary epithelial cells, suggesting the involvement of NF-κB in the development of mammary glands [90]. In triple negative breast cancer cells, NF-κB has been shown to exhibit trans-activity as it binds upstream of CD44 promoter and induces tumor cell proliferation [91]. Transformed cells grow continuously, circumventing cell cycle check, and NF-κB signaling indirectly controls cancer cell cycle progression by upregulating the expression of cyclin D, a regulator of G1 checkpoint control [92]. Furthermore, it has been shown that Rac1, a small GTPase, induces lung cancer cell proliferation, possibly through NF-κB activation [93].

Inhibition of Apoptosis

NF-κB transcriptionally regulates the expression of several anti-apoptotic genes such as cIAP1/2, c-FLIP, Bcl-2 Bfl-1/A1 and Bcl-X$_L$ which help in cancer cells survival, circumvention of programmed cell death pathway and mitigation of cytotoxic effects of anticancer therapies [94, 95]. Tumors with constitutive NF-κB activation are highly resistant towards anti-cancer therapies and, in this scenario, inhibition of NF-κB could be helpful in sensitizing cancer cells against therapeutic drugs [95]. Pro-inflammatory cytokine TNF-α activates NF-κB and suppresses apoptosis by upregulating the levels of anti-apoptotic proteins [96]. Additionally, TNF-α also generates ROS which might be lethal for the malignant cells; therefore, NF-κB, as a counter defense mechanism, upregulates the expression of anti-oxidant genes (manganese-superoxide dismutase and ferritin heavy chain) to combat deleterious effect of ROS [96, 97]. Generally, NF-κB is thought to be an anti-apoptotic molecule but there are some evidences which advocate the pro-apoptotic nature of NF-κB signaling pathway in a context dependent manner. In neuroblastoma, activation of NF-κB by doxorubicin exhibits more profound killing of N-type neuroblastoma cells without the activation of caspases [98]. Another *in vitro* study suggested that leukemia cells treated with etoposide activate NF-κB which, subsequently, induces apoptosis [99]. These unusual behaviors of NF-κB suggest that stimuli used in these studies activated alternate signaling pathways thereby inducing apoptosis, with possibly no relevance of activated NF-κB. In any case, the underlying mechanisms of such unusual NF-κB activity remain unclear, and need to be explored further.

Migration and Invasion

Some of the cytokines/chemokines expressed aberrantly in an NF-κB-dependent manner possess chemotactic potential which can accelerate tumor cell migration and angiogenesis. One such NF-κB-dependent cytokine, IL-8, has been associated with angiogenesis [100, 101] and we have shown that IL-8 derived from gemcitabine treated pancreatic cancer cells promotes the growth of endothelial cells with enhanced invasiveness [101]. We postulated that interactions between tumor cells and surrounding endothelial cells through IL-8 might help in building a route for cancer cells to migrate towards growth favorable niche. Another angiogenic molecule, vascular endothelial growth factor (VEGF), found also to be under the control of NF-κB, induces angiogenesis, and inhibition of NF-κB by small molecule inhibitor reduces the expression of VEGF [102]. NF-κB binding sites are present in the promoter regions of matrix metalloproteinases (MMPs) like MMP-2, -3, and -9, which degrade extracellular matrix and help in the tumor cell invasion [103]. Similarly, it has been reported that constitutive activation of NF-κB in pancreatic cancer cells facilitated upregulation of urokinase-type plasminogen activator (uPA) which may be involved in tumor invasion and metastasis [104]. Moreover, we have also observed that gemcitabine-induced ROS upregulates NF-κB in pancreatic cancer cells and these cells exhibit great invasive property [105]. Besides these findings, there are some studies which stated a contradictory role of NF-κB in angiogenesis. Swiatkowska and co-workers reported that TNF-α-mediated ROS generation activated NF-κB, leading to the up-regulation of plasminogen activator inhibitor-1 (PAI-1), a serine protease inhibitor which suppressed the activity of tissue plasminogen activator (tPA), urokinase, and also inhibited VEGF-VEGFR-2 signaling axis [106, 107].

NF-κB as a Target for Cancer Chemoprevention

Chemoprevention is described as the use of pharmacological agents to suppress the growth of cancer, and reverse the process of tumorigenesis in its earliest stages. Genetic mutations and epigenetic alterations play crucial role in the development of malignancies. Therefore, chemoprevention can only be achieved by understanding the underlying molecular mechanisms of carcinogenesis and the mechanism(s) of action of these agents [108]. Broadly, cancer chemopreventive agents can be grouped into four categories such as diet associated agents, hormonal, medications, and vaccines [109]. Phytochemicals derived from our dietary products control the growth of tumor cells both *in vitro* and *in vivo* by exhibiting various modes of action. They may act as antioxidants, pro-oxidants, and anti-inflammatory or antagonize tumor's metabolism [109]. Moreover, chemopreventive agents derived from hormones are more effective against steroid-related cancers like breast and prostate cancer. Anti-inflammatory drugs, like aspirin, also act as chemopreventive agents in gastrointestinal cancer [110, 111]. Another prescribed drug for type II diabetes, metformin, was found to be associated with reduced risk of various cancers [112-114]. Chemopreventive property of cholesterol reducing drugs, statins, is evident in their inhibition of cancer growth, induction of apoptosis and inhibition of angiogenesis [114-116]. Virus or bacterial infection-associated cancers such as hepatocellular carcinoma, cervical cancer and stomach cancer can be inhibited by using vaccines [117, 118].

Cancer chemopreventive response can be achieved by addressing the two most important aspects - specificity of the agents and the therapeutic efficacy [119]. NF-κB signaling pathway is extensively studied in many human malignancies [120] where it is associated with tumor cell proliferation, inflammation, defects in apoptotic pathway, etc. [121]. Inhibition of NF-κB by various natural or chemically synthesized molecules obstructs cancer growth by suppressing tumor cells' proliferation, and reducing the expression of cell survival molecules with enhanced level of apoptosis *in vitro* as well as *in vivo*. Various studies have reported that active ingredients from our diets or from natural sources, such as, epigallocatechin-3-gallate (EGCG), curcumin, resveratrol, honokiol and isoflavones are capable of inhibiting NF-κB [12, 122].

Epigallocatechin-3-gallate is a polyphenol present in green tea with diverse beneficial properties [123]. Epidemiologic data have suggested that intake of green tea decreases cancer incidence [124] and mounting evidence has advocated anti-proliferative property of EGCG in many cancer model systems [125, 126]. Treatments with EGCG suppress NF-κB activation by inhibiting p65 localization or by blocking degradation of IκB-α [127]. EGCG-treated myeloma cells display low levels of p65 mRNA and phosphorylated IκB-α. Further, EGCG, along with proteasome inhibitor, bortezomib synergistically induces apoptosis [128]. EGCG possesses both anti and pro-oxidant behavior; therefore, it can be used for cancer prevention [129].

Another and perhaps one of the most extensively researched phytochemical, curcumin, is used as an anti-proliferative agent against various cancers. Anticancer property of curcumin is attributed to its anti-oxidant role that inhibits free radicals-induced lipid peroxidation and damage to DNA [130]. In recent years, it has been shown that curcumin treatment induces reactive oxygen species (ROS) mediated killing of cancer cells [131, 132]. Curcumin inhibits growth of cutaneous T cell lymphoma by downregulating IKK which leads to the inhibition of NF-κB [132]. Curcumin also inhibits TNF-α-mediated activation of NF-κB in human myeloid cells [133]. Like curcumin, resveratrol also displays anti-tumor activities against many cancers. Resveratrol is a phytoalexin, present in red grapes, peanuts and blueberries [134]. Resveratrol was used in oriental medicines for treating different human diseases like cardiovascular, antiviral neuroprotective and antiaging [135]. Resveratrol serves as anti-inflammatory agent by suppressing cyclooxygenase-2 activity and NF-κB activation. Resveratrol scavenges ROS and induces apoptosis in cancer cells. Moreover, resveratrol also potentiates the cytotoxic effect of other anticancer agents, cytokines, or radiation [135, 136]. Thus, resveratrol has the ability to inhibit or delay the development of many cancers. In recent years, another plant (*Magnolia* spp.) derived phytochemical, honokiol, has garnered interest as a novel therapeutic agent against cancer [137, 138]. Honokiol targets multiple signaling pathways such as NF-κB, m-TOR, STAT3 and EGFR. These pathways are deregulated in cancer and display crucial role in cancer initiation and progression [139, 140]. Recently, it has been shown in a preclinical study that honokiol effectively reverses the effect of 7,12-Dimethylbenz[a]anthracene (DMBA) by inhibiting NF-κB [141]. Moreover, another study by Chilampalli and co-workers reported chemopreventive role of honokiol in UV-induced skin cancer [142], and our own work revealed a suppression of tumor-stromal crosstalk by honokiol in a pancreatic cancer model [143]. These findings suggest that honokiol could be a novel chemopreventive or therapeutic agent. An extensive amount of literature is available to support the NF-κB-modulating action of several chemopreventive agents [144-147], and such action is believed to be crucial for their anticancer properties.

NF-κB as a Target for Cancer Therapy

NF-κB signaling pathway is linked with cancer development, progression, as well as with therapy resistance [12, 148]. Therefore, suppression of NF-κB and or signaling associated with its activation is an attractive molecular target for developing therapies against cancer. Inhibition of NF-κB pathway can be achieved by developing inhibitors against various steps of NF-κB activation pathway, such as IKK inhibitors, proteasome inhibitors, anti-oxidants, and cell permeable peptides [149-151]. Furthermore, the transcriptional activity of NF-κB can be reduced by inhibiting NF-κB binding to DNA, using decoy oligodeoxynucleotides, and inhibitor (glucocorticoids), which block the transactivation of NF-κB or silence NF-κB gene resulting in the blockade of NF-κB signaling pathway [151-154].

IKK Inhibitors

Inhibition of NF-κB activation could be achieved by inhibiting kinase activity of IKK with the specific inhibitors of IKKα and/or IKKβ. Irreversible inhibitors, BAY 11-7082 and BAY 11-7085, block phosphorylation of IκBα, suppress proteasomal degradation of IκBα and inhibit NF-κB translocation to the nucleus. *In vitro* study suggested that BAY 11-7082 treatment blocks NF-κB activation and induces mitochondrial mediated apoptosis. Treatment with BAY 11-7082 downregulated the expression of cyclin A and CDK-2 and induced cell cycle arrest [155]. Another study suggested that BAY 11-7082 induces robust cell death in primary adult leukemic T cells, as compared to normal peripheral blood mononuclear cells (PBMCs), by inhibiting the activation of NF-κB [156]. BAY 11-7085 treatment potentiates the efficacy of cisplatin in ovarian cancer by downregulating the expression of X-linked inhibitor of apoptosis protein (XIAP) and suppressing tumor cell invasiveness [157]. Pham and colleagues suggested that BAY 11-7082 treatment blocked constitutive activation of NF-κB, and induced apoptosis in non-Hodgkin mantle cell lymphoma by downregulating expression of Bcl-X$_L$ and Bfl/A1 [158]. In pre-clinical study, it has been shown that colon cancer cells-bearing mice treated with BAY 11-7085 fail to develop tumor, as compared to vehicle treated mice [159]. Co-administration of BAY11-7085 with histone deacetylase inhibitor, suberoylanilide hydroxamic acid (SAHA), reduces NF-κB in non-small cell lung cancer cells and induces enhanced apoptosis, as compared to single treatment [160].

Proteasome Inhibitors

Intracellular proteins can be degraded by ubiquitin proteasome pathway and it has been found that proteasome inhibitors possess anticancer property. These inhibitors are able to inhibit the degradation of IκB, resulting in cytoplasmic retention of NF-κB. Experimental evidences have suggested that the 26S proteasome inhibitor, bortezomib, exhibits anti-tumor activity against wide variety of cancers [161, 162]. In the year 2008, U.S. FDA approved bortezomib for the treatment of multiple myeloma (MM) patients. Combined treatment of bortezomib and irinotecan has been shown to inhibit NF-κB activation, and induce apoptosis in pancreatic cancer cells [163]. Additionally, it has been found that bortezomib enhances therapeutic efficiency of gemcitabine in the mouse model of pancreatic cancer [164]. Another proteasome inhibitor, MG132, has been shown to block NF-κB activation resulting in diminished resistance against gemcitabine [165]. It has been also shown that MG132, along with anthracycline idarubicin, inhibited constitutive NF-κB and induced apoptosis in leukemic stem cells, without affecting normal hematopoietic stem cells [166]. However, studies have also shown that bortezomib/MG132 treatment could activate NF-κB in MM and endometrial cancer, which may suggest that anti-tumor activity of these inhibitors are not associated just with the inhibition of NF-κB [167].

Protein Phosphatases

Phosphatases, through their ability to counterbalance kinase activity, can be used to inhibit NF-κB activation. Protein phosphatase 2A (PP2A), a serine/threonine phosphatase, blocks the activity of IKKβ. Anti-leukemic drug, cytarabine, inhibits NF-κB by dephosphorylating p65 through PP2A and B [168]. PP2C family member, PPM1A, has been reported to dephosphorylate RelA, and inhibit NF-κB activity and cancer invasiveness [169]. Another member of PP2C family, WIP1, dephosphorylates p65 subunit of NF-κB, thus inhibiting NF-κB activation, and, consequently, the mice lacking WIP1 exhibit enhanced inflammation [170].

Blockage of NF-κB Nuclear Translocation

Inhibition of NF-κB activation can be achieved by using small cell-permeable peptides. One such peptide, SN50, competes with NF-κB for the nuclear translocation, blocking its nuclear translocation. SN50 also induces the differentiation of glioma stem-like cells and reduces tumorigenicity in the mouse model of brain cancer [171]. However, SN50's action is not NF-κB-specific, and it blocks nuclear translocation of other transcription factors as well [172]. In order to address this limitation, either SN50 can be modified to increase specificity or alternate specific inhibitors of NF-κB can be developed. Another inhibitor of NF-κB translocation, dehydroxymethylepoxyquinomicin, has been shown to block nuclear translocation of NF-κB in multiple myeloma cells in an IκB-α-independent manner [173].

Acetylation Inhibition

Post translation modifications (PTMs) such as acetylation and deacetylation play an important role in controlling proteins activity and function [174]. PTMs regulate NF-κB activity either by activating or inhibiting its transcriptional activity. In the nucleus, lysine (K) residues of p65, a subunit of NF-κB, undergo acetylation at multiple sites. The transcriptional activity of NF-κB increases due to the acetylation of lysine residue at positions 221 and 310. Moreover, acetylation at K122 and K123 suppresses DNA binding affinity of p65 which may lead to nuclear export of p65 [175, 176]. In recent years, many compounds derived from natural sources have been shown to suppress NF-κB activity by inhibiting acetylation, and these molecules may possess anti-tumor activity. Gallic acid, derived from oak bark or gallnuts, inhibits NF-κB acetylation and activation resulting in low levels of inflammatory molecules, which forms the basis of its anticancer activity against colon and lung cancer [177, 178].

Apart from the above mentioned methods, a number of other methods have been tested to inhibit NF-κB. For example, the use of decoy oligodeoxynucleotides (ODNs) has been demonstrated to reduce DNA binding capability of NF-κB [179], Kawamura et al. have shown that administration of ODN significantly reduces cachexia in a colon cancer mice model [180]. Moreover, deleterious effects of NF-κB in cancer pathology can be minimized by the use of various anti-oxidants. An advantage of using anti-oxidants is that they can inhibit both inducible and constitutive activation of NF-κB reported in a variety of cancer models [109, 181, 182]. Additionally, small molecules synthesized by microorganisms (fungi, bacteria and virus) also inhibit NF-κB activation and, one such small molecule, gliotoxin, produced by *Aspergillus fumigatus*, prevents IκB degradation [183] and exhibits anti-tumor role in breast cancer [184].

Conclusions and Perspectives

NF-κB affects immune response by regulating the production of pro-inflammatory cytokines and through the recruitment of immune cells. NF-κB signaling pathway is activated to help against pathogenic infections and other stress. The central role of NF-κB signaling in onset as well as progression of cancer is well known. NF-κB is a key transcription factor that functions as a point of convergence for several oncogenic signaling pathways. It has been implicated in the development of resistance to various therapies, as well as in the metastases of primary tumors to distant sites. All this makes NF-κB as an extremely attractive target for anticancer therapy. However, before targeted therapies against NF-κB become a reality, several key challenges need to be addressed. Even though activated NF-κB signaling has been reported in a number of cancer models and clinical samples, it needs to be recognized that activation of NF-κB signaling is not universal. Further, since there are two distinct pathways for NF-κB activation, the benefits of inhibiting one pathway over the other also need to be carefully considered. This calls for a more rigorous genetic screening of cancer patients, and underlines the importance of personalized cancer therapy to identify patients who are mostly like to benefit from anti-NF-κB therapy. This is critical because NF-κB is important for normal physiological functioning. In particular, its role in innate immunity is well known. Any targeted therapy against NF-κB is bound to adversely affect the immune system. Thus, the benefits vs. adverse effects of specific targeting of NF-κB need to be carefully weighed. It is possible that specific targeting of key upstream or downstream factor(s) might circumvent the possibility of immunosuppression, and might be a better approach. However, this needs to be clearly established through detailed mechanistic studies. Despite the challenges of targeting NF-κB in clinics, it is apparent that an overwhelmingly large amount of pre-clinical data is supportive of continued efforts to evolve and refine strategies to target NF-κB with minimum to no adverse effects.

Conflicts of Interest

No potential conflict of interest to disclose.

Acknowledgements

The authors would like to thank Dr. Sanjeev K. Srivastava for reviewing the manuscript. This work is financially supported by NIH/NCI [R01CA175772, and U01CA185490 (to APS)] and USA-MCI.

References

1. Coussens, L.M., Z. Werb. Inflammation and cancer. Nature, 2002. 420: 860-67.
2. Rakoff-Nahoum, S. Why cancer and inflammation? Yale J Biol Med, 2006. 79: 123-30.
3. Landskron, G., M. De la Fuente, P. Thuwajit, C. Thuwajit, M.A. Hermoso. Chronic inflammation and cytokines in the tumor microenvironment. J Immunol Res, 2014. 149185.
4. Lawrence, T. The nuclear factor NF-kappaB pathway in inflammation. Cold Spring Harb Perspect Biol, 2009. 1: a001651.
5. Karin, M. Nuclear factor-kappaB in cancer development and progression. Nature, 2006. 441: 431-36.
6. Sen, R., D. Baltimore. Inducibility of kappa immunoglobulin enhancer-binding protein NF-kappa B by a posttranslational mechanism. Cell, 1986. 47: 921-28.
7. Aggarwal, B.B. Nuclear factor-kappaB: the enemy within. Cancer Cell, 2004. 6: 203-08.
8. Zhang, Q., M.J. Lenardo, D. Baltimore. 30 years of NF-kappaB: a blossoming of relevance to human pathobiology. Cell, 2017. 168: 37-57.
9. Sun, S.C. The non-canonical NF-kappaB pathway in immunity and inflammation. Nat Rev Immunol, 2017. 17: 545-558
10. Pahl, H.L. Activators and target genes of Rel/NF-kappaB transcription factors. Oncogene, 1999. 18: 6853-66.
11. Hayden, M.S., S. Ghosh. NF-kappaB in immunobiology. Cell Res, 2011. 21: 223-44.
12. Baud, V., M. Karin. Is NF-kappaB a good target for cancer therapy? Hopes and pitfalls. Nat Rev Drug Discov, 2009. 8: 33-40.
13. Balkwill, F., A. Mantovani. Inflammation and cancer: back to Virchow? Lancet, 2001. 357: 539-45.
14. Kundu, J.K., Y.J. Surh. Inflammation: gearing the journey to cancer. Mutat Res, 2008. 659: 15-30.
15. Algul, H., M. Treiber, M. Lesina, R.M. Schmid. Mechanisms of disease: chronic inflammation and cancer in the pancreas—a potential role for pancreatic stellate cells. Nat Clin Pract Gastroenterol Hepatol, 2007. 4: 454-62.
16. Ardies, C.M. Inflammation as cause for scar cancers of the lung. Integr Cancer Ther, 2003. 2: 238-46.
17. Karampitsakos, T., V. Tzilas, R. Tringidou, P. Steiropoulos, V. Aidinis, S.A. Papiris, D. Bouros, A. Tzouvelekis. Lung cancer in patients with idiopathic pulmonary fibrosis. Pulm Pharmacol Ther, 2017. 45: 1-10.

18. Michaud, D.S. Chronic inflammation and bladder cancer. Urol Oncol, 2007. 25: 260-68.
19. Bianchi, C., T. Bianchi. Malignant mesothelioma: global incidence and relationship with asbestos. Ind Health, 2007. 45: 379-87.
20. Arif, S., A. Blanes, S.J. Diaz-Cano. Hashimoto's thyroiditis shares features with early papillary thyroid carcinoma. Histopathology, 2002. 41: 357-62.
21. Elkahwaji, J.E., W. Zhong, W.J. Hopkins, W. Bushman. Chronic bacterial infection and inflammation incite reactive hyperplasia in a mouse model of chronic prostatitis. Prostate, 2007. 67: 14-21.
22. Khatoon, J., R.P. Rai, K.N. Prasad. Role of Helicobacter pylori in gastric cancer: updates. World J Gastrointest Oncol, 2016. 8: 147-58.
23. Xing, T., R. Camacho Salazar, Y.H. Chen. Animal models for studying epithelial barriers in neonatal necrotizing enterocolitis, inflammatory bowel disease and colorectal cancer. Tissue Barriers, 2017. e1356901.
24. Douglas, J.L., J.K. Gustin, B. Dezube, J.L. Pantanowitz, A.V. Moses. Kaposi's sarcoma: a model of both malignancy and chronic inflammation. Panminerva Med, 2007. 49: 119-38.
25. Schottenfeld, D., J. Beebe-Dimmer. Chronic inflammation: a common and important factor in the pathogenesis of neoplasia. CA Cancer J Clin, 2006. 56: 69-83.
26. Polk, D.B., R.M. Peek, Jr. Helicobacter pylori: gastric cancer and beyond. Nat Rev Cancer, 2010. 10: 403-14.
27. Jurjus, A., A. Eid, S. Al Kattar, M.N. Zeenny, A. Gerges-Geagea, H. Haydar, et al. Inflammatory bowel disease, colorectal cancer and type 2 diabetes mellitus: the links. BBA Clin, 2016. 5: 16-24.
28. Colotta, F., P. Allavena, A. Sica, C. Garlanda, A. Mantovani. Cancer-related inflammation, the seventh hallmark of cancer: links to genetic instability. Carcinogenesis, 2009. 30: 1073-81.
29. Takai, A., T. Toyoshima, M. Uemura, Y. Kitawaki, H. Marusawa, H. Hiai, et al. A novel mouse model of hepatocarcinogenesis triggered by AID causing deleterious p53 mutations. Oncogene, 2009. 28: 469-78.
30. Mishra, A., S. Liu, G.H. Sams, D.P. Curphey, R. Santhanam, L.J. Rush, et al. Aberrant overexpression of IL-15 initiates large granular lymphocyte leukemia through chromosomal instability and DNA hypermethylation. Cancer Cell, 2012. 22: 645-55.
31. Elinav, E., R. Nowarski, C.A. Thaiss, B. Hu, C. Jin, R.A. Flavell. Inflammation-induced cancer: crosstalk between tumours, immune cells and microorganisms. Nat Rev Cancer, 2013. 13: 759-71.
32. Schetter, A.J., N.H. Heegaard, C.C. Harris. Inflammation and cancer: interweaving microRNA, free radical, cytokine and p53 pathways. Carcinogenesis, 2010. 31: 37-49.
33. Grivennikov, S., E. Karin, J. Terzic, D. Mucida, G.Y. Yu, S. Vallabhapurapu, et al. IL-6 and Stat3 are required for survival of intestinal epithelial cells and development of colitis-associated cancer. Cancer Cell, 2009. 15: 103-13.
34. Pikarsky, E., R.M. Porat, I. Stein, R. Abramovitch, S. Amit, S. Kasem, et al. NF-kappaB functions as a tumour promoter in inflammation-associated cancer. Nature, 2004. 431: 461-66.
35. Greten, F.R., L. Eckmann, T.F. Greten, J.M. Park, Z.W. Li, L.J. Egan, et al. IKKbeta links inflammation and tumorigenesis in a mouse model of colitis-associated cancer. Cell, 2004. 118: 285-96.

36. Voronov, E., D.S. Shouval, Y. Krelin, E. Cagnano, D. Benharroch, Y. Iwakura, et al. IL-1 is required for tumor invasiveness and angiogenesis. Proc Natl Acad Sci USA, 2003. 100: 2645-50.
37. Bates, R.C., A.M. Mercurio. Tumor necrosis factor-alpha stimulates the epithelial-to-mesenchymal transition of human colonic organoids. Mol Biol Cell, 2003. 14: 1790-1800.
38. Wu, Y., J. Deng, P.G. Rychahou, S. Qiu, B.M. Evers, B.P. Zhou. Stabilization of snail by NF-kappaB is required for inflammation-induced cell migration and invasion. Cancer Cell, 2009. 15: 416-28.
39. Gocheva, V., H.W. Wang, B.B. Gadea, T. Shree, K.E. Hunter, A.L. Garfall, et al. IL-4 induces cathepsin protease activity in tumor-associated macrophages to promote cancer growth and invasion. Genes Dev, 2010. 24: 241-55.
40. Kitamura, T., K. Kometani, H. Hashida, A. Matsunaga, H. Miyoshi, H. Hosogi, et al. SMAD4-deficient intestinal tumors recruit CCR1+ myeloid cells that promote invasion. Nat Genet, 2007. 39: 467-75.
41. Qian, B.Z., J. Li, H. Zhang, T. Kitamura, J. Zhang, L.R. Campion, et al. CCL2 recruits inflammatory monocytes to facilitate breast-tumour metastasis. Nature, 2011. 475: 222-25.
42. Sun, S.C., S.C. Ley. New insights into NF-kappaB regulation and function. Trends Immunol, 2008. 29: 469-78.
43. M.S. Hayden, S. Ghosh. Shared principles in NF-kappaB signaling. Cell, 2008. 132: 344-62.
44. Ghosh, S., M.S. Hayden. New regulators of NF-kappaB in inflammation. Nat Rev Immunol, 2008. 8: 837-48.
45. Ruland, J. Return to homeostasis: downregulation of NF-kappaB responses. Nat Immunol, 2011. 12: 709-14.
46. Mahoney, D.J., H.H. Cheung, R.L. Mrad, S. Plenchette, C. Simard, E. Enwere, et al. Both cIAP1 and cIAP2 regulate TNFalpha-mediated NF-kappaB activation. Proc Natl Acad Sci USA, 2008. 105: 11778-83.
47. Ghosh, S., M. Karin. Missing pieces in the NF-kappaB puzzle. Cell, 2002. 109(Suppl): S81-96.
48. Bonizzi, G., M. Karin. The two NF-kappaB activation pathways and their role in innate and adaptive immunity. Trends Immunol, 2004. 25: 280-88.
49. Oeckinghaus, A., M.S. Hayden, S. Ghosh. Crosstalk in NF-kappaB signaling pathways. Nat Immunol, 2011. 12: 695-708.
50. Mitchell, S., J. Vargas, A. Hoffmann. Signaling via the NFkappaB system. Wiley Interdiscip Rev Syst Biol Med, 2016. 8: 227-41.
51. Wertz, I.E., V.M. Dixit. Signaling to NF-kappaB: regulation by ubiquitination. Cold Spring Harb Perspect Biol, 2010. 2: a003350.
52. Compagno, M., W.K. Lim, A. Grunn, S.V. Nandula, M. Brahmachary, Q. Shen, et al. Mutations of multiple genes cause deregulation of NF-kappaB in diffuse large B-cell lymphoma. Nature, 2009. 459: 717-21.
53. Kato, M., M. Sanada, I. Kato, Y. Sato, J. Takita, K. Takeuchi, et al. Frequent inactivation of A20 through gene mutation in B-cell lymphomas. Rinsho Ketsueki, 2011. 52: 313-19.
54. Enesa, K., M. Zakkar, H. Chaudhury, A. Luong le, L. Rawlinson, J.C. Mason, et al. NF-kappaB suppression by the deubiquitinating enzyme Cezanne: a novel negative feedback loop in pro-inflammatory signaling. J Biol Chem, 2008. 283: 7036-45.
55. Kovalenko, A., C. Chable-Bessia, G. Cantarella, A. Israel, D. Wallach, G.

Courtois. The tumour suppressor CYLD negatively regulates NF-kappaB signalling by deubiquitination. Nature, 2003. 424: 801-5.
56. Cildir, G., K.C. Low, V. Tergaonkar. Noncanonical NF-kappaB signaling in health and disease. Trends Mol Med, 2016. 22: 414-29.
57. Coope, H.J., P.G. Atkinson, B. Huhse, M. Belich, J. Janzen, M.J. Holman, et al. CD40 regulates the processing of NF-kappaB2 p100 to p52. EMBO J, 2002. 21: 5375-85.
58. Claudio, E., K. Brown, S. Park, H. Wang, U. Siebenlist. BAFF-induced NEMO-independent processing of NF-kappaB2 in maturing B cells. Nat Immunol, 2002. 3: 958-65.
59. Ramakrishnan, P., W. Wang, D. Wallach. Receptor-specific signaling for both the alternative and the canonical NF-kappaB activation pathways by NF-kappaB-inducing kinase. Immunity, 2004. 21: 477-89.
60. Nonaka, M., R. Horie, K. Itoh, T. Watanabe, N. Yamamoto, S. Yamaoka. Aberrant NF-kappaB2/p52 expression in Hodgkin/Reed-Sternberg cells and CD30-transformed rat fibroblasts. Oncogene, 2005. 24: 3976-86.
61. Murray, S.E., F. Polesso, A.M. Rowe, S. Basak, Y. Koguchi, K.G. Toren, et al. NF-kappaB-inducing kinase plays an essential T cell-intrinsic role in graft-versus-host disease and lethal autoimmunity in mice. J Clin Invest, 2011. 121: 4775-86.
62. Novack, D.V., L. Yin, A. Hagen-Stapleton, R.D. Schreiber, D.V. Goeddel, F.P. Ross, et al. The IkappaB function of NF-kappaB2 p100 controls stimulated osteoclastogenesis. J Exp Med, 2003. 198: 771-81.
63. Dejardin, E., N.M. Droin, M. Delhase, E. Haas, Y. Cao, C. Makris, et al. The lymphotoxin-beta receptor induces different patterns of gene expression via two NF-kappaB pathways. Immunity, 2002. 17: 525-35.
64. Saitoh, T., M. Nakayama, H. Nakano, H. Yagita, N. Yamamoto, S. Yamaoka. TWEAK induces NF-kappaB2 p100 processing and long lasting NF-kappaB activation. J Biol Chem, 2003. 278: 36005-12.
65. Jin, J., H. Hu, H.S. Li, J. Yu, Y. Xiao, G.C. Brittain, et al. Noncanonical NF-kappaB pathway controls the production of type I interferons in antiviral innate immunity. Immunity, 2014. 40: 342-54.
66. Ruckle, A., E. Haasbach, I. Julkunen, O. Planz, C. Ehrhardt, S. Ludwig. The NS1 protein of influenza A virus blocks RIG-I-mediated activation of the noncanonical NF-kappaB pathway and p52/RelB-dependent gene expression in lung epithelial cells. J Virol, 2012. 86: 10211-17.
67. Matta, H., P.M. Chaudhary. Activation of alternative NF-kappaB pathway by human herpes virus 8-encoded Fas-associated death domain-like IL-1 beta-converting enzyme inhibitory protein (vFLIP). Proc Natl Acad Sci USA, 2004. 101: 9399-9404.
68. Xiao, G., M.E. Cvijic, A. Fong, E.W. Harhaj, M.T. Uhlik, M. Waterfield, et al. Retroviral oncoprotein tax induces processing of NF-kappaB2/p100 in T cells: evidence for the involvement of IKKalpha. EMBO J, 2001. 20: 6805-15.
69. Ge, J., H. Xu, T. Li, Y. Zhou, Z. Zhang, S. Li, et al. A Legionella type IV effector activates the NF-kappaB pathway by phosphorylating the IkappaB family of inhibitors. Proc Natl Acad Sci USA, 2009. 106: 13725-30.
70. Luftig, M., T. Yasui, V. Soni, M.S. Kang, N. Jacobson, E. Cahir-McFarland, et al. Epstein-Barr virus latent infection membrane protein 1 TRAF-binding site induces NIK/IKK alpha-dependent noncanonical NF-kappaB activation. Proc Natl Acad Sci USA, 2004. 101: 141-46.
71. Sun, S.C. The noncanonical NF-kappaB pathway. Immunol Rev, 2012. 246: 125-40.

72. Vallabhapurapu, S., A. Matsuzawa, W. Zhang, P.H. Tseng, J.J. Keats, H. Wang, et al. Nonredundant and complementary functions of TRAF2 and TRAF3 in a ubiquitination cascade that activates NIK-dependent alternative NF-kappaB signaling. Nat Immunol, 2008. 9: 1364-70.
73. Zarnegar, B.J., Y. Wang, D.J. Mahoney, P.W. Dempsey, H.H. Cheung, J. He, et al. Noncanonical NF-kappaB activation requires coordinated assembly of a regulatory complex of the adaptors cIAP1, cIAP2, TRAF2 and TRAF3 and the kinase NIK. Nat Immunol, 2008. 9: 1371-78.
74. Razani, B., B. Zarnegar, A.J. Ytterberg, T. Shiba, P.W. Dempsey, C.F. Ware, et al. Negative feedback in noncanonical NF-kappaB signaling modulates NIK stability through IKKalpha-mediated phosphorylation. Sci Signal, 2010. 3: ra41.
75. Jin, J., Y. Xiao, J.H. Chang, J. Yu, H. Hu, R. Starr, et al. The kinase TBK1 controls IgA class switching by negatively regulating noncanonical NF-kappaB signaling. Nat Immunol, 2012. 13: 1101-09.
76. Hu, H., G.C. Brittain, J.H. Chang, N. Puebla-Osorio, J. Jin, A. Zal, et al. OTUD7B controls non-canonical NF-kappaB activation through deubiquitination of TRAF3. Nature, 2013. 494: 371-74.
77. Gray, C.M., C. Remouchamps, K.A. McCorkell, L.A. Solt, E. Dejardin, J.S. Orange, et al. Noncanonical NF-kappaB signaling is limited by classical NF-kappaB activity. Sci Signal, 2014. 7: ra13.
78. Guo, H., J.B. Callaway, J.P. Ting. Inflammasomes: mechanism of action, role in disease, and therapeutics. Nat Med, 2015. 21: 677-87.
79. Latz, E., T.S. Xiao, A. Stutz. Activation and regulation of the inflammasomes. Nat Rev Immunol, 2013. 13: 397-411.
80. Boaru, S.G., E. Borkham-Kamphorst, E. Van de Leur, E. Lehnen, C. Liedtke, R. Weiskirchen. NLRP3 inflammasome expression is driven by NF-kappaB in cultured hepatocytes. Biochem Biophys Res Commun, 2015. 458: 700-06.
81. Zhong, Z., A. Umemura, E. Sanchez-Lopez, S. Liang, S. Shalapour, J. Wong, et al. NF-kappaB restricts inflammasome activation via elimination of damaged mitochondria. Cell, 2016. 164: 896-910.
82. Gambhir, S., D. Vyas, M. Hollis, A. Aekka, A. Vyas. Nuclear factor kappa B role in inflammation associated gastrointestinal malignancies. World J Gastroenterol, 2015. 21: 3174-83.
83. Wang, N., H. Liang, K. Zen. Molecular mechanisms that influence the macrophage m1-m2 polarization balance. Front Immunol, 2014. 5: 614.
84. Porta, C., M. Rimoldi, G. Raes, L. Brys, P. Ghezzi, D. Di Liberto, et al. Tolerance and M2 (alternative) macrophage polarization are related processes orchestrated by p50 nuclear factor kappaB. Proc Natl Acad Sci USA, 2009. 106: 14978-83.
85. Aronica, M.A., A.L. Mora, D.B. Mitchell, P.W. Finn, J.E. Johnson, J.R. Sheller, et al. Preferential role for NF-kappaB/Rel signaling in the type 1 but not type 2 T cell-dependent immune response in vivo. J Immunol, 1999. 163: 5116-24.
86. Park, S.H., G. Cho, S.G. Park. NF-kappaB Activation in T Helper 17 Cell Differentiation. Immune Netw, 2014. 14: 14-20.
87. Baker, R.G., M.S. Hayden, S. Ghosh. NF-kappaB, inflammation, and metabolic disease. Cell Metab, 2011. 13: 11-22.
88. Rocha, V.Z., P. Libby. Obesity, inflammation, and atherosclerosis. Nat Rev Cardiol, 2009. 6: 399-409.
89. Chaturvedi, M.M., B. Sung, V.R. Yadav, R. Kannappan, B.B. Aggarwal. NF-kappaB addiction and its role in cancer: 'one size does not fit all'. Oncogene, 2011. 30: 1615-30.

90. Brantley, D.M., C.L. Chen, R.S. Muraoka, P.B. Bushdid, J.L. Bradberry, F. Kittrell, et al. Nuclear factor-kappaB (NF-kappaB) regulates proliferation and branching in mouse mammary epithelium. Mol Biol Cell, 2001. 12: 1445-55.
91. Smith, S.M., Y.L. Lyu, L. Cai. NF-kappaB affects proliferation and invasiveness of breast cancer cells by regulating CD44 expression. PLoS One, 2014. 9: e106966.
92. Hinz, M., D. Krappmann, A. Eichten, A. Heder, C. Scheidereit, M. Strauss. NF-kappaB function in growth control: regulation of cyclin D1 expression and G0/G1-to-S-phase transition. Mol Cell Biol, 1999. 19: 2690-98.
93. Gastonguay, A., T. Berg, A.D. Hauser, N. Schuld, E. Lorimer, C.L. Williams. The role of Rac1 in the regulation of NF-kappaB activity, cell proliferation, and cell migration in non-small cell lung carcinoma. Cancer Biol Ther, 2012. 13: 647-56.
94. Karin, M., Y. Cao, F.R. Greten, Z.W. Li. NF-kappaB in cancer: from innocent bystander to major culprit. Nat Rev Cancer, 2002. 2: 301-10.
95. Wang, C.Y., J.C. Cusack, Jr., R. Liu, A.S. Baldwin, Jr. Control of inducible chemoresistance: enhanced anti-tumor therapy through increased apoptosis by inhibition of NF-kappaB. Nat Med, 1999. 5: 412-17.
96. Nakano, H., A. Nakajima, S. Sakon-Komazawa, J.H. Piao, X. Xue, K. Okumura. Reactive oxygen species mediate crosstalk between NF-kappaB and JNK. Cell Death Differ, 2006. 13: 730-37.
97. Kamata, H., S. Honda, S. Maeda, L. Chang, H. Hirata, M. Karin. Reactive oxygen species promote TNFalpha-induced death and sustained JNK activation by inhibiting MAP kinase phosphatases. Cell, 2005. 120: 649-61.
98. Bian, X., L.M. McAllister-Lucas, F. Shao, K.R. Schumacher, Z. Feng, A.G. Porter, et al. NF-kappa B activation mediates doxorubicin-induced cell death in N-type neuroblastoma cells. J Biol Chem, 2001. 276: 48921-29.
99. Bessho, R., K. Matsubara, M. Kubota, K. Kuwakado, H. Hirota, Y. Wakazono, et al. Pyrrolidine dithiocarbamate, a potent inhibitor of nuclear factor kappa B (NF-kappa B) activation, prevents apoptosis in human promyelocytic leukemia HL-60 cells and thymocytes. Biochem Pharmacol, 1994. 48: 1883-89.
100. Koch, A.E., P.J. Polverini, S.L. Kunkel, L.A. Harlow, L.A. DiPietro, V.M. Elner, et al. Interleukin-8 as a macrophage-derived mediator of angiogenesis. Science, 1992. 258: 1798-1801.
101. Khan, M.A., S.K. Srivastava, A. Bhardwaj, S. Singh, S. Arora, H. Zubair, et al. Gemcitabine triggers angiogenesis-promoting molecular signals in pancreatic cancer cells: therapeutic implications. Oncotarget, 2015. 6: 39140-50.
102. Xu, C., G. Shen, C. Chen, C. Gelinas, A.N. Kong. Suppression of NF-kappaB and NF-kappaB-regulated gene expression by sulforaphane and PEITC through IkappaBalpha, IKK pathway in human prostate cancer PC-3 cells. Oncogene, 2005. 24: 4486-95.
103. Bond, M., R.P. Fabunmi, A.H. Baker, A.C. Newby. Synergistic upregulation of metalloproteinase-9 by growth factors and inflammatory cytokines: an absolute requirement for transcription factor NF-kappaB. FEBS Lett, 1998. 435: 29-34.
104. Wang, W., J.L. Abbruzzese, D.B. Evans, P.J. Chiao. Overexpression of urokinase-type plasminogen activator in pancreatic adenocarcinoma is regulated by constitutively activated RelA. Oncogene, 1999. 18: 4554-63.
105. Arora, S., A. Bhardwaj, S. Singh, S.K. Srivastava, S. McClellan, C.S. Nirodi, et al. An undesired effect of chemotherapy: gemcitabine promotes pancreatic cancer cell invasiveness through reactive oxygen species-dependent, nuclear

factor kappaB- and hypoxia-inducible factor 1alpha-mediated up-regulation of CXCR4. J Biol Chem, 2013. 288: 21197-22207.
106. Swiatkowska, M., J. Szemraj, C.S. Cierniewski. Induction of PAI-1 expression by tumor necrosis factor alpha in endothelial cells is mediated by its responsive element located in the 4G/5G site. FEBS J, 2005. 272: 5821-31.
107. Wu, J., T.L. Strawn, M. Luo, L. Wang, R. Li, M. Ren, et al. Plasminogen activator inhibitor-1 inhibits angiogenic signaling by uncoupling vascular endothelial growth factor receptor-2-alphaVbeta3 integrin cross talk. Arterioscler Thromb Vasc Biol, 2015. 35: 111-20.
108. Sporn, M.B., N. Suh. Chemoprevention of cancer. Carcinogenesis, 2000. 21: 525-30.
109. Zubair, H., S. Azim, A. Ahmad, M.A. Khan, G.K. Patel, S. Singh, et al. Cancer chemoprevention by phytochemicals: nature's healing touch. Molecules, 2017. 22: E395.
110. Thorat, M.A., J. Cuzick. Role of aspirin in cancer prevention. Curr Oncol Rep, 2013. 15: 533-40.
111. Drew, D.A., Y. Cao, A.T. Chan. Aspirin and colorectal cancer: the promise of precision chemoprevention. Nat Rev Cancer, 2016. 16: 173-86.
112. Higurashi, T., K. Hosono, H. Takahashi, Y. Komiya, S. Umezawa, E. Sakai, et al. Metformin for chemoprevention of metachronous colorectal adenoma or polyps in post-polypectomy patients without diabetes: a multicentre double-blind, placebo-controlled, randomised phase 3 trial. Lancet Oncol, 2016. 17: 475-83.
113. Evans, J.M., L.A. Donnelly, A.M. Emslie-Smith, D.R. Alessi, A.D. Morris. Metformin and reduced risk of cancer in diabetic patients. BMJ, 2005. 330: 1304-05.
114. Gronich, N., G. Rennert. Beyond aspirin-cancer prevention with statins, metformin and bisphosphonates. Nat Rev Clin Oncol, 2013. 10: 625-42.
115. Chan, K.K., A.M. Oza, L.L. Siu. The statins as anticancer agents. Clin Cancer Res, 2003. 9: 10-19.
116. Nielsen, S.F., B.G. Nordestgaard, S.E. Bojesen. Statin use and reduced cancer-related mortality. N Engl J Med, 2012. 367: 1792-1802.
117. De Flora, S., P. Bonanni. The prevention of infection-associated cancers. Carcinogenesis, 2011. 32: 787-95.
118. Guo, C., M.H. Manjili, J.R. Subjeck, D. Sarkar, P.B. Fisher, X.Y. Wang. Therapeutic cancer vaccines: past, present, and future. Adv Cancer Res, 2013. 119: 421-75.
119. Sporn, M.B., N. Suh. Chemoprevention: an essential approach to controlling cancer. Nat Rev Cancer, 2002. 2: 537-43.
120. Darnell, Jr., J.E. Transcription factors as targets for cancer therapy. Nat Rev Cancer, 2002. 2: 740-49.
121. Ghosh, S., M.S. Hayden. Celebrating 25 years of NF-kappaB research. Immunol Rev, 2012. 246: 5-13.
122. Sarkar, F.H., Y. Li. NF-kappaB: a potential target for cancer chemoprevention and therapy. Front Biosci, 2008. 13: 2950-59.
123. Zhong, L., J. Hu, W. Shu, B. Gao, S. Xiong. Epigallocatechin-3-gallate opposes HBV-induced incomplete autophagy by enhancing lysosomal acidification, which is unfavorable for HBV replication. Cell Death Dis, 2015. 6: e1770.
124. Johnson, J.J., H.H. Bailey, H. Mukhtar. Green tea polyphenols for prostate cancer chemoprevention: a translational perspective. Phytomedicine, 2010. 17: 3-13.

125. Shankar, S., S. Ganapathy, S.R. Hingorani, R.K. Srivastava. EGCG inhibits growth, invasion, angiogenesis and metastasis of pancreatic cancer. Front Biosci, 2008. 13: 440-52.
126. Sanna, V., C.K. Singh, R. Jashari, V.M. Adhami, J.C. Chamcheu, I. Rady, et al. Targeted nanoparticles encapsulating (-)-epigallocatechin-3-gallate for prostate cancer prevention and therapy. Sci Rep, 2017. 7: 41573.
127. Li, Y.J., S.L. Wu, S.M. Lu, F. Chen, Y. Guo, S.M. Gan, et al. (-)-Epigallocatechin-3-gallate inhibits nasopharyngeal cancer stem cell self-renewal and migration and reverses the epithelial-mesenchymal transition via NF-kappaB p65 inactivation. Tumour Biol, 2015. 36: 2747-61.
128. Wang, Q., J. Li, J. Gu, B. Huang, Y. Zhao, D. Zheng, et al. Potentiation of (-)-epigallocatechin-3-gallate-induced apoptosis by bortezomib in multiple myeloma cells. Acta Biochim Biophys Sin (Shanghai), 2009. 41: 1018-26.
129. Tao, L., S.C. Forester, J.D. Lambert. The role of the mitochondrial oxidative stress in the cytotoxic effects of the green tea catechin, (-)-epigallocatechin-3-gallate, in oral cells. Mol Nutr Food Res, 2014. 58: 665-76.
130. Olivera, A., T.W. Moore, F. Hu, A.P. Brown, A. Sun, D.C. Liotta, et al. Inhibition of the NF-kappaB signaling pathway by the curcumin analog, 3,5-Bis(2-pyridinylmethylidene)-4-piperidone (EF31): anti-inflammatory and anti-cancer properties. Int Immunopharmacol, 2012. 12: 368-77.
131. Kang, J., J. Chen, Y. Shi, J. Jia, Y. Zhang. Curcumin-induced histone hypoacetylation: the role of reactive oxygen species. Biochem Pharmacol, 2005. 69: 1205-13.
132. Khan, M.A., S. Gahlot, S. Majumdar. Oxidative stress induced by curcumin promotes the death of cutaneous T-cell lymphoma (HuT-78) by disrupting the function of several molecular targets. Mol Cancer Ther, 2012. 11: 1873-83.
133. Singh, S., B.B. Aggarwal. Activation of transcription factor NF-kappaB is suppressed by curcumin (diferuloylmethane) corrected. J Biol Chem, 1995. 270: 24995-25000.
134. Ahmad, A., S. Farhan Asad, S. Singh, S.M. Hadi. DNA breakage by resveratrol and Cu(II): reaction mechanism and bacteriophage inactivation. Cancer Lett, 2000. 154: 29-37.
135. Shankar, S., G. Singh, R.K. Srivastava. Chemoprevention by resveratrol: molecular mechanisms and therapeutic potential. Front Biosci, 2007. 12: 4839-54.
136. Bishayee, A., T. Politis, A.S. Darvesh. Resveratrol in the chemoprevention and treatment of hepatocellular carcinoma. Cancer Treat Rev, 2010. 36: 43-53.
137. Prasad, R., T. Singh, S.K. Katiyar. Honokiol inhibits ultraviolet radiation-induced immunosuppression through inhibition of ultraviolet-induced inflammation and DNA hypermethylation in mouse skin. Sci Rep, 2017. 7: 1657.
138. Averett, C., S. Arora, H. Zubair, S. Singh, A. Bhardwaj, A.P. Singh. Molecular targets of honokiol: a promising phytochemical for effective cancer management. Enzymes, 2014. 36: 175-93.
139. Arora, S., S. Singh, G.A. Piazza, C.M. Contreras, J. Panyam, A.P. Singh. Honokiol: a novel natural agent for cancer prevention and therapy. Curr Mol Med, 2012. 12: 1244-52.
140. Arora, S., A. Bhardwaj, S.K. Srivastava, S. Singh, S. McClellan, B. Wang, et al. Honokiol arrests cell cycle, induces apoptosis, and potentiates the cytotoxic effect of gemcitabine in human pancreatic cancer cells. PLoS One, 2011. 6: e21573.

141. Wang, Z., X. Zhang. Chemopreventive activity of honokiol against 7,12-dimethylbenz[a]anthracene-induced mammary cancer in female Sprague Dawley rats. Front Pharmacol, 2017. 8: 320.
142. Chilampalli, S., X. Zhang, H. Fahmy, R.S. Kaushik, D. Zeman, M.B. Hildreth, et al. Chemopreventive effects of honokiol on UVB-induced skin cancer development. Anticancer Res, 2010. 30: 777-83.
143. Averett, C., A. Bhardwaj, S. Arora, S.K. Srivastava, M.A. Khan, A. Ahmad, et al. Honokiol suppresses pancreatic tumor growth, metastasis and desmoplasia by interfering with tumor-stromal cross-talk. Carcinogenesis, 2016. 37: 1052-61.
144. Gerhauser, C. Cancer chemoprevention and nutriepigenetics: state of the art and future challenges. Top Curr Chem, 2013. 329: 73-132.
145. Luqman, S., J.M. Pezzuto. NFkappaB: a promising target for natural products in cancer chemoprevention. Phytother Res, 2010. 24: 949-63.
146. Ahmad, A., K.R. Ginnebaugh, Y. Li, S.B. Padhye, F.H. Sarkar. Molecular targets of naturopathy in cancer research: bridge to modern medicine. Nutrients, 2015. 7: 321-34.
147. Sung, B., S. Prasad, V.R. Yadav, B.B. Aggarwal. Cancer cell signaling pathways targeted by spice-derived nutraceuticals. Nutr Cancer, 2012. 64: 173-97.
148. Melisi, D., P.J. Chiao. NF-kappaB as a target for cancer therapy. Expert Opin Ther Targets, 2007. 11: 133-44.
149. Schreck, R., B. Meier, D.N. Mannel, W. Droge, P.A. Baeuerle. Dithiocarbamates as potent inhibitors of nuclear factor kappa B activation in intact cells. J Exp Med, 1992. 175: 1181-94.
150. Cho, S., Y. Urata, T. Iida, S. Goto, M. Yamaguchi, K. Sumikawa, et al. Glutathione downregulates the phosphorylation of I kappa B: autoloop regulation of the NF-kappaB-mediated expression of NF-kappaB subunits by TNF-alpha in mouse vascular endothelial cells. Biochem Biophys Res Commun, 1998. 253: 104-08.
151. Lee, C.H., Y.T. Jeon, S.H. Kim, Y.S. Song. NF-kappaB as a potential molecular target for cancer therapy. Biofactors, 2007. 29: 19-35.
152. D'Acquisto, F., A. Ialenti, A. Ianaro, R. Di Vaio, R. Carnuccio. Local administration of transcription factor decoy oligonucleotides to nuclear factor-kappaB prevents carrageenin-induced inflammation in rat hind paw. Gene Ther, 2000. 7: 1731-37.
153. Higgins, K.A., J.R. Perez, T.A. Coleman, K. Dorshkind, W.A. McComas, U.M. Sarmiento, et al. Antisense inhibition of the p65 subunit of NF-kappaB blocks tumorigenicity and causes tumor regression. Proc Natl Acad Sci USA, 1993. 90: 9901-05.
154. McKay, L.I., J.A. Cidlowski. Cross-talk between nuclear factor-kappaB and the steroid hormone receptors: mechanisms of mutual antagonism. Mol Endocrinol, 1998. 12: 45-56.
155. Chen, L., Y. Ruan, X. Wang, L. Min, Z. Shen, Y. Sun, et al. Bay 11-7082, a nuclear factor-kappaB inhibitor, induces apoptosis and S phase arrest in gastric cancer cells. J Gastroenterol, 2014. 49: 864-74.
156. Mori, N., Y. Yamada, S. Ikeda, Y. Yamasaki, K. Tsukasaki, Y. Tanaka, et al. Bay 11-7082 inhibits transcription factor NF-kappaB and induces apoptosis of HTLV-I-infected T-cell lines and primary adult T-cell leukemia cells. Blood, 2002. 100: 1828-34.
157. Mabuchi, S., M. Ohmichi, Y. Nishio, T. Hayasaka, A. Kimura, T. Ohta, et al. Inhibition of NFkappaB increases the efficacy of cisplatin in in vitro and in vivo ovarian cancer models. J Biol Chem, 2004. 279: 23477-85.

158. Pham, L.V., A.T. Tamayo, L.C. Yoshimura, P. Lo, R.J. Ford. Inhibition of constitutive NF-kappaB activation in mantle cell lymphoma B cells leads to induction of cell cycle arrest and apoptosis. J Immunol, 2003. 171: 88-95.
159. Scaife, C.L., J. Kuang, J.C. Wills, D.B. Trowbridge, P. Gray, B.M. Manning, et al. Nuclear factor kappaB inhibitors induce adhesion-dependent colon cancer apoptosis: implications for metastasis. Cancer Res, 2002. 62: 6870-78.
160. Rundall, B.K., C.E. Denlinger, D.R. Jones. Combined histone deacetylase and NF-kappaB inhibition sensitizes non-small cell lung cancer to cell death. Surgery, 2004. 136: 416-25.
161. Adams, J., V.J. Palombella, E.A. Sausville, J. Johnson, A. Destree, D.D. Lazarus, et al. Proteasome inhibitors: a novel class of potent and effective antitumor agents. Cancer Res, 1999. 59: 2615-22.
162. Adams, J. The proteasome: a suitable antineoplastic target. Nat Rev Cancer, 2004. 4: 349-60.
163. Shah, S.A., M.W. Potter, T.P. McDade, R. Ricciardi, R.A. Perugini, P.J. Elliott, et al. 26S proteasome inhibition induces apoptosis and limits growth of human pancreatic cancer. J Cell Biochem, 2001. 82: 110-22.
164. Bold, R.J., S. Virudachalam, D.J. McConkey. Chemosensitization of pancreatic cancer by inhibition of the 26S proteasome. J Surg Res, 2001. 100: 11-17.
165. Arlt, A., A. Gehrz, S. Muerkoster, J. Vorndamm, M.L. Kruse, U.R. Folsch, et al. Role of NF-kappaB and Akt/PI3K in the resistance of pancreatic carcinoma cell lines against gemcitabine-induced cell death. Oncogene, 2003. 22: 3243-51.
166. Guzman, M.L., C.F. Swiderski, D.S. Howard, B.A. Grimes, R.M. Rossi, S.J. Szilvassy, et al. Preferential induction of apoptosis for primary human leukemic stem cells. Proc Natl Acad Sci USA, 2002. 99: 16220-25.
167. Hideshima, T., H. Ikeda, D. Chauhan, Y. Okawa, N. Raje, K. Podar, et al. Bortezomib induces canonical nuclear factor-kappaB activation in multiple myeloma cells. Blood, 2009. 114: 1046-52.
168. Sreenivasan, Y., A. Sarkar, S.K. Manna. Mechanism of cytosine arabinoside-mediated apoptosis: role of Rel A (p65) dephosphorylation. Oncogene, 2003. 22: 4356-69.
169. Lu, X., H. An, R. Jin, M. Zou, Y. Guo, P.F. Su, et al. PPM1A is a RelA phosphatase with tumor suppressor-like activity. Oncogene, 2014. 33: 2918-27.
170. Chew, J., S. Biswas, S. Shreeram, M. Humaidi, E.T. Wong, M.K. Dhillion, et al. WIP1 phosphatase is a negative regulator of NF-kappaB signalling. Nat Cell Biol, 2009. 11: 659-66.
171. Zhang, L., X. Ren, Y. Cheng, X. Liu, J.E. Allen, Y. Zhang, et al. The NFkappaB inhibitor, SN50, induces differentiation of glioma stem cells and suppresses their oncogenic phenotype. Cancer Biol Ther, 2014. 15: 602-11.
172. Tafani, M., B. Pucci, A. Russo, L. Schito, L. Pellegrini, G.A. Perrone, et al. Modulators of HIF1alpha and NFκB in cancer treatment: is it a rational approach for controlling malignant progression? Front Pharmacol, 2013. 4: 13.
173. Tatetsu, H., Y. Okuno, M. Nakamura, F. Matsuno, T. Sonoki, I. Taniguchi, et al. Dehydroxymethylepoxyquinomicin, a novel nuclear factor-kappaB inhibitor, induces apoptosis in multiple myeloma cells in an IkappaBalpha-independent manner. Mol Cancer Ther, 2005. 4: 1114-20.
174. Gray, S.G., B.T. The. Histone acetylation/deacetylation and cancer: an "open" and "shut" case. Curr Mol Med, 2001. 1: 401-29.
175. Kiernan, R., V. Bres, R.W. Ng, M.P. Coudart, S. El Messaoudi, C. Sardet, et al. Post-activation turn-off of NF-kappa B-dependent transcription is regulated by acetylation of p65. J Biol Chem, 2003. 278: 2758-66.

176. Chen, L.F., W.C. Greene. Shaping the nuclear action of NF-kappaB. Nat Rev Mol Cell Biol, 2004. 5: 392-401.
177. Al-Halabi, R., M. Bou Chedid, R. Abou Merhi, H. El-Hajj, H. Zahr, R. Schneider-Stock, et al. Gallotannin inhibits NFkB signaling and growth of human colon cancer xenografts. Cancer Biol Ther, 2011. 12: 59-68.
178. Choi, K.C., Y.H. Lee, M.G. Jung, S.H. Kwon, M.J. Kim, W.J. Jun, et al. Gallic acid suppresses lipopolysaccharide-induced nuclear factor-kappaB signaling by preventing RelA acetylation in A549 lung cancer cells. Mol Cancer Res, 2009. 7: 2011-21.
179. Tomita, N., T. Ogihara, R. Morishita. Transcription factors as molecular targets: molecular mechanisms of decoy ODN and their design. Curr Drug Targets, 2003. 4: 603-08.
180. Kawamura, I., R. Morishita, N. Tomita, E. Lacey, M. Aketa, S. Tsujimoto, et al. Intratumoral injection of oligonucleotides to the NF kappa B binding site inhibits cachexia in a mouse tumor model. Gene Ther, 1999. 6: 91-97.
181. Bubici, C., S. Papa, K. Dean, G. Franzoso. Mutual cross-talk between reactive oxygen species and nuclear factor-kappa B: molecular basis and biological significance. Oncogene, 2006. 25: 6731-48.
182. Hayakawa, M., H. Miyashita, I. Sakamoto, M. Kitagawa, H. Tanaka, H. Yasuda, et al. Evidence that reactive oxygen species do not mediate NF-kappaB activation. EMBO J, 2003. 22: 3356-66.
183. Pahl, H.L., B. Krauss, K. Schulze-Osthoff, T. Decker, E.B. Traenckner, M. Vogt, et al. The immunosuppressive fungal metabolite gliotoxin specifically inhibits transcription factor NF-kappaB. J Exp Med, 1996. 183: 1829-40.
184. Vigushin, D.M., N. Mirsaidi, G. Brooke, C. Sun, P. Pace, L. Inman, et al. Gliotoxin is a dual inhibitor of farnesyltransferase and geranylgeranyltransferase I with antitumor activity against breast cancer in vivo. Med Oncol, 2004. 21: 21-30.

CHAPTER 3

The S100A7/8/9 Proteins: Novel Biomarker and Therapeutic Targets for Solid Tumor Stroma

Sanjay Mishra[1], Dinesh Ahirwar[1], Mohd W. Nasser[2]* and Ramesh K. Ganju[1]*

[1] Department of Pathology, The Ohio State University Medical Center, Columbus, Ohio, USA
[2] Department of Biochemistry and Molecular Biology, University of Nebraska Medical Center, Omaha, NE, USA

Introduction

The association between carcinogenesis and inflammation has long been recognized and chronic inflammation is considered as one of the hallmarks of cancer [1]. The very initial stage of cancer development is abrupt, persistent and chronic inflammation. Chronic inflammatory diseases often increase the risk of cancer development [2-4]. In addition, non-steroid anti-inflammatory drugs (NSAIDs), especially aspirin, have been shown to benefit different types of cancer patients [1, 5-8]. Chronic inflammation causes release of reactive oxygen species (ROS), which can induce DNA damage and initiate malignant transformation [2]. The proteins of S100 family are known to be expressed at the site of inflammation and abundantly present in various inflammatory diseases including cancer. Based on their inflammatory nature, S100 proteins have been analyzed for their ability to serve as biomarkers in cancers. Various comparative and functional genomics studies have reported that S100 genes are differentially expressed in cancer cells [9, 10]. Among various S100 family genes, S100A7, S100A8 and S100A9 (S100A8/A9) have emerged as critical genes involved in inflammatory processes. S100A8/A9 proteins are often produced together [11]. An elevated level of S100A7 and S100A8/A9 is often found in inflammatory diseases such as psoriasis, arthritis, diabetes, and inflammatory bowel disease [12-14]. In fact, strong up-regulation of these proteins has also been observed in many tumors, including gastric, esophageal, colon, pancreatic, bladder, ovarian, thyroid,

*Corresponding author: Wasim.nasser@unmc.edu; ramesh.ganju@osumc.edu

breast and skin cancers [15, 16]. In this chapter, we will describe the altered expression, biological functions and therapeutic targeting of S100A7/8/9 in different types of human cancers.

Role of S100A7 in Different Human Malignancies

Psoriasin (S100A7) was first identified as an 11.4 kDa cytoplasmic protein and reported as a new member of the S100 gene family which is located within the S100 gene cluster on 1q21 chromosomal locus. This secreted protein shares the typical calcium binding domains and is extracted from the skin of psoriasis patients. It has been reported that S100A7 can be expressed by normal cultured cells and also by malignant cells. This report suggested its abnormal role in cell cycle progression and differentiation [17]. Currently, several reports supported the deregulated role of this protein in different human malignancies apart from its role in inflammatory skin disease. In this section, we have described the differential expressions and functional significance of S100A7 in different types of human cancers.

S100A7 in Lung Cancer

The overexpression of S100A7 protein has been reported in lung squamous cell carcinoma. It has been reported that the knock down of S100A7 gene inhibited the proliferation of NCI-H520 cells. The inhibition of S100A7 has also been found to reduce the *in-vivo* tumor growth and metastasis by suppressing the phosphorylation of NF-κB in lung cancer [18]. Hu et al. have shown higher expression of mRNA level of S100A7 in the squamous cell carcinoma compared to paired normal and non-paired normal tissues. They have also shown that the higher expression of S100A7 is associated with poorer prognosis of squamous cell lung carcinoma. They have found that the overexpression of S100A7 in SK-MES-1 lung cancer cells lead to an increased *in-vitro* growth and invasiveness. Their group has also shown that the downregulation of S100A7 resulted in decreased cell growth and invasion with increased cell adhesiveness [19]. Their study thus strongly suggested the role of S100A7 in modulating the lung cancer cells growth and invasion. One study also analyzed the expression of S100A7 in 150 squamous cell carcinoma subjects (53 well, 51 moderately, and 46 poorly differentiated samples) and 159 adenocarcinoma subjects (49 well, 52 moderately, and 58 poorly differentiated samples) [20]. In this study, S100A7 has been reported as squamous cell carcinoma marker with a specificity and sensitivity of 86.8% and 70.7%, respectively. It has also been reported that the specific high expression of both S100A7 mRNA and protein in squamous cell carcinomas, adenosquamous carcinomas and large cell lung carcinomas without any positive staining in adenocarcinomas or paired non-cancerous lung tissues [21]. They have reported the weak staining of S100A7 in the inflammatory cells of benign lung diseases. In this study, the increased level of S100A7 protein was also found in the sera of squamous cell carcinoma patients

[21]. The most important finding is that elevated S100A7 protein could be detected in the sera of patients with squamous cell carcinomas. Furthermore, S100A7 overexpression was also associated with trans-differentiation of lung adenocarcinoma to squamous carcinoma by the Hippo-YAP pathway in lung cancer cells [22]. In this study, they have also reported an increased expression of DNp63 with reduced expressions of thyroid transcription factor 1 (TTF1) and aspartic proteinase napsin (napsin A) proteins in lung cancer cells. Aforementioned studies strongly suggest that S100A7 could be used a novel biomarker for lung squamous cell carcinoma. However, further studies are warranted to establish the role of S100A7 as a novel biomarker and target of lung SCC.

Interestingly, Zhang et al. identified the increased expression of S100A7 in H226Br lung cancer cells (the brain metastatic cell line of NCI-H226) by using two-dimensional electrophoresis (2-DE) followed by a tandem mass spectrometer with a matrix-assisted laser desorption/ionization (MALDI) [23]. They found the positive staining of S100A7 in 8/21 (38%) primary lung squamous cell carcinoma and in 3/5 (60%) matched brain metastasis tissues, without any positive staining in the brain metastases tissue from the primary adenocarcinoma samples, the primary brain tumors, all local positive lymph nodes from the primary non-small cell lung carcinoma and non-cancer brain tissues. This study strongly suggests that increased S100A7 expression is associated with squamous cell carcinoma metastasis to brain. Since lung cancer is the leading cause of brain metastasis, this study has high clinical relevance for predicting brain metastasis.

S100A7 in Cervical and Ovarian Cancers

S100A7 is a small molecular weight EF-hand calcium-binding protein and its role in cell proliferation, migration, invasion and tumor metastasis has been well defined. Although its role in cervical and ovarian cancers has not yet been fully elucidated, there are some reports which suggest its role in cervical and ovarian cancers. It has been shown that S100A7 showed increased immunohistochemical staining in cervical cancer tissues as compared to adjacent normal tissues [24]. In this study, they have reported the increased expressions of S100A7 in high grade cervical intraepithelial neoplasm (CIN) as compared to cervical cancer. They have also shown that S100A7 expression is significantly associated with tumor grade and lymph node metastasis. The overexpression of S100A7 in cervical cancer cells also showed increased migration, invasion and metastasis without affecting the proliferating ability of cervical cancer cells. In this study, authors have found that S100A7 mediated increased cervical cancer cells migration and metastasis by RAGE-ERK signaling pathway. The overexpression of S100A7 were also found to induce the epithelial-mesenchymal transition (EMT) in cervical cancer cells, which is one of the most confounding factor for invasion and metastasis in different human malignancies. Kong et al. have shown that S100A7 expression is regulated by Hippo-YAP signaling pathway in well differentiated cervical cells (HCC94) [25]. Their studies also reported the increased expression of

S100A7 and YAP phosphorylation with decreased nuclear accumulation in dense and suspension culture of cervical cancer cells. In this study, they have explored the transcriptional regulation of S100A7 in cervical cancer cells and reported that S100A7 is repressed by YAP at transcriptional level by TEAD1 protein. They have found the weak expression of S100A7 in poorly differentiated SiHa cervical cancer cells in both conditions (suspension cell culture condition or activation of the Hippo signaling pathway which have important role in cervical cancer pathogenesis). In their studies, they have also reported that S100A7 expression positively correlates with pYAP-S127 and negatively correlates with nuclear YAP in low malignant but not in advanced cervical tissue and lingual SCC tissue samples.

Gagnon et al. have also identified the autoantibodies against S100A7 in the blood plasma of ovarian cancer patients using proteomics, immunology and ELISA-bases approaches [26]. For this study, they developed innovative technique called two-dimensional differential gel electrophoresis (2D-DGE) analysis of immunoprecipitated tumor antigens. Here, they have identified the novel circulating autoantibodies generated against the S100A7 protein in the plasma of ovarian cancer patients. Moreover, they have found the increased expression of S100A7 mRNA and proteins in malignant ovarian cancer tissues as compared to benign and normal tissues by qRT-PCR and IHC, respectively. They also reported the significantly increased level of anti-S100A7 antibody in blood plasma of early and late-stage ovarian cancer patients (n=130) as compared to normal and benign gynecologic subjects. Recently, it has also been shown that the increased expression of S100A7 in epithelial ovarian cancer (EOC) is associated with chemoresistance and metastasis by regulating the MAPK signaling [27]. In this study, it has been shown that S100A7 is upregulated in EOC cell lines (Caov3 and SKOV3) and EOC derived tumor tissues and showed increased chemoresistance to cisplatin treatment. They have also found that the miR-330-5p can target 3′UTR of S100A7 mRNA transcripts and suppresses its expression in EOC. The overexpression of miR-330-5p was found to inhibit the EOC cells proliferation, migration and invasion by downregulating the S100A7 and thus it might be used as potential therapeutic regimen to target S100A7 in different cancers including EOC.

S100A7 in Melanoma and Skin Cancer

The deregulated expression of S100A7 has been reported in several tumors of epithelial origins but little is known about its expression in melanocyte-derived tumors of neuroectodermal origin. Petersson et al. have analyzed the differential expression of S100 proteins by using SAGE Genie informatics [28]. In this study, they have not found any significant changes in differential expression of S100A4, S100A7, S100A8, S100A9 and S100A11 in melanocytes and melanocytic lesions. But, one study reported by another group has identified the increased level of S100A7 in the urine of patients with melanoma (77%, 24 out of 31) as compared to healthy subjects (41%, seven

Table 1: The timeline showing key events about S100A7 protein discovery and functions

1991-1995	• Molecular cloning and expression analysis of Psoriasin (S100A7) • Binding with calcium and expression analysis in breast cancer • Localization to 1q21-q22 and amino acid sequencing identified
1996-2000	• Novel role in cancer (bladder and breast carcinomas) • 3-Dimensional structural and crystal structure analysis • Genomic and mutational analysis • Interaction with E-FABP protein
2001-2005	• S100A7 as a transglutaminase substrate • Genomic and mutational analysis. • Role in ER(-) invasive breast cancer • Antimicrobial role against E. coli infection • Interaction with RanBPM and Jab1
2006-2010	• Identification of mouse S100A15 ortholog and role in lung cancer • Modulation of immune response genes and neutrophils activation • Reciprocal regulation between S100A7 and p-catenin signaling • Chemotactic activity of S100A7 through RAGE • Structural and functional characterization of its triple mutant form
2011-2015	• Role of ICAM-1 signaling in S100A7 expression • Interaction of S100A7 with integrin b6 subunit • Role of S100A7 in inflammation induced mammary tumorigenesis • miR-29b in regulation of S100A7 in breast cancer • Novel role of S100A7/RAGE in breast tumor microenvironment • Function of disulphide-reduced S100A7 in apoptosis
2016-2018	• Role of cholesterol in IL-17A signaling in psoriasis • Role of Hippo-YAP pathway in S100A7 signaling • Role of S100A7 in systemic sclerosis and atherosclerosis • Development of S100A7 neutralizing monoclonal antibodies

out of 17) and patients with different types of cancer (53%, 39 out of 73) [29]. But, till date no study has been reported about its role in metastasis and angiogenesis in melanoma.

Although there are increasing evidences about the role of S100 family proteins in different cancers and internal diseases, not much is known about their distribution in normal skin and in sweat gland tumors. It has been reported that S100A7 showed positive staining in the acrosyringium, ductal, and secretory portions of the eccrine gland and in the inner layer of the apocrine gland [30]. The intense positive staining of S100A7 was reported in the inner layer of the acrosyringium and eccrine ducts and also in Syringoma, Eccrine poroma, Extramammary Paget's disease (EMPD) and Syringocystadenoma papilliferum. Alowami et al. have also reported the essential role of S100A7 in skin cancer progression and they have identified the abnormal expressions of S100A7 in abnormal keratinocytes in actinic keratosis, *in-situ* and invasive squamous cell carcinoma [31]. They observed weak expressions of S100A7 in the basal epidermal layer or in superficial or invasive basal cell carcinoma and highly increased expression in squamous carcinoma *in-situ*. In this study, they also reported the low expression of

S100A7 in both matched and unmatched invasive squamous cell carcinoma. It has also been found that the expression of S100A7 in abnormal squamous lesions was positively correlated with mitotic count in all of these abnormal pathophysiological conditions. Moubayed et al. also reported the increased expressions of S100A7 in patients with precancerous skin lesions, squamous cell carcinoma and basal cell carcinoma as compared to normal skin samples by using *in-situ* hybridization and IHC analysis [32]. One study also observed increased expressions of S100A7 in carcinoma *in-situ*, and in keratoacanthoma and differentiated squamous cell carcinoma [33]. On the other hand, in normal epithelium, S100A7 was found to express in the superficial, differentiated region of the epithelium instead of basal region. It has also been shown that the overexpression of S100A7 in epidermal cancer cell line (A431) caused increased cell proliferation *in-vitro* and *in-vivo* tumor growth in squamous cell carcinoma of skin [34]. In this study, they also demonstrated the decreased expressions of GATA-3, caspase-14 and three squamous differentiation markers, TG-1, involucrin and keratin-1 in A431 cells after overexpression of S100A7 gene. Here, authors delineated the important role of S100A7 in squamous cell carcinoma by the regulation of squamous cell carcinoma cells dedifferentiation in skin cancer. Recently, two more studies also explored the reciprocal negative feedback signaling of S100A7 and YAP signaling in A431 cells [35, 36]. In one study, it has been reported that S100A7 expression is repressed by YAP by the Hippo signaling pathway in A431 cells [36]. In another study, authors have reported that S100A7 attenuated the expression and activity of YAP by the p65/NFκB-mediated suppression of ΔNp63 [35]. Wolf et al. have also identified the potential role of S100A7 in regulating different immune cells in skin inflammation by using novel doxycycline-regulated skin specific mS100a7a15 transgenic mouse model [37]. In this study, they have observed increased infiltration and elevated concentrations of T helper 1 (T(H)1) and T(H)17 proinflammatory cytokines, which have been associated with psoriasis. These effects were found to be mediated by the mS100a7a15 binding to the receptor of advanced glycation end products (RAGE) and caused inflammation by acting as a chemoattractant for leukocytes [37]. So, these results provide a novel preclinical psoriatic mouse model to study the effects of S100A7A15-RAGE axis in regulating the innate immune system in inflammation-induced pathogenesis of psoriasis.

S100A7 and Oral, Head & Neck Cancer

The abnormal expression of S100A7 is also reported in oral cancers and Kaur et al. have identified the differential expression of S100A7 in tissue samples of oral lesions with dysplasia and oral cancers as compared to normal oral tissues by using proteomics approach [38]. In this study, authors have used IHC analysis of S100A7 expressions and correlated it with p16 expression and HPV 16/18 status. Here, Kaplan-Meier survival analysis showed low oral cancer-free survival (OCFS) of 68.6 months in patients showing strong positive cytoplasmic staining of S100A7 as compared to subjects with weak or no cytoplasmic immunostaining (mean OCFS = 122.8 months). The other statistical analysis (Multivariate Cox regression) also showed increased cytoplasmic S100A7 staining as the most potential biomarker linked with cancer development in dysplastic lesions. Furthermore, one study also revealed the aberrant expression of S100A7 in oral squamous cell carcinoma (OSCC) by using proteomics approach [39]. In this study, they have validated the increased expression of S100A7 in OSCC tissues samples by using qRT-PCR and immunofluorescence. In this study, authors have found the significant correlation of increased mRNA ratios between malignant and normal cases with early UICC stage, tumor grades and lymph node metastasis. They have also reported the increased expression of S100A7 in well-differentiated tumor samples as compared to poorly and moderately differentiated tumor samples. In a very elegant study by Meyer's group, it has been reported that S100A7 forms complex with beta-catenin and it negatively regulates beta-catenin signaling in SCC of oral cavity. They have also shown that S100A7 expression is increased in the early tumor stages and lost in the later stages of SCC of oral cavity. These studies suggest that S100A7 plays tumor suppressive role in SCC of oral cavity [40].

S100A7 in Breast Cancer

The tumor promoting activity of S100A7 has been reported in breast cancer by regulating different oncogenic and tumor-suppressive signaling cascades. Rhee et al. also reported the altered expression of S100A7 during the transformation process and breast cancer progression using the whole gene profiling [41]. It has been also reported that the mRNA transcripts of S100A7 gene differentially expressed in between *in situ* ductal carcinoma and invasive breast carcinoma tumor tissues samples [17]. In this study, authors have reported the high expression of S100A7 in ductal carcinoma *in situ*. Emberley et al. also reported the novel interacting partner (RanBPM) of S100A7 by the yeast two-hybrid assay and co-immunoprecipitation and correlated their expressions with clinicopathological features *in-vivo* in breast cancer [42]. They showed the increased expressions of S100A7 and RanBPM mRNAs in breast cancer cell lines and tumor tissues as compared to normal cells and tissue samples. They also reported the association of increased S100A7/RanBPM ratio with both ER negative and PR negative breast cancer subtypes, and inflammatory cell infiltrates within the breast tumor. This same group also reported another novel binding partner (c-jun activation-domain binding protein 1, also known as Jab1) of S100A7 by using the yeast two-hybrid assay and co-immunoprecipitation in breast cancer cells and reported its novel role in breast cancer pathogenesis [43]. The increased expression of S100A7 in breast cancer cells showed enhanced nuclear translocation of Jab1 with increased expressions of AP-1 and HIF-1-dependent genes. Simultaneously, it also showed decreased expression of the cell-cycle inhibitor p27(Kip1) in S100A7 overexpressing breast cancer cells. The authors revealed that the S100A7 increased expression and interaction with Jab1 was also linked with changes in cellular functions and thus associated with increased tumor growth and metastasis *in-vitro* and *in-vivo* studies.

Differential Role of S100A7 in Breast Cancer

Our lab demonstrated the clinical significance of S100A7 expressions in ERα(+) breast cancer and have found that the overexpression of S100A7 has tumor-suppressive effects in ERα(+) breast cancer. The overexpression of S100A7 in ERα(+) breast cancer cells (MCF7 and T47D) showed reduced proliferation, migration and wound healing capabilities [44]. It was also found that the S100A7 overexpression in MCF-7 cells showed decreased breast tumor growth in *in-vivo* xenograftmouse model as compared to control group. The in-depth detail analysis showed that tumor suppressive effects of S100A7 in ERα(+) breast cancer cells mediated through the fine regulation of the β-catenin/TCF4 pathway and increased interaction of β-catenin and E-cadherin in S100A7-overexpressed ERα(+) breast cancer cells. We have also found that the overexpression of S100A7 in ERα(+) breast cancer cells inhibited the expressions and activation of β-catenin, p-GSK3β, TCF4, cyclin D1, and c-myc in ERα(+) breast cancer cells. Tumors derived from mice model injected with S100A7-overexressing MCF-7 cell line displayed decreased activation of the β-catenin/TCF4 signaling pathway. Another lab has also demonstrated that the negative influence of S100A7 on cancer growth is mediated by histone demethylase LSD1 [45]. We also observed that the S100A7 differentially modulates EGF-induced migration and invasion of ERα (+) and (-) subtypes [46]. It has been reported by Watson's group that long term incubation of MCF7 ERα(+) cells with oncostatin M reduced the expression of ERα. These studies further demonstrated that both oncostatin M and IL-6 increased S100A7 and Stat3 activation and thereby these cytokines enhance breast cancer tumorigenesis [47]. We have also shown that the epidermal growth factor (EGF) induces S100A7 expression in breast cancer cells (MCF-7 and MDA-MB-468) [48]. It was found that S100A7 down regulation inhibits the EFG-induced migration of breast cancer cells through decrease phosphorylation of EGFR (Tyr1173) and HER2 (Tyr1248) in S100A7 knock-down cell lines as compared to scramble control. Furthermore, the decreased phosphorylation of Src (Tyr416) and p-SHP2 (Tyr542) was observed in S100A7 knock-down cell lines as compared to scrambled control. These S100A7 knock-down cells also showed decreased angiogenesis and breast tumor-mediated osteoclastic resorption in an intra-tibial bone injection SCID mouse model. The downregulation of S100A7 was found to decrease the osteoclast size and number as compared to the scrambled control [48]. In addition, we observed that the S100A7 differentially modulates EGF-induced migration and invasion of ERα (+) and (-) subtypes [46]. The overexpression of S100A7 in MDA-MB-231 cells (ERα negative) activated matrix metalloproteinase-9 (MMP-9) secretion during invasion by increased activation of NF-κB signaling. The overexpression of S100A7 in MCF-7 cells

(ERα positive) reduced the MMP-9 secretion during migration and invasion by inhibiting the Rac-1 signaling pathway. In addition, the overexpression of S100A7 in MDA-MB-231 cells showed increased metastasis as compared to scrambled control in *in-vivo* nude mice model. In tissue microarray analysis, the increased expression of S100A7 was observed in ERα(-) metastatic carcinoma, mainly in lymph node regions. Additionally, we have also shown that the overexpression of S100A7 increased the expressions of miR-29b in ERα(+) breast cancer cells while inhibiting its expression in ERα(-) breast cancer cells [49]. This differential effect of S100A7 expression in ER subtypes of breast cancer was also supported by the gene expression dataset of TCGA invasive breast cancer dataset. In this study, it was demonstrated that the differential regulation of miR-29b in different subtypes of ER breast cancer subtypes was regulated the modulation of NF-κB and p53 activation and regulation of PI3K p85α and CDC42. The use of exogenous miR-29b or miR-29b-Decoy in these differentially expressed S100A7-breast cancer cells showed potent effects of cancer cells proliferation and tumor growth in nude mice model. Thus, for the first time we have shown the differential regulation of S100A7 depending on the ERα status in breast cancer cells via distinct modulation of NF-κB/miR-29b/p53 signaling axis.

S100A7 Modulates Breast Cancer Microenvironment

Our research group has also shown that the S100A7 increases the mammary tumorigenesis through the activation of inflammatory signaling pathways in invasive estrogen receptor (ER)α-negative breast cancers [50]. In this study, we demonstrated the role of S100A7 breast cancer growth or metastasis using our novel human S100A7 or its murine homologue mS100a7a15. In this study, we have reported that the overexpression of S100A7 or its murine homologue mS100a7a15 increased the breast cancer cells proliferation by increased expression of different proinflammatory signaling molecules in ERα-negative breast cancer cells. It was found that the orthotopic implantation of MVT-1 breast cancer cells into the mammary fat pad of these doxyclycline-induced bi-transgenic mice model (MMTV-mS100a7a15 mice) increased the tumor growth and subsequent metastasis. The overexpression of mS100a7a15 in the mammary gland of these mice model showed increased ductal hyperplasia with increased expressions of diverse signaling molecules associated with proliferation, tissue remodeling, and inflammation and macrophage recruitment. Here, we have reported that the lung tissues of these mice model showed increased infiltration of tumor-associated macrophages (TAM) with enhanced expression of prometastatic genes. The *in-vivo* depletion of TAM also suppressed the effects on tumor growth, angiogenesis and metastasis mediated through activation of mS100a7a15 in these mice model. Moreover, it was also found that the treatment of soluble hS100A7 or mS100a7a15 increased the chemotaxis of macrophages *via* activation of RAGE receptors in ERα(-) breast cancer cells. Thus, this work revealed a potential mouse model of S100A7 overexpression which can be used in future to explore the tumor promoting activity of S100A7 in

ERα(-) breast cancer by also modulating the tumor microenvironment through the activation of inflammatory signaling cascades. We have also reported that RAGE, a multifunctional receptor of S100A7 protein, was found to be overexpressed in highly aggressive triple-negative breast cancer (TNBC) tissues and cells [51]. In this we have observed that the RAGE-deficient mouse model showed decreased tumor growth. We have also reported that the RAGE/S100A7 conditioned the breast tumor microenvironment by causing the recruitment of MMP9-positive tumor-associated macrophages. These studies strongly suggest that S100A7 enhances TNBC growth and metastasis through activation of RAGE receptor. Interestingly, an elegant study has shown through integrative genomic analysis that chromosome 1q21.3 amplification occurs in recurrent breast tumors. This locus contains S100 family proteins and amplified region of the chromosome has shown increased expression of S100A7 and S100A8/9 proteins in the recurrent breast cancer cells. These studies suggest that S100A7 may be a valuable biomarker for relapse and metastatic breast cancer [52].

S100A7 as a Therapeutic Target in Breast Cancer

It was also found that the antibody-based neutralization of RAGE showed decreased metastasis in an established model of lung metastasis. Here, we have found that the binding of S100A7 to RAGE activated the ERK-NF-κB signaling and causes increased cell migration. To study the effects of S100A7 *in-vivo*, we have also developed a transgenic breast cancer mouse model of mS100a7a15 and it was observed that the treatment of either soluble RAGE or RAGE neutralizing antibody significantly reduced the breast tumor progression and subsequent metastasis [53]. Recently, several neutralizing antibodies of S100A7 have been generated and studies have shown that blocking S100A7-RAGE axis is essential for inhibiting metastasis in preclinical mouse models. In this study, authors have identified the potential role of S100A7 in the formation of a proinflammatory and proangiogenic tumor microenvironment which supports the tumor progression and metastasis. Moreover, they have also observed the important contribution of S100A7 in the formation of the pre-metastatic niche which favors the metastasis to distant organs. Overall, majority of the preclinical data presented in the literature strongly suggest that S100A7 plays crucial role in enhancing TNBC growth and metastasis via modulating tumor microenvironment and serve as novel biomarker for early detection as well as novel target for therapy (Fig. 1).

The Role of S100A8/A9 in Different Malignancies

Among various S100 family genes, S100A8 and S100A9 (S100A8/A9) have emerged as critical genes involved in tumor progression and metastasis. S100A8 and S100A9 exist as homodimers as well as anti-parallel heterodimers of S100A8/A9, also known as calprotectin and Ca^{2+}-induced tetramers [12-

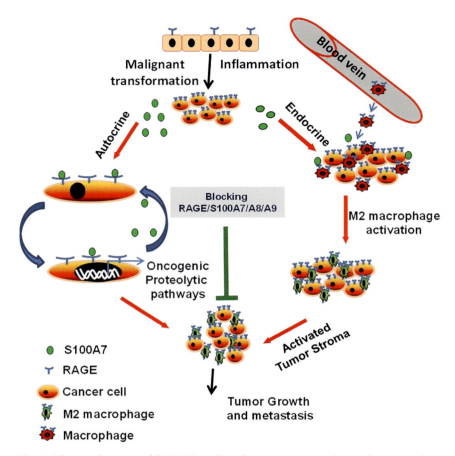

Fig. 1: The mechanism of S100A7 mediated tumor progression and metastasis.

14, 54-56]. S100A8/A9 are expressed at higher amount in aggressive high grade tumors or estrogen receptor (ER) negative compared to low grade or ER positive breast tumor samples, respectively [11, 57, 58]. Higher expression of S100A9 predicts poor patient survival in breast cancer [58]. These findings may help to identify biologically aggressive tumors and, thus, patients who might benefit from more intensive adjuvant therapy.

S100A8/A9 and Breast Tumor

An interesting study specifically evaluated S100 family proteins in two different cohorts of breast cancer patients. In both cohorts, the expression of S100A8 and S100A9 mRNA level was elevated in high-grade compared to low-grade tumors, in estrogen receptor-negative compared to estrogen receptor-positive tumors and in basal-type compared to non-basal types breast cancer [11]. The expression levels of S100A8 and S100A9 were higher in human epidermal growth factor receptor 2 (Her2)-amplified and basal-like BC [57]. Mechanical studies suggest that S100A8/A9 acts by binding to receptor for advanced glycation end-products (RAGE) [59]. Upon binding to RAGE, S100A8/A9 promotes the migration, invasion and metastasis of human breast cancer cells through actin polymerization and epithelial-mesenchymal transition by stabilizing Snail transcription factor [59, 60]. A differential role of extracellular and intracellular S100A8 and S100A9 has been observed. In contrast to extracellular S100A8/A9 enhancing tumor cell proliferation, intracellular-expressed S100A8/A9 inhibits cell proliferation [60]. Intracellular S100A8/A9 promotes an epithelial-like phenotype through the induction of epithelial markers, including E-Cadherin, integrin alpha-5 and Zona Occludens 1 (ZO-1) and negatively regulate oncogenic signaling molecule Focal Adhesion Kinase-1 (FAK-1) signaling [60]. These results suggest that depending on the intracellular or extracellular localization of S100A8/A9, these molecules may inhibit or support tumor growth, respectively. The messenger (m)RNA levels of S100A8 and S100A9 were inversely correlated with ESR1 and GATA3 expression [57]. Poor overall survival was associated with high expression levels of S100A9, but not with high expression levels of S100A8 in BC [57]. Elevated expression of S100A8 in breast epithelial as well as stromal compartments was significantly associated with poor breast cancer outcomes, regardless of estrogen receptor status [61, 62]. The S100 family proteins have also been found to contribute to metastatic niche formation into distant organs that favors tumor cell seeding and proliferation. The expression of S100A8 increases many folds into the common sites of metastasis including the lungs in time dependent manner after tumor implantation. This increased expression of S100A8 directly correlates with the number of RAGE-positive myeloid-derived suppressor cells (MDSCs) recruited to the lungs of tumor bearing mice [63, 64]. Another study conducted by Hiratsuka and colleagues found similar role of S100A8/A9 in metastatic niche formation. Implantation of primary tumor induces expression of S100A8/A9 in the lungs and recruitment of MDSCs [65]. The tumor cells have been shown to secrete various growth factors and signaling factors to induce S100A8 expression. Tumor cells derived factors, including macrophage inhibitory factor (MIF), vascular endothelial growth factor (VEGF) and tumor necrosis factor-α (TNF-α) are known to induce S100A8 expression [61]. The MDSCs recruited to the metastatic niche express reduced level of immune-activating co-stimulatory molecules and are immune-suppressive in nature [61, 63]. In addition to supporting tumor growth by activating oncogenic pathways, recruiting stromal cells and suppressing anti-tumor immunity, S100A8 higher expression also contributes to the chemoresistance in breast cancer suggesting the diverse role played by S100A8/A9 in breast cancer progression, metastasis and drug resistance [66].

S100A8/A9 and Colon Cancer

S100A8 and S100A9 proteins are also overexpressed in colon cancer samples. S100A8 and S100A9 levels, analyzed by various methods including mRNA-

PCR, immunohistochemistry and western blot, were elevated in more than 50% of Colorectal cancer (CRC) tissues and their expression in tumor cells was associated with differentiation, tumor stage and lymph node metastasis. Additional studies showed that recombinant S100A8/A9 increases levels of β-catenin and its target genes c-myc and MMP7. Furthermore, β-catenin knockdown partially abolished the effects of recombinant S100A8 and S100A9 proteins to promote the viability and migration of CRC cells. These observations suggest that S100A8/A9 enhances the progression of CRC through upregulating β-catenin mediated signaling pathways and gene expression [67]. In addition to tumor cells, MDSCs also secrete S100A8/A9, which interacts with RAGE and carboxylated glycans present on the membrane of colon tumor cells and promote activation of MAPK and NF-κB signaling pathways [68]. S100A8/A9 activated tumor cells upregulate various cytokines/chemokines and pro-tumor factors involved in angiogenesis, tumor migration, and formation of metastatic niche [68]. *In vivo* experiments using mice lacking S100A9 showed significantly reduced tumor incidence, growth and metastasis, reduced chemokine levels, and reduced infiltration of CD11b(+)Gr1(+) cells within tumors and pre-metastatic organs [68]. Another study analyzing the role of S100A8 and S100A9 in chemically-induced colitis-associated cancer mouse model found that S100A8/A9 act as chemoattractant to recruit macrophages and these cells promote proliferation and invasion of colon cancer cells by activating the Akt1-Smad5-Id3 signaling axis [69]. Based on pro-tumor functions of S100A8/A9 reported by various studies, anti-S100A9 antibody was developed to evaluate the possibilities of targeting S100A9-mediated signaling pathway as a therapeutic opportunity. Administration of a neutralizing anti-S100a9 antibody significantly inhibited chemically-induced colitis by reducing the infiltration of innate immunity cells and production of pro-inflammatory cytokines [70].

S100A8/A9 and Prostate Cancer

To analyze the potential of S100A8/A9 as prognostic biomarkers in PCa, the researchers have analyzed S100A8/A9 levels in benign prostate hyperplasia (BPH) and PCa patient samples. It was found that both S100A8 and S100A9 were highly expressed in patients with aggressive disease and shorter recurrence-free time. S100A8/A9 levels correlated positively with expression levels obtained from tissue staining [71]. Using a doxycycline regulated S100A8/A9 overexpression system, it was observed that increased expression of S100A8/A9 in prostate cancer (PCa) cells enhances their ability to grow as tumors in mice and invade surrounding tissue by recruiting immune cells, especially neutrophils [72]. In addition, intra-cardiac injection of S100A8/A9 overexpressing prostate cancer cells resulted in higher colonization of tumor cells into the lungs compared to control cells [72]. Oncogenic molecule, such as Prostaglandin E2 (PGE2), has been shown to induce the expression of S100A8 and S100A9 in prostate cancer cells [73]. PGE2 is a member of prostaglandin lipids, which are lipid mediators and are known to be involved in diverse physiological and pathological processes, including pain, inflammation, renal functions and cancer progression [74]. Transcription factor binding sequence analysis revealed that PGE2 induced overexpression of S100A8 by enhancing the binding of CCAAT/enhancer-binding-protein-beta transcription factor to S100A8 promotor [73]. In addition to lipid mediators, hypoxia also induces expression of S100A8/A9 in prostate cancer cells [75]. Furthermore, overexpression of hypoxia-inducing factor-1α (HIF-1α) increased mRNA expression as well as secretion [75]. Chromatin immunoprecipitation experiments confirmed the binding of HIF-1α to S100A8 and S100A9 promotors. A direct correlation of S100A8/A9 expression with HIF-1 expression was also observed in prostate cancer patient samples [75]. These studies suggest that various tumor inducing factors activate different transcription factors to enhance the expression of S100A8/A9 which results in enhanced tumor cell proliferation, recruitment of immune-suppressive cells, and increased production of cytokine/chemokine which all together support tumor progression.

S100A8/A9 and Squamous Cell Carcinoma

S100A8/A9 proteins were also reported to be overexpressed in cutaneous squamous cell carcinoma (SCC) patient samples compared to pre-malignant skin tumors [76]. Ectopic expression of S100A8/A9 in SCC12 cells enhanced proliferation and motility of these cells. When implanted subcutaneously into mice, S100A8/A9 overexpressing SCC12 cells showed enhanced tumor growth [76]. Another study supported SCC promotive role of S100A8/A9 and RAGE. The authors reported that SCC samples have upregulated RAGE and S100A8/A9 compared to normal epidermis and exogenous S100A8/A9 induced SCC keratinocytes *in vitro* [77]. Exogenous S100A8/A9 activates P38/SAPK/JNK/ERK1/2 signaling pathway to enhance the proliferation of SCC keratinocytes [77]. To identify molecular changes involved in SCC initiation, global genomic DNA methylation was performed and it was observed that S100A8 CpG loci was hypomethylated in SCC tumors compared to adjacent normal tissue [78]. Further analysis showed that S100A8 hypomethylation was associated with reduced overall survival [78]. However, conflicting functions of S100A8/A9 have been reported in SCC. Khammanivong et al. have observed that S100A8/A9 is downregulated in squamous cell carcinomas of the cervix, esophagus, and the head and neck (HNSCC) [79]. Furthermore, ectopic expression of S100A8/A9 suppressed SCC *in vitro* and *in vivo* [79]. In addition, another study suggested protective role of S100A9 against SCC using S100A9 knockout mouse model and chemically induced SCC model [80].

S100A8/A9 and Urinary Bladder Cancer

In search of minimal invasive biomarkers for cancer, the global proteomic analysis of sera from urinary bladder cancer (UBC) patients revealed that S100A8 and S100A9 are tumor-associated proteins [81]. Increased expression of S100A8 was associated with invasive tumors and increased expression of S100A9 was linked to high grade bladder tumors [81]. Higher expression of S100A8/A9 was also reported to predict poor recurrence-free survival in UBC patients [81]. Using global proteomic approach, another study identified that S100A8 was highly expressed in deceased leukemia patient samples and its high expression predicts worse patient survival [82]. Proteomic analysis of bile samples from gallbladder cancer patients revealed elevated levels of S100A8 compared to benign gallbladder diseases [83]. Higher levels of S100A8/A9 have also been observed in various other types of cancers, including oral squamous cell carcinoma, and renal cell carcinoma [84, 85]. Furthermore, S100A9 positivity was associated with disease progression after chemotherapy [86]. Another report suggests no association between serum S100A8/A9 levels and UBC [87].

S100A8/A9 and Melanoma

In melanoma patients, immune-based antibody drug ipilimumab treatment response has been shown to be positively associated with increase in eosinophil count and negatively associated with elevated amounts of MDSC and monocytes as compared with basal levels and with responding patients. It was also observed that ipilimumab treatment in non-responders resulted in elevated serum concentrations of S100A8/A9 and HMGB1 that attract and activate MDSCs [88]. In another study, S100A8 levels were found to be associated with clinical response rate in acute myeloblastic leukemia (AML). Knockdown of S100A8 increases autophagy and chemotherapy-induced cytotoxicity in AML cells. In addition, S100A8 directly regulates autophagy protein Beclin 1 and dissociates Beclin1-Bcl-2 complex [89, 90]. In search for novel S100 protein receptors, by using affinity isolation-mass spectrometry, cell surface glycoprotein EMMPRIN/BASIGIN (CD147/BSG) was identified as specific receptor to S100A9, which does not bind to S100A8. It was found that besides RAGE, S100A9 interacts with EMMPRIN to activate TNF receptor-associated factor TRAF2 and induce cytokines and metalloproteases expression [91]. Immunohistochemistry analysis revealed that EMMPRIN was expressed at the edge of melanoma lesion and adjacent epidermis and co-localize with S100A9 [91]. Similarly, ALCAM and MCAM molecules were identified as novel receptors for S100A8/A9 and induce melanoma progression by activating ROS production and NF-κB activation [92]. ß-1,3-galactosyl-O-glycosyl-glycoprotein ß-1,6-N-acetylglucosaminyltransferase 3 (GCNT3) has been shown to be overexpressed in metastatic melanomas. Molecular inhibition of GCNT3 resulted in reduced migration and invasion of melanoma cells, mediated by loss of S100A8/A9 signaling [93].

Pancreatic Cancer

S100A8 and S100A9 overexpression was also associated with poor recurrence-free and overall survival in pancreatic ductal adenocarcinoma (PDAC) [94]. Similar to other cancers, S100A8 and S100A9 secreted from

PDAC cells recruit immunosuppressive MDSCs. These recruited MDSCs also express low levels of cytotoxic T lymphocyte associated protein-4 (CTLA-4) on the surface [95]. Another study identified S100A8/A9 complex activating immunosuppressive molecule programmed death-ligand 1 (PD-L1) and suppressing cytotoxic T lymphocyte antigen-4 (CTLA-4) on MDSCs [96]. In contrast to these studies, a significantly reduced level of S100A9 was found in PDAC patients, which was also correlated with a reduced overall survival [97].

S100A8/A9 and Lung Cancer

The overexpression of S100A8/A9 has also been appreciated in lung cancer. Recently, higher expression of S100A8 and S100A9 was reported in 71.2% and 76.9% non-small cell lung cancer patient samples, respectively, in comparison to adjacent normal [98]. The higher expression of S100A8/A9 was also associated with degree of tumor differentiation [98]. S100A8/A9 plays a crucial role in the formation of pre-metastatic niche. Tumor secreted factors (TSFs) such as VEGF-A, TGF-β and TNF-α may induce expression of S100A8/A9 in pre-metastatic endothelial cells and macrophages [65]. S100A8/A9 secreted by lung stromal cells promotes recruitment of CD11b+ myeloid cells to establish a pre-metastatic niche in the lungs, to which tumor cells migrate [65]. Subsequently, it was shown that S100A8 induces the expression of serum amyloid A 3 (SAA3) in pre-metastatic lungs, followed by

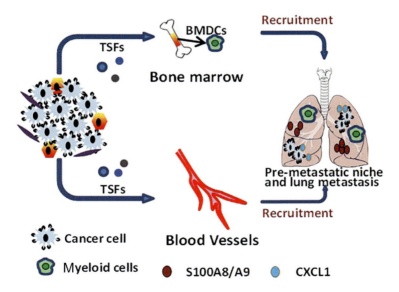

Fig. 2: S100A8/A9 creates pre-metastatic niche. The cancer cells secret tumor secreted factors (TSFs) such as VEGF-A, TNF-α and TGF-β, which stimulate expression of S100A8/A9 in the pre-metastatic organs to recruit CD11b+ myeloid cells. S100A8/A9 also induces CXCL1 to tumor cells to seed on pre-metastatic niche. BMDCs: bone marrow-derived cells.

activation of NF-κβ signaling, resulting in accumulation of CD11b+ myeloid cells [99]. In addition, S100A8/A9 can also induce the expression of CXCL1 in pre-metastatic lungs, which has been shown to attract tumor cells [68]. The role of S100A8/A9 in pre-metastatic niche formation has been summarized in Fig. 2.

Conclusion

The process of tumor progression is extremely complex involving various molecular and cellular components. Often, tumor cells secrete various proteins and signaling molecules to activate self or other cells surrounding tumor mass. Among various small molecules secreted by tumor cells, S100 protein family members including S100A7, S100A8 and S100A9 have been found to regulate various processes of tumor initiation, progression and metastasis. These molecules enhance tumor cell growth by activating cell proliferation pathways. In addition, S100A7, A8/A9 proteins are found to act as chemoattractant for immune cells and recruit immunosuppressive MDSCs and macrophages to the TME. These recruited cells negatively regulate T-cell activity and secrete various angiogenesis molecules, MMPs and immune suppressive molecules to promote tumor progression and metastasis. The possible role of S100A7, A8/A9 in different cancers, especially breast cancer, has been illustrated in Figs 1 and 2. Based on their pro-tumor functions in different type of cancers, various strategies have been developed to evaluate the therapeutic potential of targeting S100A7, A8/A9 against different cancers. Various biomarker studies have shown these molecules as predictor of poor prognosis in different cancers. Although sufficient data is available showing S100A7, A8/A9 act as pro-tumor molecules, few studies suggest that these molecules may also act as suppressor of tumor growth in a context dependent manner. Therefore, more studies are required to establish the functions of S100A7, A8/A9 in different cancers, including breast cancer. The pre-clinical studies also suggest that S100A7, A8/A9 may hold potential to be developed as biomarker for early detection and targeted therapies for solid tumors.

Acknowledgment

The work is supported, in part, by NIH grant R01CA109527 and DoD breast cancer breakthrough award W81XWH1910088 to RKG.

References

1. Crusz, S.M., F.R. Balkwill. Inflammation and cancer: advances and new agents. Nature Reviews: Clinical Oncology, 2015. 12(10): 584-96.

2. Coussens, L.M., Z. Werb. Inflammation and cancer. Nature, 2002. 420(6917): 860-7.
3. Grivennikov, S.I., F.R. Greten, M. Karin. Immunity, inflammation, and cancer. Cell, 2010. 140(6): 883-99.
4. Lotze, M., R. Herberman. Cancer as a chronic inflammatory disease: role of immunotherapy. *In*: D. Morgan, U. Forssmann, M. Nakada (eds.). Cancer and Inflammation: Progress in Inflammation Research. 2004. Birkhäuser, Basel.
5. Dovizio, M., A. Bruno, S. Tacconelli, P. Patrignani. Mode of action of aspirin as a chemopreventive agent. Recent Results in Cancer Res. 2013. 191: 39-65.
6. Algra, A.M., P.M. Rothwell. Effects of regular aspirin on long-term cancer incidence and metastasis: a systematic comparison of evidence from observational studies versus randomised trials. The Lancet. Oncology, 2012. 13(5): 518-27.
7. Rothwell, P.M., J.F. Price, F.G. Fowkes, A. Zanchetti, M.C. Roncaglioni, G. Tognoni, et al. Short-term effects of daily aspirin on cancer incidence, mortality, and non-vascular death: analysis of the time course of risks and benefits in 51 randomised controlled trials. Lancet, 2012. 379(9826): 1602-12.
8. Rothwell, P.M., M. Wilson, J.F. Price, J.F. Belch, T.W. Meade, Z. Mehta. Effect of daily aspirin on risk of cancer metastasis: a study of incident cancers during randomised controlled trials. Lancet, 2012. 379(9826): 1591-601.
9. Schafer, B.W., C.W. Heizmann. The S100 family of EF-hand calcium-binding proteins: functions and pathology. Trends Biochem Sci, 1996. 21(4): 134-40.
10. Carlsson, H., S. Petersson, C. Enerback. Cluster analysis of S100 gene expression and genes correlating to psoriasin (S100A7) expression at different stages of breast cancer development. Int J Oncol, 2005. 27(6): 1473-81.
11. McKiernan, E., E.W. McDermott, D. Evoy, J. Crown, M.J. Duffy. The role of S100 genes in breast cancer progression. Tumour Biol, 2011. 32(3): 441-50.
12. Vogl, T., A. Stratis, V. Wixler, T. Voller, S. Thurainayagam, S.K. Jorch, et al. Autoinhibitory regulation of S100A8/S100A9 alarmin activity locally restricts sterile inflammation. J Clin Invest, 2018. 128(5): 1852-66.
13. Roth, J., M. Goebeler, C. Sorg. S100A8 and S100A9 in inflammatory diseases. Lancet, 2001. 357(9261): 1041.
14. Kraakman, M.J., M.K. Lee, A. Al-Sharea, D. Dragoljevic, T.J. Barrett, E. Montenont et al. Neutrophil-derived S100 calcium-binding proteins A8/A9 promote reticulated thrombocytosis and atherogenesis in diabetes. J Clin Invest, 2017. 127(6): 2133-47.
15. Gebhardt, C., J. Nemeth, P. Angel, J. Hess. S100A8 and S100A9 in inflammation and cancer. Biochem Pharmacol, 2006. 72(11): 1622-31.
16. Salama, I., P.S. Malone, F. Mihaimeed, J.L. Jones. A review of the S100 proteins in cancer. Eur J Surg Oncol, 2008. 34(4): 357-64.
17. Leygue, E., L. Snell, T. Hiller, H. Dotzlaw, K. Hole, L.C. Murphy, et al. Differential expression of psoriasin messenger RNA between in situ and invasive human breast carcinoma. Cancer Res, 1996. 56(20): 4606-09.
18. Liu, G., Q. Wu, X. Song, J. Zhang. Knockdown of S100A7 reduces lung squamous cell carcinoma cell growth in vitro and in vivo. Int J Clin Exp Pathol, 2014. 7(11): 8279-89.
19. Hu, M., L. Ye, F. Ruge, X. Zhi, L. Zhang, W.G. Jiang. The clinical significance of Psoriasin for non-small cell lung cancer patients and its biological impact on lung cancer cell functions. BMC Cancer, 2012. 12: 588.
20. Tsuta, K., Y. Tanabe, A. Yoshida, F. Takahashi, A.M. Maeshima, H. Asamura, et al. Utility of 10 immunohistochemical markers including novel markers

(desmocollin-3, glypican 3, S100A2, S100A7, and Sox-2) for differential diagnosis of squamous cell carcinoma from adenocarcinoma of the Lung. J Thorac Oncol, 2011. 6(7): 1190-99.
21. Zhang, H., Q. Zhao, Y. Chen, Y. Wang, S. Gao, Y. Mao, et al. Selective expression of S100A7 in lung squamous cell carcinomas and large cell carcinomas but not in adenocarcinomas and small cell carcinomas. Thorax, 2008. 63(4): 352-9.
22. Wang, R., Y. Li, E. Hu, F. Kong, J. Wang, J. Liu, et al. S100A7 promotes lung adenocarcinoma to squamous carcinoma transdifferentiation, and its expression is differentially regulated by the Hippo-YAP pathway in lung cancer cells. Oncotarget, 2017. 8(15): 24804-14.
23. Zhang, H., Y. Wang, Y. Chen, S. Sun, N. Li, D. Lv, et al. Identification and validation of S100A7 associated with lung squamous cell carcinoma metastasis to brain. Lung Cancer, 2007. 57(1): 37-45.
24. Tian, T., X. Li, Z. Hua, J. Ma, X. Wu, Z. Liu, et al. S100A7 promotes the migration, invasion and metastasis of human cervical cancer cells through epithelial-mesenchymal transition. Oncotarget, 2017. 8(15): 24964-77.
25. Kong, F., Y. Li, E. Hu, R. Wang, J. Wang, J. Liu, et al. The characteristic of S100A7 induction by the Hippo-YAP pathway in cervical and glossopharyngeal squamous cell carcinoma. PloS One, 2016. 11(12): e0167080.
26. Gagnon, A., J.H. Kim, J.O. Schorge, B. Ye, B. Liu, K. Hasselblatt, et al. Use of a combination of approaches to identify and validate relevant tumor-associated antigens and their corresponding autoantibodies in ovarian cancer patients. Clin Cancer Res, 2008. 14(3): 764-71.
27. Lin, M., B. Xia, L. Qin, H. Chen, G. Lou. S100A7 regulates ovarian cancer cell metastasis and chemoresistance through MAPK signaling and is targeted by miR-330-5p. DNA Cell Biol, 2018.
28. Petersson, S., E. Shubbar, L. Enerback, C. Enerback. Expression patterns of S100 proteins in melanocytes and melanocytic lesions. Melanoma Res, 2009. 19(4): 215-25.
29. Brouard, M.C., J.H. Saurat, G. Ghanem, G. Siegentaler. Urinary excretion of epidermal-type fatty acid-binding protein and S100A7 protein in patients with cutaneous melanoma. Melanoma Res, 2002. 12(6): 627-31.
30. Zhu, L., S. Okano, M. Takahara, T. Chiba, Y. Tu, Y. Oda et al. Expression of S100 protein family members in normal skin and sweat gland tumors. J Dermatol Sci, 2013. 70(3): 211-9.
31. Alowami, S., G. Qing, E. Emberley, L. Snell, P.H. Watson. Psoriasin (S100A7) expression is altered during skin tumorigenesis. BMC Dermatol, 2003. 3: 1.
32. Moubayed, N., M. Weichenthal, J. Harder, E. Wandel, M. Sticherling, R. Glaser. Psoriasin (S100A7) is significantly up-regulated in human epithelial skin tumours. J Cancer Res Clin Oncol, 2007. 133(4): 253-61.
33. Martinsson, H., M. Yhr, C. Enerback. Expression patterns of S100A7 (psoriasin) and S100A9 (calgranulin-B) in keratinocyte differentiation. Exp Dermatol, 2005. 14(3): 161-8.
34. Li, T., Z. Qi, F. Kong, Y. Li, R. Wang, W. Zhang, et al. S100A7 acts as a dual regulator in promoting proliferation and suppressing squamous differentiation through GATA-3/caspase-14 pathway in A431 cells. Exp Dermatol, 2015. 24(5): 342-8.
35. Li, Y., F. Kong, Q. Shao, R. Wang, E. Hu, J. Liu, et al. YAP expression and activity are suppressed by S100A7 via p65/NF-kappaB-mediated repression of DeltaNp63. Mol Cancer Res, 2017. 15(12): 1752-63.
36. Li, Y., F. Kong, J. Wang, E. Hu, R. Wang, J. Liu et al. S100A7 induction is

repressed by YAP via the Hippo pathway in A431 cells. Oncotarget, 2016. 7(25): 38133-42.
37. Wolf, R., F. Mascia, A. Dharamsi, O.M. Howard, C. Cataisson, V. Bliskovski et al. Gene from a psoriasis susceptibility locus primes the skin for inflammation. Sci Transl Med, 2010. 2(61): 61ra90.
38. Kaur, J., A. Matta, I. Kak, G. Srivastava, J. Assi, I. Leong, I. Witterick, et al. S100A7 overexpression is a predictive marker for high risk of malignant transformation in oral dysplasia. Int J Cancer, 2014. 134(6): 1379-88.
39. Kesting, M.R., H. Sudhoff, R.J. Hasler, M. Nieberler, C. Pautke, K.D. Wolff, et al. Psoriasin (S100A7) up-regulation in oral squamous cell carcinoma and its relation to clinicopathologic features. Oral Oncol, 2009. 45(8): 731-6.
40. Zhou, G., T.X. Xie, M. Zhao, S.A. Jasser, M.N. Younes, D. Sano, et al. Reciprocal negative regulation between S100A7/psoriasin and beta-catenin signaling plays an important role in tumor progression of squamous cell carcinoma of oral cavity. Oncogene, 2008. 27(25): 3527-38.
41. Rhee, D.K., S.H. Park, Y.K. Jang. Molecular signatures associated with transformation and progression to breast cancer in the isogenic MCF10 model. Genomics, 2008. 92(6): 419-28.
42. Emberley, E.D., R.D. Gietz, J.D. Campbell, K.T. HayGlass, L.C. Murphy, P.H. Watson. RanBPM interacts with psoriasin in vitro and their expression correlates with specific clinical features in vivo in breast cancer. BMC Cancer, 2002. 2: 28.
43. Emberley, E.D., Y. Niu, E. Leygue, L. Tomes, R.D. Gietz, L.C. Murphy, P.H. Watson. Psoriasin interacts with Jab1 and influences breast cancer progression. Cancer Res, 2003. 63(8): 1954-61.
44. Deol, Y.S., M.W. Nasser, L. Yu, X. Zou, R.K. Ganju. Tumor-suppressive effects of psoriasin (S100A7) are mediated through the beta-catenin/T cell factor 4 protein pathway in estrogen receptor-positive breast cancer cells. J Biol Chem, 2011. 286(52): 44845-54.
45. Yu, S.E., Y.K. Jang. The histone demethylase LSD1 is required for estrogen-dependent S100A7 gene expression in human breast cancer cells. Biochem Biophys Res Commun, 2012. 427(2): 336-42.
46. Sneh, A., Y.S. Deol, A. Ganju, K. Shilo, T.J. Rosol, M.W. Nasser, et al. Differential role of psoriasin (S100A7) in estrogen receptor alpha positive and negative breast cancer cells occur through actin remodeling. Breast Cancer Res Treat, 2013. 138(3): 727-39.
47. West, N.R., P.H. Watson. S100A7 (psoriasin) is induced by the proinflammatory cytokines oncostatin-M and interleukin-6 in human breast cancer. Oncogene, 2010. 29(14): 2083-92.
48. Paruchuri, V., A. Prasad, K. McHugh, H.K. Bhat, K. Polyak, R.K. Ganju. S100A7-downregulation inhibits epidermal growth factor-induced signaling in breast cancer cells and blocks osteoclast formation. PLoS One, 2008. 3(3): e1741.
49. Zhao, H., T. Wilkie, Y. Deol, A. Sneh, A. Ganju, M. Basree, et al. miR-29b defines the pro-/anti-proliferative effects of S100A7 in breast cancer. Mol Cancer, 2015. 14(1): 11.
50. Nasser, M.W., Z. Qamri, Y.S. Deol, J. Ravi, C.A. Powell, P. Trikha, et al. S100A7 enhances mammary tumorigenesis through upregulation of inflammatory pathways. Cancer Res, 2012. 72(3): 604-15.
51. Nasser, M.W., N.A. Wani, D.K. Ahirwar, C.A. Powell, J. Ravi, M. Elbaz, et al. RAGE mediates S100A7-induced breast cancer growth and metastasis by modulating the tumor microenvironment. Cancer Res, 2015. 75(6): 974-85.

52. Goh, J.Y., M. Feng, W. Wang, G. Oguz, S. Yatim, P.L. Lee, et al. Chromosome 1q21.3 amplification is a trackable biomarker and actionable target for breast cancer recurrence. Nat Med, 2017. 23(11): 1319-30.
53. Nasser, M.W., M. Elbaz, D.K. Ahirwar, R.K. Ganju. Conditioning solid tumor microenvironment through inflammatory chemokines and S100 family proteins. Cancer Lett, 2015. 365(1): 11-22.
54. Padilla, L., S. Dakhel, J. Adan, M. Masa, J.M. Martinez, L. Roque, et al. S100A7: from mechanism to cancer therapy. Oncogene, 2017. 36: 6749-61.
55. Donato, R., S100: a multigenic family of calcium-modulated proteins of the EF-hand type with intracellular and extracellular functional roles. Int J Biochem Cell Biol, 2001. 33(7): 637-68.
56. Heizmann, C.W., G. Fritz, B.W. Schafer. S100 proteins: structure, functions and pathology. Front Biosci, 2002. 7: d1356-68.
57. Leukert, N., T. Vogl, K. Strupat, R. Reichelt, C. Sorg, J. Roth. Calcium-dependent tetramer formation of S100A8 and S100A9 is essential for biological activity. J Mol Biol, 2006. 359(4): 961-72.
58. Bao, Y.I., A. Wang, J. Mo. S100A8/A9 is associated with estrogen receptor loss in breast cancer. Oncol Lett, 2016. 11(3): 1936-42.
59. Bergenfelz, C., A. Gaber, R. Allaoui, M. Mehmeti, K. Jirstrom, T. Leanderson, et al. S100A9 expressed in ER(-)PgR(-) breast cancers induces inflammatory cytokines and is associated with an impaired overall survival. Br J Cancer, 2015. 113(8): 1234-43.
60. Yin, C., H. Li, B. Zhang, Y. Liu, G. Lu, S. Lu et al. RAGE-binding S100A8/A9 promotes the migration and invasion of human breast cancer cells through actin polymerization and epithelial-mesenchymal transition. Breast Cancer Res Treat, 2013. 142(2): 297-309.
61. Cormier, K., J. Harquail, R.J. Ouellette, P.A. Tessier, R. Guerrette, G.A. Robichaud. Intracellular expression of inflammatory proteins S100A8 and S100A9 leads to epithelial-mesenchymal transition and attenuated aggressivity of breast cancer cells. Anticancer Agents Med Chem, 2014. 14(1): 35-45.
62. Drews-Elger, K., E. Iorns, A. Dias, P. Miller, T.M. Ward, S. Dean, et al. Infiltrating S100A8+ myeloid cells promote metastatic spread of human breast cancer and predict poor clinical outcome. Breast Cancer Res Treat, 2014. 148(1): 41-59.
63. Miller, P., K.M. Kidwell, D. Thomas, M. Sabel, J.M. Rae, D.F. Hayes, et al. Elevated S100A8 protein expression in breast cancer cells and breast tumor stroma is prognostic of poor disease outcome. Breast Cancer Res Treat, 2017. 166(1): 85-94.
64. Vrakas, C.N., R.M. O'Sullivan, S.E. Evans, D.A. Ingram, C.B. Jones, T. Phuong, et al. The measure of DAMPs and a role for S100A8 in recruiting suppressor cells in breast cancer lung metastasis. Immunol Invest, 2015. 44(2): 174-88.
65. Eisenblaetter, M., F. Flores-Borja, J.J. Lee, C. Wefers, H. Smith, R. Hueting, et al. Visualization of tumor-immune interaction – target-specific imaging of S100A8/A9 reveals pre-metastatic niche establishment. Theranostics, 2017. 7(9): 2392-2401.
66. Hiratsuka, S., A. Watanabe, H. Aburatani, Y. Maru. Tumour-mediated upregulation of chemoattractants and recruitment of myeloid cells predetermines lung metastasis. Nat Cell Biol, 2006. 8(12): 1369-75.
67. Li, Y.H., H.T. Liu, J. Xu, A.Y. Xing, J. Zhang, Y.W. Wang, et al. The value of detection of S100A8 and ASAH1 in predicting the chemotherapy response for breast cancer patients. Hum Pathol, 2018. 74: 156-63.
68. Duan, L., R. Wu, L. Ye, H. Wang, X. Yang, Y. Zhang, X. Chen, et al. S100A8 and S100A9 are associated with colorectal carcinoma progression and contribute

to colorectal carcinoma cell survival and migration via Wnt/beta-catenin pathway. PloS One, 2013. 8(4): e62092.
69. Ichikawa, M., R. Williams, L. Wang, T. Vogl, G. Srikrishna. S100A8/A9 activate key genes and pathways in colon tumor progression. Mol Cancer Res, 2011. 9(2): 133-48.
70. Zhang, X., L. Wei, J. Wang, Z. Qin, Y. Lu, X. Zheng, et al. Suppression colitis and colitis-associated colon cancer by anti-S100a9 antibody in mice. Front Immunol, 2017. 8: 1774.
71. Yun, S.J., C. Yan, P. Jeong, H.W. Kang, Y.H. Kim, E.A. Kim, et al. Comparison of mRNA, protein, and urinary nucleic acid levels of S100A8 and S100A9 between prostate cancer and BPH. Ann Surg Oncol, 2015. 22(7): 2439-45.
72. Grebhardt, S., K. Muller-Decker, F. Bestvater, M. Hershfinkel, D. Mayer. Impact of S100A8/A9 expression on prostate cancer progression in vitro and in vivo. J Cell Physiol, 2014. 229(5): 661-71.
73. Miao, L., S. Grebhardt, J. Shi, I. Peipe, J. Zhang, D. Mayer. Prostaglandin E2 stimulates S100A8 expression by activating protein kinase A and CCAAT/enhancer-binding-protein-beta in prostate cancer cells. Int J Biochem Cell Biol, 2012. 44(11): 1919-28.
74. Smyth, E.M., T. Grosser, M. Wang, Y. Yu, G.A. FitzGerald. Prostanoids in health and disease. J Lipid Res, 2009. 50(Suppl): S423-8.
75. Grebhardt, S., C. Veltkamp, P. Strobel, D. Mayer. Hypoxia and HIF-1 increase S100A8 and S100A9 expression in prostate cancer.Int J Cancer, 2012. 131(12): 2785-94.
76. Choi, D.K., Z.J. Li, I.K. Chang, M.K. Yeo, J.M. Kim, K.C. Sohn et al. Clinicopathological roles of S100A8 and S100A9 in cutaneous squamous cell carcinoma in vivo and in vitro. Arch Dermatol Res, 2014. 306(5): 489-96.
77. Iotzova-Weiss, G., P.J. Dziunycz, S.N. Freiberger, S. Lauchli, J. Hafner, T. Vogl, et al. S100A8/A9 stimulates keratinocyte proliferation in the development of squamous cell carcinoma of the skin via the receptor for advanced glycation-end products. PloS One, 2015. 10(3): e0120971.
78. Liu, K., Y. Zhang, C. Zhang, Q. Zhang, J. Li, F. Xiao, et al. Methylation of S100A8 is a promising diagnosis and prognostic marker in hepatocellular carcinoma. Oncotarget, 2016. 7(35): 56798-810.
79. Khammanivong, A., B.S. Sorenson, K.F. Ross, E.B. Dickerson, R. Hasina, M.W. Lingen, et al. Involvement of calprotectin (S100A8/A9) in molecular pathways associated with HNSCC. Oncotarget, 2016. 7(12): 14029-47.
80. McNeill, E., N. Hogg. S100A9 has a protective role in inflammation-induced skin carcinogenesis. Int J Cancer, 2014. 135(4): 798-808.
81. Minami, S., Y. Sato, T. Matsumoto, T. Kageyama, Y. Kawashima, K. Yoshio, et al. Proteomic study of sera from patients with bladder cancer: usefulness of S100A8 and S100A9 proteins. Cancer Genomics Proteomics, 2010. 7(4): 181-9.
82. Nicolas, E., C. Ramus, S. Berthier, M. Arlotto, A. Bouamrani, C. Lefebvre, et al. Expression of S100A8 in leukemic cells predicts poor survival in de novo AML patients. Leukemia, 2011. 25(1): 57-65.
83. Wang, W., K.X. Ai, Z. Yuan, X.Y. Huang, H.Z. Zhang. Different expression of S100A8 in malignant and benign gallbladder diseases. Dig Dis Sci, 2013. 58(1): 150-62.
84. Driemel, O., N. Escher, G. Ernst, C. Melle, F. von Eggeling. S100A8 cellular distribution in normal epithelium, hyperplasia, dysplasia and squamous cell carcinoma and its concentration in serum. Anal Quant Cytol Histol, 2010. 32(4): 219-24.

85. Zhang, L., H. Jiang, G. Xu, H. Wen, B. Gu, J. Liu, et al. Proteins S100A8 and S100A9 are potential biomarkers for renal cell carcinoma in the early stages: results from a proteomic study integrated with bioinformatics analysis. Mol Med Rep, 2015. 11(6): 4093-4100.
86. Basso, D., D. Bozzato, A. Padoan, S. Moz, C.F. Zambon, P. Fogar, et al. Inflammation and pancreatic cancer: molecular and functional interactions between S100A8, S100A9, NT-S100A8 and TGFbeta1. Cell Commun Signal, 2014. 12: 20.
87. Yasar, O., T. Akcay, C. Obek, F.A. Turegun. Significance of S100A8, S100A9 and calprotectin levels in bladder cancer. Scand J Clin Lab Invest, 2017. 77(6): 437-41.
88. Gebhardt, C., A. Sevko, H. Jiang, R. Lichtenberger, M. Reith, K. Tarnanidis, et al. Myeloid cells and related chronic inflammatory factors as novel predictive markers in melanoma treatment with Ipilimumab. Clin Cancer Res, 2015. 21(24): 5453-59.
89. Yang, L., M. Yang, H. Zhang, Z. Wang, Y. Yu, M. Xie, et al. S100A8-targeting siRNA enhances arsenic trioxide-induced myeloid leukemia cell death by down-regulating autophagy. Int J Mol Med, 2012. 29(1): 65-72.
90. Yang, M., P. Zeng, R. Kang, Y. Yu, L. Yang, D. Tang, et al. S100A8 contributes to drug resistance by promoting autophagy in leukemia cells. PloS One, 2014. 9(5): e97242.
91. Hibino, T., M. Sakaguchi, S. Miyamoto, M. Yamamoto, A. Motoyama, J. Hosoi, et al. S100A9 is a novel ligand of EMMPRIN that promotes melanoma metastasis. Cancer Res, 2013. 73(1): 172-83.
92. Ruma, I.M., E.W. Putranto, E. Kondo, H. Murata, M. Watanabe, P. Huang, et al. MCAM, as a novel receptor for S100A8/A9, mediates progression of malignant melanoma through prominent activation of NF-kappaB and ROS formation upon ligand binding. Clin Exp Metastasis, 2016. 33(6): 609-27.
93. Sumardika, I.W., C. Youyi, E. Kondo, Y. Inoue, I.M.W. Ruma, H. Murata, et al. Beta-1,3-galactosyl-O-glycosyl-glycoprotein beta-1,6-N-acetylglucosaminyltransferase 3 increases MCAM stability, which enhances S100A8/A9-mediated cancer motility. Oncology Res, 2018. 26(3): 431-44.
94. Chen, K.T., P.D. Kim, K.A. Jones, K. Devarajan, B.B. Patel, J.P. Hoffman, et al. Potential prognostic biomarkers of pancreatic cancer. Pancreas, 2014. 43(1): 22-7.
95. Basso, D., P. Fogar, M. Plebani. The S100A8/A9 complex reduces CTLA4 expression by immature myeloid cells: implications for pancreatic cancer-driven immunosuppression. Oncoimmunology, 2013. 2(6): e24441.
96. Basso, D., P. Fogar, M. Falconi, E. Fadi, C. Sperti, C. Frasson, et al. Pancreatic tumors and immature immunosuppressive myeloid cells in blood and spleen: role of inhibitory co-stimulatory molecules PDL1 and CTLA4. An in vivo and in vitro study. PloS One, 2013. 8(1): e54824.
97. Moz, S., D. Basso, A. Padoan, D. Bozzato, P. Fogar, C.F. Zambon, et al. Blood expression of matrix metalloproteinases 8 and 9 and of their inducers S100A8 and S100A9 supports diagnosis and prognosis of PDAC-associated diabetes mellitus. Clin Chim Acta, 2016. 456: 24-30.
98. Huang, H., Q. Huang, T. Tang, L. Gu, J. Du, Z. Li, et al. Clinical significance of calcium-binding protein S100A8 and S100A9 expression in non-small cell lung cancer. Thorac Cancer, 2018. 9(7): 800-4.
99. Hiratsuka, S., A. Watanabe, Y. Sakurai, S. Akashi-Takamura, S. Ishibashi, K. Miyake, et al. The S100A8-serum amyloid A3-TLR4 paracrine cascade establishes a pre-metastatic phase. Nat Cell Biol, 2008. 10(11): 1349-55.

CHAPTER 4

Nuclear Receptor Coactivators: Mechanism and Therapeutic Targeting in Cancer

Andrew Cannon[#], Christopher Thompson[#], Rakesh Bhatia and Sushil Kumar[*]

Department of Biochemistry and Molecular Biology, University of Nebraska Medical Center, Omaha, Nebraska, USA

Introduction

General Structure and Function of Nuclear Receptors Coactivators

Nuclear receptor coactivators contain highly conserved sequences that facilitate interaction with and recruitment of other nuclear proteins comprising the complexes necessary for chromatin remodeling that allow the promotion or inhibition of gene transcriptional activity. These structures generally include N-terminal located basic helix-loop-helix (bHLH) domain and two Per/Arnt/Sim (PAS-A and -B) domains, three centrally located leucine-rich sequences (LXXLL or FXXLF) known as the nuclear receptor (NR)-box, and two activation domains: AD1/CID (CREB-binding protein interacting domain) and AD2 situated at the C-terminal end. Through these domains, NCOAs form massive complexes with ligand-bound nuclear receptors, histone-modifying proteins (acetyl- or methyltransferases), other coactivators/corepressors/coregulators, and DNA-binding transcription factors. The N-terminal domains of NCOAs are conserved regions and have been experimentally shown to mediate dimerization of transcription factors with chromatin (bHLH) [1-3] and stabilize these transient interactions while facilitating interactions with the α-helical LXXLL NR-box domains of other proteins (PAS) [2, 4-6]. AD1/CID is also conserved across the paralogs and known to bind p300/CREB-binding protein while AD2 mediates the binding of histone methyltransferases [7]. The NR-box domain contains α-helical

---

[*]Corresponding author: skumar@unmc.edu
[#] Equal contribution

LXXLL or FXXLF sequence that binds activated nuclear receptors. The three principal structural domains of NCOAs are conserved with some slight differences. Functionality, similarly, is preserved throughout the NCOA family, but minor differences have been described. For example, in contrast to NCOA2, NCOA1 and NCOA3 possess moderate intrinsic acetyltransferase properties. The sequence and location of binding sites on NCOAs are predominately conserved; however, some differences have been observed. For instance, the interaction of NCOA2 and NCOA3 with aryl hydrocarbon receptor (AHR) requires the α-helix domain of aryl hydrocarbon receptor (AHR) bind the N-terminally located bHLH-PAS domain in NCOA2 and 3, but co-immunoprecipitation assays showed that NCOA1 has a unique sequence next to C-terminally located AD1/CID domain which can bind AHR [1, 8]. To add to this complexity, there is a high degree of variability in the presence of certain conserved domains in individual nuclear receptor coactivators. For example, a significant portion of known coactivators lacks NR-Box domains and interacts with nuclear receptors indirectly or through utilization of alternate interaction domains. This variability in the presence of conserved domains allows functional diversification of nuclear receptor coactivators, thereby conferring unique capabilities and limitations to an NR coactivator. Finally, NR-coactivators are tightly regulated by post-translational modifications (PTMs) including phosphorylation, acetylation, ubiquitinylation, and sumoylation. In some cases, these PTMs act as a molecular on/off switch, while in other cases, they tune the interaction of the NCOAs with binding partners and by extension the coactivator activity. Importantly, these PTMs allow cells to control NCOA activity in response to specific cellular conditions. Collectively, NCOAs have been shown to have diverse cellular actions and mediate the effects of ligand binding with numerous nuclear steroid receptors, including thyroid hormone receptor (THR), estrogen receptor (ERα and ERβ), progesterone receptor (PR), androgen receptor (AR), glucocorticoid receptor (GR), mineralocorticoid receptor (MR), retinoic acid receptor (RXR), peroxisome proliferator-activated receptors (PPAR), vitamin D receptor (VDR), and others [9-11], via their NR-box domains.

Nuclear Receptor Coactivators in Cancer

Given their ability to potentiate the function of various nuclear receptors, several nuclear receptor coactivators have been shown to play critical roles in the development and progression of cancer. While most of the work regarding the function of nuclear receptor coactivators in cancer has been conducted in the setting of hormone sensitive cancers, a significant body of literature exists for cancers that are not classically associated with hormone sensitivity. In the following section, we address the role of several nuclear receptor co-activator in hormone sensitive and hormone insensitive cancers (Table 1).

Table 1: Summary of nuclear receptor coactivators, their overarching function in each cancer, and interacting partners in each disease setting

NCOA	Cancer	Functions in cancer	Nuclear receptor partners
NCOA1	Breast	Angiogenesis, invasion and metastasis	HIF-1α, AP-1
	Prostate	Androgen receptor transcriptional activity, proliferation and invasion	ROR-γ, AR
	Colorectal	Cell invasion	Insufficient data
	Hepatocellular	Proliferation and diethylnitrosamine induced carcinogenesis	β-catenin
	Breast	Cell survival and proliferation	Insufficient data
	Prostate	Accelerated carcinogenesis, metastasis, insensitivity to androgen deprivation	AR
NCOA2	Colorectal	Promotion of apoptosis and suppression of cell growth	Insufficient data
	Hepatocellular	Suppression of tumorigenesis and cell proliferation	Insufficient data
	Breast	Promotion of cell proliferation and tumorigenesis	ER, PR, E2F1
	Prostate	Cell invasion, migration, and metastasis	ROR-γ, AP-1
	Colorectal	Cell proliferation, invasion, and promotion of carcinogen induced carcinogenesis	Insufficient data
NCOA3	Hepatocellular	Cell proliferation, invasion, and migration	Insufficient data
	Pancreatic	Promotion of EGFR signaling and mucin expression	RAR
	AML	Promote cell differentiation	VDR
	CML	Resistance to TRAIL induced apoptosis	Insufficient data
	NSCLC	Promote EGFR signaling	Insufficient data

(Contd.)

Table 1: (*Contd.*)

NCOA	Cancer	Functions in cancer	Nuclear receptor partners
NCOA4	Prostate	Transcriptional coactivation of AR, promote anchorage independent growth resistance to androgen deprivation	Aryl hydrocarbon receptor, PPARγ, AR
	Thyroid	Fusion with RET	Insufficient data
	Breast	Promotion of metastasis and mitochondrial biogenesis	PPARγ (presumed)
PGC1-α	Ovary	Drug resistance	PPARγ (presumed)
	Hepatocellular	Mitochondrial biogenesis, invasion and metastasis	PPARγ (presumed)
	Renal	Chemosensitization	PPARγ (presumed)
	Breast	Adhesion independent conditions, anti-estrogen resistance	AP-1, BRCA1
FHL2	Prostate	Prostate cancer cell invasion,	Androgen receptor
	Colorectal	Tumorigenesis, invasion, metastasis, EMT	Sp1, Snail1, KLF8, Smad2/3/4

Nuclear Receptor Coactivator 1 (NCOA1, SRC1)

Hormone Sensitive Cancers

In breast cancer, studies have revealed several mechanisms by which NCOA1 contributes to disease progression, and therapy resistance/failure. The formation of new tumor vasculature is critical for both sustaining the high metabolic demands of rapidly growing tumors as well as for the initiation of the invasion/metastasis cascade. In several mouse models, the expression of NCOA1 was associated with microvessel density (MVD) as assessed by CD31 staining and qRT-PCR expression of CD31 [12]. Knockout and overexpression of NCOA1 in these models resulted in decrease and increase in MVD, respectively [12]. NCOA1 mediated this effect by enhancing transcription of HIF-1alpha and AP-1 gene target, VEGFa. In patient samples, the association of NCOA1 expression with MVD was validated and high NCOA1 expression and high MVD was associated with decreased overall survival (OS) [12]. Consistent with these findings of more aggressive disease in patients expressing high levels of NCOA1, a second study found increased numbers of circulating tumor cells as well as increased involvement of the lung by metastatic lesions (normalized to primary tumor volume) in a Tg(Neu x hNCOA1) mouse model [13]. As in the previous study, NCOA1 acted as a coactivator of AP-1 transcriptional complex to augment CSF-1 expression [13]. Subsequent knockdown of CSF-1 abrogated lung colonization confirming the involvement of this cytokine in the metastatic process [13]. The combined effect of high NCOA1 and high CSF-1 expression resulted in a statistically significant decrease in OS of breast cancer patients compared to the low NCOA1 low CSF-1 expression cohort [13]. Additional studies of the role of NCOA1 in the breast cancer metastasis showed that knockdown of NCOA1 in mouse-derived breast cancer cells lines attenuated the migratory capacity of breast cancer cells through downregulation of AP-1-mediated expression of integrin A5 and disruption of focal adhesion kinase signaling [14].

Because of the reported role of NCOA1 in breast cancer progression, several studies have attempted to use NCOA1 expression as a prognostic marker. When NCOA1 is applied as prognostic marker in a cohort of 285 ER+ patients receiving tamoxifen, NCOA1 expression was significantly higher in patients who survived longer than six months without disease progression; however, this association was not significant in COX regression analysis of progression-free survival (PFS) [15]. Interestingly, when NCOA1 was examined as a differential prognostic marker between breast cancer patients with a history of smoking (n=50) and those without (n=48), high NCOA1 expression was associated with increased risk of mortality and disease recurrence in the smoking cohort only [16]. Cumulatively, these results suggest that NCOA1 may differentially modulate the response of breast cancer cells to various stimuli; in the setting of estrogen receptor inhibition, NCOA1 may compete with NCOA3, a known contributor to tamoxifen resistance, and minimize NCOA3-mediated tamoxifen resistance. In contrast, NCOA1 itself may mediate the effects of smoking induced signaling through Aryl-hydrocarbon nuclear receptors and contribute to more aggressive disease in smokers.

Similar to breast cancer, NCOA1 has also been studied in prostate cancer. In this setting, NCOA1 knockdown attenuated the proliferation of androgen receptor (AR) positive cell lines MDA-PCa and LNCaP, but not AR-negative cell line PC3 [17]. Further, NCOA1 knockdown significantly decreased AR+ and AR– cell line migration and invasion. Gene expression analysis by cDNA microarray showed that NCOA1 knockdown in AR+ MDA-PCa cells resulted in differential regulation of 433 genes compared to knockdown in PC3 cells, which modulate expression of only 41 genes, suggesting that the transcriptional signatures associated with NCOA1 in prostate cancer cells are dependent on AR [17]. This result is consistent with another study which showed that NCOA1 is partially responsible for the transcriptional activation of the AR locus in cooperation with ROR-γ [18]. Among the genes differentially regulated upon NCOA1 knockdown in MDA-PCa cells, PRKD1 was identified as being highly upregulated after NCOA1 knockdown. To further demonstrate the role of PRKD1 in inhibition of prostate cancer cell migration, the knockdown of PRKD1 in NCOA1 KD cells significantly rescued the migratory phenotype of AR+ prostate cancer cells [17]. From these two studies, it is possible that NCOA1 is a key regulator of AR transcriptional activity and/or the transcription of AR itself; however, it remains difficult to delineate the contribution of these mechanisms leading to the phenotype observe upon NCOA1 KD. In sum, the confluence of these results suggests that NCOA1 may be involved in the regulation of malignant cell behavior in the setting of hormone sensitive malignancy largely through differential regulation of gene expression. However, the contribution of NCOA1's in these pathologies is complicated by their cooperation with transcription factors which, inevitably make the function of NCOA1 highly context dependent. Notably, this context dependence emerges in breast cancer patient cohorts who are exposed to different treatments and/or environmental stimuli and in prostate cancer where the expression of AR may play a critical role in determining the effect of NCOA1 expression.

Hormone Insensitive Cancers

In contrast to hormone sensitive breast and prostate cancers in which the function of NCOA1 have been extensively studied, the literature regarding NCOA1 expression in hormone insensitive cancers is distributed across several cancer types. In the colon, NCOA1 overexpression is detected in around 80% of colorectal cancer (CRC) patients [19] while physiological expression of NCOA1 is limited to mucosal cells lining the crypts [20]. However, comprehensive studies on the consequences of NCOA1 expression in colorectal cancer is limited and constrained to a single publication by Meerson and Yehuda [21]. In this study, miR-4443 expression was induced downstream of MEK1/2 and cEBP signaling following leptin or insulin treatment of colorectal carcinoma cell line, HCT-116. Using an *in silico* and *in vitro* approach, miR-4443 was predicted and shown to target the 3' UTR of NCOA1. Despite not achieving statistical significance in that study, invasion assay using HCT-116 cells, transfected with miR-4443 seem to support that KD of NCOA1 may interfere with the invasive potential of colorectal cancer cells. These data are suggestive of a larger mechanism by which obesity, and associated reduced leptin secretion, may interfere with anti-proliferative and anti-invasive signals and permit higher NCOA1 expression in colorectal tissues, contributing to the increased risk of colorectal cancer.

While the role of NCOA1 in gastric cancer (GC) has not been explored extensively, a study by Frycz et al. using next-generation sequencing of GC patient tumor samples identified that tumors arising from the gastric cardia had significantly reduced expression of NCOA1 compared to surrounding normal tissue [22]. When comparing the expression of hormone receptors in matched gastric cardia tissue samples, expression of NCOA1, NCOR1, estrogen receptor 2 (ESRβ), androgen receptor (AR), steroid dehydrogenase (HSD3B1) and steroid sulfatase (STS) were significantly decreased in CRC samples; aromatase was significantly increased; but no significant differences were observed for ESRα, PELP1, CREBBP, or NR2F1. The significant decrease in STS and aromatase levels suggested that despite no detection of major estrogen synthesis inhibition, the inability of desulfation of STS substrates and reduced NCOA1 may be inhibiting estradiol-protective effects in the gastric cardia. However, this study did not address the role of reduced NCOR1 expression, which reduces the associated transcriptional repression exhibited by this nuclear protein.

Of the hormone insensitive cancers, the role of NCOA1 in hepatocellular carcinoma (HCC) has been the most investigated. Ma and group identified expression of microRNAs in HCC tumor samples and further explored the role of miR-105 in HCC [23]. Expression analysis revealed that miR-105 had lower expression in 134 HCC samples compared to normal tissues. *In silico* analysis of potential miR-105 targets predicted NCOA1 as a likely candidate. Survival and progression analyses demonstrated that high miR-105 expression, which was inversely correlated with NCOA1, was associated with longer overall survival and progression-free survival. NCOA1 silencing and miR-105 inhibition confirmed that NCOA1 expression is negatively regulated by miR-105, correlating to better HCC prognosis. Consistent with Ma and team's finding on the involvement of NCOA1 in HCC, Tong et al. [11] sought to explore NCOA1 in HCC mechanistically, and showed that 20 of 40 HCC-patient tumor samples had higher NCOA1 expression compared to matched normal tissue by immunoassay and a 3-fold higher expression by qRT-PCR analysis. Further, expression of NCOA1 was positively correlated with expression of proliferating cell nuclear antigen (PCNA), suggesting that NCOA1 expression may be involved in proliferation and progression of HCC. Tong and colleagues used an shRNA to KD NCOA1 in HCC cell line,

HepG2 and observed decreased cell proliferation, colony formation and DNA synthesis (measured by percentage EdU incorporation). Exploring the causes of cell cycle arrest, it was observed that NCOA1-overexpressing patient-derived HCC samples had almost a 2.5-fold increase in c-Myc expression, and HCC cell line transfected with NCOA1-targeting shRNA had a decrease in c-Myc expression compared to control cells. Concurrent transfection of shNCOA1 and HA-cMyc was able to rescue cell proliferation independent of NCOA1 expression. Lithium chloride-treated NCOA1 knockout mice showed significantly lower expression of c-Myc than wild-type mice, and co-immunoprecipitation detected NCOA1 and β-catenin interaction, indicating that the Wnt/β-catenin pathway cooperates with NCOA1 to induce c-Myc expression. *In vivo*, intraperitoneal injection of diethylnitrosamine produced fewer liver lesions in NCOA1 knockout mice compared to wild-type. The findings from this work clearly demonstrate the part NCOA1 plays in c-Myc expression downstream of Wnt/β-catenin signaling in the growth and progression of HCC.

Nuclear Receptor Coactivator 2 (NCOA2, GRIP1, SRC-2, TIF2)

Hormone Sensitive Cancers

In the breast cancer, comparatively little is known regarding the function of NCOA2. Despite this, NCOA2 appears to have a unique function in cancer cells. In both HeLa and MCF7 cells, NCOA2 is activated via phosphorylation at serine 736, downstream of the MAP kinase pathway. This phosphorylation event in turn augments the ability of NCOA2 to act as a coactivator at estrogen response elements [24, 25]. Despite these interesting and highly suggestive findings, the functional significance of phosphorylated NCOA2 has not been directly studied. In fact, only one study has investigated the functional significance of NCOA2 in breast cancer. This study investigated the distinct functions of p160 family proteins in breast cancer and tamoxifen resistance. Results demonstrated that knockdown of NCOA2 did not sensitize LCC2, a tamoxifen insensitive cell line, to tamoxifen *in vitro* [26]. Further analysis, however, revealed an increase in apoptosis and decreased proliferation of LCC2 cells in culture [26]. Cell cycle analysis revealed that NCOA2 knockdown resulted in G1/G0 arrest [26]. However, it is difficult to draw conclusions on the function of NCOA2 in breast cancer due to the paucity of functional studies, particularly in relation to the ability of NCOA2 to act as a coactivator of ER.

In contrast to breast cancer, NCOA2 has been studied extensively in prostate cancer. Analysis of alteration in AR related pathways in 281 prostate cancer patients identified that a copy number gain of 8q13.3 spans NCOA2 and was associated with increased expression of NCOA2 transcripts [27]. Overall, 17% of patients had gains in the NCOA2 genomic region and there was an increased association of this genomic change with metastasis; 8% of primary tumors and 37% of metastatic lesions showed increased expression of NCOA2 [27]. Additionally, three somatic mutations were observed in NCOA2 and these were predicted to be functionally significant. Because of the androgen sensitive nature of prostate cancer, the authors examined the role of NCOA2 in the regulation of AR transcriptional activity. In these studies, overexpression of NCOA2 primed prostate cancer cells to respond to decreased levels of androgen, and increase the overall response to androgen [27]. The identification of NCOA2 as a critical mediator of prostate cancer progression in human disease led to intensive functional study of NCOA2 in mouse models. Here, overexpression of NCOA2 (NCOA2^{OE}) in prostate epithelium drove the accumulation of prostatic intraepithelial neoplasms (PIN) and accelerated the development of prostatic adenocarcinoma in heterozygous PTEN deletion (PTEN$^{+/-}$) [28]. Further PTEN$^{+/-}$ NCOA2^{OE} mice demonstrated a marked increase in the number of lung and lymph node metastases compared to PTEN$^{+/-}$ mice. In human samples, both the expression of NCOA2 and the expression of NCOA2 signature genes were significantly elevated in metastatic samples as compared to primary tumors [28]. Interestingly, another study showed that NCOA2 knockdown in prostate cancer cell lines resulted in decreased fatty acid content associated with decreased utilization of glutamine for fatty acid synthesis and downregulation of genes involved in fatty acid synthesis (i.e. FASN) [29]. This study went on to demonstrate that the reprogrammed use of glutamine by NCOA2 augmented the ability of prostate cancer cells to survive in clonogenic and soft agar assays [29]. These findings were validated *in vivo* using orthotopic injection models; tumors derived from NCOA2 knockdown cells were smaller in volume and demonstrated decreased metastases. Additionally, NCOA2 KD tumors had decreased cellular glutamate and oleic acid concentrations [29]. Finally, in a prostate cancer patient cohort, researchers observed increased expression of both NCOA2 and FASN in metastatic prostate cancer tissue as well as increased concentrations of glutamate and oleic acid suggesting that the observations from the mouse model are consistent with human disease [29]. Because of the association of metastatic disease and castration resistance in human prostate cancer, further study focused on the contribution of NCOA2 to castration resistance. Here, castrated PTEN$^{+/-}$ mice showed substantial upregulation of NCOA2 as compared to uncastrated PTEN$^{+/-}$ mice. In cell lines as well as in prostate-specific deletion of NCOA2 in PTEN$^{+/-}$ mice (PTEN$^{+/-}$ NCOA2$^{-/-}$), loss of NCOA2 stunted the progression of PIN lesions to adenocarcinoma and decreased the capacity of prostate cancer cells to proliferate in the absence of androgen. These changes in disease response to loss of androgen signaling were attributed to NCOA2-mediated activation of the PI3K/AKT pathway. Further, in humans, high expression of NCOA2 was significantly associated with decreased time to biochemical recurrence (PSA) and disease-specific death [28]. Finally, as was the case in breast cancer, where NCOA2 was found to act as a bridge between activation of growth factor signaling and transcriptional activity of ER, NCOA2 was found to be downstream of EGF signaling through phosphorylation at S736 and in turn augmented transcriptional activity of AR [30]. While the functional consequences of this NCOA2 activity in both breast and prostate cancer remain to be explored, it may very well serve as a mediator of anti-hormone therapy resistance. Overall, NCOA2 has been shown to be a key mediator of the emergence of aggressive disease in prostate cancer through acting both as a transcriptional coactivator of AR, and metabolic reprogramming of cancer cells to support a malignant phenotype. While similar extensive studies of NCOA2 have not been conducted in breast cancer, this nuclear receptor coactivator may play critical roles in breast cancer.

Hormone Insensitive Cancers

Expression of NCOA2 in CRC has been investigated in a study [31], which identified a fusion transcript between LACTB2 and NCOA2 from CRC-patient tumor samples, which was absent in normal colonocytes. Albeit its expression was detected in only six of 99 tumor samples with no expression in adjacent normal samples. The study by Yu et al. demonstrate by RT-PCR and IHC that NCOA2 expression is downregulated in CRC tumors compared to normal colon tissue and its CRC-specific actions, in contrast to many other malignancies, is tumor suppressive. The group was able to demonstrate using two minimal NCOA2-expressing CRC cell lines, HCT116 and SW1116, that transfection of NCOA2 was able to abrogate colony formations, induce apoptosis, and reduce both migration and invasion of cancer cells. Additionally, overexpressing NCOA2 substantially reduced the volume and weight of murine flank xenograft tumors. Given that NCOA2-mediated transcription increases Frzb, a regulator of the Wnt/β-catenin pathway, Yu and colleagues proposed that NCOA2 downregulation in CRC cells might facilitate increased Wnt signaling, and Wnt-luciferase assay confirmed that NCOA2 transfection reduced β-catenin translational activity. Gene microarray further confirmed that NCOA2 transfection resulted in a 3-fold or greater reduction in 38% of Wnt-associated genes compared to vector-only transfected cells. Notably, this study overlooked the need to correlate NCOA2 expression in CRC-patients to disease characteristics such as progression rates or overall survival.

The function of NCOA2 in HCC is not adequately addressed; however, Suresh and colleagues identified a possible mechanism by which NCOA2 acts to suppress tumor formation [32]. With almost half of all human HCC cases exhibiting overexpression of MYC, the group conducted a *Sleeping Beauty*-mediated transposon mutagenesis and discovered that loss of NCOA2 with MYC overexpression increased the risk of tumorigenesis [33]. In a mouse model, they reveal that NCOA2 binds the proximal promoters of Vegf, Fgf1, and Masp1, with expression of all three upregulated in shNCOA2 KD. Conversely, NCOA2 KD decreased the expression of Shp, Dkk4, and Cadm4 in HCC cell line, Huh7. The KD of these three genes increased proliferation and tumor volume from orthotopic implantation. Overexpression of Shp, Dkk4, Cadm4, and Thrsp decreased cell proliferation and tumor volume, but only Shp and Cadm4 demonstrated reductions of orthotopic tumor burden in the setting of NCOA2 KD. These data together suggest a plausible mechanism by which NCOA2 acts with nuclear partners to repress liver tumorigenesis.

Nuclear Receptor Coactivator 3 (NCOA3, RAC3, SRC-3, AIB1, p/CIP)

Hormone Sensitive Cancers

Perhaps the most well studied nuclear receptor coactivator in the hormone-sensitive cancer is NCOA3. Initially, interest in NCOA3 in cancer began with the identification of NCOA3 as being upregulated in breast cancer [34]. Analysis of an unselected cohort of 105 primary breast tumors found that the NCOA3 locus was amplified in 10% of breast cancers and that 64% of tumors overexpressed NCOA3 [34]. Subsequent investigation of the function of NCOA3 in breast cancer revealed a myriad of roles in breast cancer development and progression. With respect to the development of breast cancer, heterozygous, whole-body knockout of NCOA3 in MMTV-neu mice resulted in decreased lateral side budding of mammary gland and significantly delayed the median time to mammary tumor development from 9 to 16 months [35]. Impressively homozygous knockout of NCOA3 prevented *neu* mediated tumor formation altogether [35]. These changes in mammary tumor development were further traced to decreased phosphorylation of neu and downstream signaling molecules, which were associated with decreased cell proliferation as assessed by PCNA staining [35]. Several additional studies support these findings. In the first of these studies, it was shown that like NCOA2, NCOA3 can act as signaling conduit from MAPK signaling to the activation of ER transcriptional activity [36]. Here it was shown that NCOA3 is phosphorylated downstream of the MAPK pathway and that this phosphorylation event results in augmented coactivator activity by NCOA3 partially through the recruitment of histone acetyltransferase p300 [36]. While the effect of NCOA3 phosphorylation on transactivation was not directly tested, it was observed that ER transcriptional activity was increased by the conditions resulting in NCOA3 phosphorylation [36]. Additionally, through distinct mechanisms, NCOA3 knockdown attenuated PI3K/AKT signaling downstream of IGF-1Rα stimulation as well as EGFR autophosphorylation and associated proliferative/prosurvival phenotypes in breast cancer cell lines [37, 38].

Similarly, the impact of NCOA3 expression on early events in breast cancer tumorigenesis was investigated through implantation models of MCFDCIS cells. In this model, inducible knockdown of NCOA3 caused a decrease in the formation and size of DCIS lesions as well as progression to invasive carcinoma. Of note, this study found that NCOA3 KD was associated with a disruption of HER2, HER3, and NTOCH leading to a decrease in the breast cancer initiating cell population [39].

While the above studies focused on the expression of full length NCOA3, similar roles in breast tumorigenesis have been shown for an NCOA3 transcript lacking exon3, also referred to as NCOA3-Δ3 that is expressed in both breast cancer cell lines as well as patient-derived tumor samples [40, 41]. The first study of NCOA3-Δ3 showed that expression of transgenic NCOA3-Δ3 in mice resulted in an increased rate of mammary epithelial cell proliferation with altered responses to growth factors and increased mammary gland weight after 13 months [42]. More interestingly, NCOA3-Δ3 expression in the setting of ERα overexpression receptor showed that compared to full length NCOA3, NCOA3-Δ3 resulted in overexpression of progesterone receptor and progesterone receptor target genes [43]. Further analysis of the function of NCOA3-Δ3 showed that its expression alters the response of breast and ovarian cancer cells to estrogens differing in natural activities [44]. These changes associated specifically with the expression of NCOA3-Δ3 suggest that they may play a unique role in augmenting the function of other hormones including progesterone in the process of breast cancer development.

In addition to the critical role of NCOA3 in the early stages of breast cancer development, several studies have reported roles for NCOA3 in other aspects of breast cancer biology. Not surprisingly, NCOA3 appears to be a critical co-activator of ER and PR in MCF7 cells [45]. In this setting, modest reduction in NCOA3 expression resulted in a statistically significant decrease in luciferase activity when expressed under estrogen and progesterone sensitive elements. Furthermore, reduction in NCOA3 expression resulted in decreased cell proliferation and attenuation of estrogen's capacity to inhibit apoptosis in MCF7 cells [45]. Consistent with this finding, implantation of NCOA3 KD or WT cells in athymic nude mice showed a marked reduction in both tumor incidence and in tumor size with NCOA3 KD [45]. Additional studies have suggested that the decreased proliferation of MCF7 with NCOA3 knockdown may be mediated by decreased expression of PLAC1 which has been shown to be critical regulator of breast cancer cell proliferation and a transcriptional target of NCOA3-ERα complex [46]. Interestingly, the interaction of NCOA3 with ERα and subsequent transcriptional activity in MCF7 has been shown to be dependent on the phosphorylation of both ERα as well as NCOA3 by CK-1δ indicating that posttranslational modification of NCOA3 may also contribute to the dysregulation of this signaling axis in breast cancer [47]. Finally, inhibition of the ERα DNA-binding domain was associated with an increase in tamoxifen sensitivity as well as loss of interaction between ERα and NCOA3 [48]. Concordantly, in 64 ER+ surgical samples, tamoxifen treatment was associated with increased NCOA3 expression but was insignificantly associated with poor clinical outcome [49]. Despite this evidence for the role of NCOA3 in ERα mediated breast cancer cell proliferation, the effects of NCOA3 on proliferation cannot strictly be attributed to its coactivation of ERα. For instance, an acetylated form of

NCOA3 has been shown to act as a coactivator of E2F1, a transcription factor required for progression through the cell cycle [50].

Because of this myriad roles in breast cancer biology, several studies have examined NCOA3 as a prognostic marker. In 2197 breast cancer patients, high NCOA3 expression was significantly associated with poor prognosis [51]. Consistent with these reports, two independent studies of 791 and 2727 breast cancer cases confirmed a SNP in the coding region of NCOA3 to be protective against the development of breast cancer [52, 53]. However, these results were not supported by a third case-control study of 172 Australian breast cancer patients though this lack of statistical significance may be due to the comparatively small sample size [54]. The fact that NCOA3 SNPs appeared to be protective against breast cancer indicates that disruption of NCOA3 function may result in a decreased propensity to develop breast cancer, but further studies are needed to fully characterize the mechanistic effect of the polymorphism in breast tissue.

The role of NCOA3 in prostate cancer has not been as extensively studied as in breast cancer. Despite this, several studies have revealed an important role of NCOA3 in prostate cancer. In the normal prostate, NCOA3 is expressed in the basal epithelium as well as in the surrounding stromal cells, but not in the luminal epithelium [55]. With oncogenesis, NCOA3 expression increases in the luminal epithelial cells. Further, NCOA3 expression increases with decreasing differentiation status. Depletion of NCOA3 in an SV40 transgene-induced model of prostate cancer impaired the progression of prostate cancer from well-differentiated to poorly-differentiated, metastatic prostate cancer resulting in an increase in survival of mice from 26.5 weeks to greater than 39 weeks after NCOA3 depletion [55]. While this study did not explore the mechanism by which NCOA3 contributes to prostate cancer progression, other studies have supported several possible mechanisms. One study demonstrated that NCOA3 and NCOA1 can function as coactivators of ROR-γ to increase the transcription of AR in metastatic, castration resistant prostate cancer thereby driving the malignant phenotype [18]. However, in the NCOA3 knockout model, loss of NCOA3 was not associated with changes in the expression of androgen receptor or NCOA1 (a possible compensatory coactivator), suggesting an alternate mechanism at least in the NCOA3 knockout TRAMP model [55]. An additional study demonstrated that NCOA3 expression is associated with increased invasion of the seminal vesicle, lymph node metastases, and biochemical recurrence. Subsequently, this study demonstrated that NCOA3 knockdown in AR-PC3 cells resulted in abrogation of PC3 cell mobility, decreased formation of lamellapodia and loss of FAK signaling and focal adhesion turnover [56]. These changes in phenotype and cellular signaling were accompanied by a decrease in AP-1 mediated expression of matrix metalloproteinase 2 and 13 [56]. These alterations in invasive phenotypes are consistent with those observed in the NCOA3$^{-/-}$ TRAMP model but require validation of a similar phenomenon in an autochthonous model of prostate cancer. Finally, other studies has demonstrated that NCOA3 is involved in the regulation of various components of the insulin-like growth factor pathway. In the first study, Zhou et al. demonstrated that inducible overexpression of NCOA3 in LNCaP and PC3 cells resulted in significantly increased cell size. This was further connected to augmented AKT and MTOR activation through an undetermined mechanism [57]. In a follow up study using a similar inducible system in PC3 cells, the same group demonstrated that NCOA3 expression upregulated the expression of insulin-like growth factors 1 and 2 as well as insulin receptor substrate 1 and 2 PIK3CA, AKT and BCL2. Subsequent analysis revealed that this upregulation occurred through transcriptional coactivation of AP-1 [58]. Despite the significance of these findings, their relation to prostate cancer progression and the phenotype seen in NCOA3$^{-/-}$ TRAMP mice remain nebulous due to a lack functional studies *in vitro* and *in vivo* validation. Irrespective of these shortcomings, the fact that NCOA3 acts as a coactivator for the AP-1 transcription factor complex is strongly supported by multiple studies and likely plays a meaningful role in the phenotype of the NCOA3$^{-/-}$ TRAMP prostate cancer model.

Hormone Insensitive Cancers

The cancer-associated functions of NCOA3 have been explored to the greatest extent in CRC. It has been shown that NCOA3 is overexpressed in about 35% of CRC patients at the protein level with the greatest expression in later stage diseases [59, 60] and about 10% of CRC patients have detectable amplification of NCOA3 transcription [59, 61]. Recent work by Li and group determined that degradation of NCOA3 in CRC cell lines was able to abrogate colony formation, reduce proliferation and inhibit migration; further NCOA3 undergoes degradation mediated by ubiquitylation requiring prerequisite phosphorylation by p38 [62]. NCOA3 has been shown to have a diverse collection of nuclear binding partners, including those normally expressed in the gut like EGFR [38] and those detected only in CRC like ERβ [63]. While NCOA3 expression in cancer cells may likely be induced, in part, by stromal signaling and has been shown to cross-talk with the FGF-signaling pathway to promote wound-induced recruitment of inflammatory immune cells, stimulating angiogenesis and inflammation [64], these latter aspects of NCOA3 have not yet been investigated within the context of malignancy-associated angiogenesis and progression. Mo and colleagues did extensive studies to elucidate the cooperative transcriptional actions of NCOA3 and Notch signaling [60]. Using NCOA3 KD in human CRC cells, RKO and murine CRC fibroblasts, CT26, they demonstrated that cellular proliferation was substantially slowed. Interestingly, NCOA3 knockdown reduced cyclins A2 and E2 and Hes1 in CRC RKO cells. Consequently, the percentage of RKO cells in G1-phase increased significantly while the percentage of cell in S-phase significantly decreased. As expected, Hes1 downregulation increased the expression of Hes1-targeted genes p27, ATOH1, and MUC2, suggesting that the NCOA3-Hes1 axis results in regulation of expression of several genes that may be important to CRC tumor establishment and progression.

Curiously, NCOA3 knockdown in mouse CRC fibroblasts, CT26 also decreased cyclin B1 expression, but the anticipated decrease of maturation-promoting factor (MPF)-complexes did not result in increased cells at the G2/M-phase transition. Interrogating the effects of the Notch intracellular domain (NICD) and its enhancer Mastermind-like Protein-1 (MAML1), Mo et al. showed that NCOA3 knockdown reduced transcription of luciferase under Hes1 promoter, which was recovered by NCOA3 transfection. NCOA3 was shown to bind to NICD and MAML1 while recruitment to the Hes1 promoter was shown to occur independent of NICD. However, ChIP-reChIP assays demonstrated that NCOA3 and NICD were simultaneously recruited to the Hes1 promoter. NICD transfection also recovered the decline in cell proliferation induced by NCOA3 KD. Murine flank implantation of shNCOA3-transfected CT26 cells produced smaller and slower growing tumors, and ki-67 staining of tumors showed that proliferative markers did not recover in tumors compared to tumors from CT26 sham-transfected cells. In vitro matrigel-invasion of NCOA3 KD RKO and CT26 cells was significantly reduced. Tail-vein injection of NCOA3 knockdown CT26 cells produced substantially fewer lung lesions, and these lesions failed to recover NCOA3 or Hes1 expression while p27 expression remained higher than control cell lesions. Again, tumor section staining for ki-67 confirmed that NCOA3 KD cells had lower proliferation than control cells. NCOA3-deficient mice also developed grossly fewer azoxymethane/dextran sulfate sodium (AOM/DSS)-induced colon lesions, while those lesions that did grow were restricted in size with the majority smaller than 2.5 mm while control tumors were mostly 2.5-5 mm. Extending these findings about NCOA3 expression in CRC, Edvardsson and team explored signaling events contributing to its overexpression. CRC is known to aberrantly express estrogen receptor beta (ERβ). Using CRC cell line, SW480, they identified several miRNAs that were downregulated by ERβ signaling. Additionally, siMYC transfection downregulated two miRNAs identified with ERβ signal: miR-17 and miR-19a. Transfection of miR-17 mimic in ERβ activated SW480 cells increased the number of live cells while transfection of a miR-17 inhibitor increased the percentage of dead cells. ERβ activation increased expression of NCOA3 by nearly 2-fold, and miR-17 inhibitor transfection of SW480 control cells increased NCOA3 by 0.5-fold; simultaneous ERβ activation and miR-17 mimic transfection resulted in reduced NCOA3 expression. Further, ERβ activation decreased cell proliferation as measured by scratch assay. This response was determined to be mediated by an increased expression of ZEB1 and downregulation of E-cadherin. These two studies suggest that NCOA3 may play a pleiotropic role in the development of CRC in a stage-specific context.

In addition to the well supported findings in CRC, NCOA3 has been reported to have a function in hematologic malignancies mostly within the confines of therapeutic intervention. One study found that GSK3 inhibition hyperphosphorylated the vitamin D receptor (VDR) and NCOA3 expression increased in acute myeloid leukemia cells [65]. After 1,25-vitamin D treatment, the interaction of NCOA3 and VDR was detected by CoIP. The resulting transcriptional cascade induced differentiation in AML cells. Conversely, a second study found that chronic myeloid leukemia (CML) overexpressed NCOA3 and its KD increased CML cell sensitivity to TNF-related apoptosis-inducing ligand (TRAIL) [66]. NCOA3 silenced CML cells showed a decrease in basal NF-κB and TRAIL-treatment induced caspase and BID activations and cleavage of PARP. Regulation of cas8 was reduced, and cytochrome c was released indicating mitochondrial damage induced by decreased expression of cFLIP. These two studies suggest that treatment of hematological cancers may be subtype-specific when considering NCOA3 targeting.

Expanding on visceral cancers, HCC are highly aggressive malignancies with high rates of multinodular advancement and morbidity and limited groups have investigated the expression of NCOA3 in these tumors. A few studies have discovered that expression is highly specific to cancer compared to normal liver and high expression correlates with shorter overall survival [67-69]. Xu and colleagues conducted mechanistic studies and saw that in three HCC cell lines, NCOA3 KD abrogated cell proliferation, migration, invasion, and colony formation [69]. Also, *in vivo* studies demonstrated that NCOA3 KD inhibited tumorigenicity. In those tumors that did develop, p21 expression was upregulated and, as expected, significantly more cells were arrested in G1 phase. NCOA3-derived xenograft tumors also showed reduced expression of PCNA and MMP9 with decreased Akt activation. These results were duplicated by treating HCC cells with a PI3K inhibitor. Interestingly, transfection of NCOA3 did not yield inverse findings.

Similarly, NCOA3 is hypothesized to be a negative prognostic factor in non-small cell lung carcinoma (NSCLC) [70-72]. In different patient cohorts, its expression is highly specific to malignant cells and the degree of expression increases in a tumor grade-specific manner with male patients shown to express higher levels of NCOA3 than female counterparts [72]. While overexpression of NCOA3 was found in 48% and amplification in 8% of patients, statistical analysis suggests that level of expression is correlated only with positive node status and shorter overall survival, most significantly in Stage III patients [71]. Lahusen et al. showed that NCOA3 KD was sufficient to abrogate EGFR signaling and subsequent cyclin D1 induction in A549 and H1975 NSCLC cells [38]. Additionally, both total and phosphorylated ERK1/2 were reduced with concurrent inhibition of pSTAT5 and pJNK. Similar to the function of NCOA3 in pancreatic cancer, EGFR phosphorylation was shown to occur in an NCOA3-dependent manner.

The molecular pathways aberrantly expressed in pancreatic cancer (PC) are complex and incompletely understood. The role that aberrant NCOA3 expression may play in these pathways of PC establishment, progression, and metastasis are yet to be wholly elucidated. Henke and colleagues observed that expression of NCOA3 in PC cell line COLO357 and two of its metastasis-educated subclones were high and linked this overexpression to elevated gene copy numbers (CNV) with 5-6 copies per cell typically detected [73]. IHC and *in situ* hybridization (ISH) for NCOA3 protein and mRNA respectively demonstrated that aberrant expression begins in high-grade PanIN lesions and peaks in adenocarcinoma. Loss of exon 3 from NCOA3 splicing (NCOA3-Δ3) generates a more active transcript, which was detected abundantly with the three PC cell lines by both IHC and ISH. In a wider PC cell line screening, Ghadimi et al. assessed CNV and discovered that 6 of 9 PC cell lines had elevated alteration of chromosome 20q12 corresponding to the location of the NCOA3 gene and FISH confirmed increased copies of NCOA3 [74]. The pathobiological effects of high NCOA3 expression were shown to correlate with higher risk of lymph node metastasis and inversely correlated with E-cadherin (a marker of epithelial-to-mesenchymal transition) [75], but NCOA3 is likely involved in other aspects of PC. The mucin (MUC) family of glycoproteins is also aberrantly and highly expressed in PC and have been linked to early malignant transformation, progression, metastasis, and therapy resistance; therefore, Kumar and coworkers investigated the role elevated NCOA3 expression in PC might have on the expression of two commonly detected MUC: MUC1 and MUC4 [76]. They showed that NCOA3 was overexpressed in six PC cell lines while expression of two other PC lines was similar to normal pancreatic cells. MUC4 expression, absent in normal pancreatic epithelium, was detected by immunostaining in 5 of the 6 NCOA3 amplified lines. NCOA3 knockdown in CD18/HPAF PC cells decreased expression of both MUC1 and MUC4 transcripts and protein. Immunostaining of PC tumors collected from rapid autopsies demonstrated that NCOA3 expression ranged from light to heavy compared to absent in normal pancreas. To explore the interaction of NCOA3 and retinoic acid (RA), a known activator of MUC4 transcription, NCOA3 KD in CD18/HPAF cells were treated with RA. Detection of MUC4 transcripts induced by RA was reduced in NCOA3 KD cells. ChIP assay confirmed that NCOA3 was being recruited to the MUC4 promoter with and without RA treatment. Further, micrococcal nuclease digestion of NCOA3 KD cells showed that NCOA3 increase the accessibility of the MUC4 promoter and its KD reduced the MUC4 expression by decreasing the accessibility of MUC4 promoter to transcriptional factors. Gene network analysis revealed a cluster of up- and down-regulated genes to NCOA3 overexpression. One of those up-regulated was FUT8, a fucosyltransferase that is known to stabilize proteins, suggesting that NCOA3 may have a secondary role of stabilizing MUC4, as FUT8 KD decreased MUC1 and MUC4 protein but had no effect on transcription. Similarly, Lahusen and team showed that NCOA3 acts to enhance EGFR signaling in PC cell line PANC-1 and its KD abrogates this signal and inhibits EGFR-mediated colony formations [38]. While EGF treatment increased phosphorylation of AKT and ERK1/2 without any abrogation by siNCOA3 KD, it did decrease phosphorylation of STAT5 and JNK (trend confirmed in secondary PC cell line COLO357). Treatment with JNK inhibitor decreased EGF-induced COLO357 absorbance by WST-1 assay, indicating that EGF is signaling at least in part via JNK activation. Immunoblot concluded that NCOA3 facilitates phosphorylation of EGFR. In summary, NCOA3 functions in PC to induce myriad biological effects due to promiscuous interactions with receptors and pathway activations.

NCOA3 is one of the most studied nuclear receptor coactivators and despite the breadth of data generated regarding its biological functions, its role in a wider range of malignancies lacks satisfactory conclusions. However, several groups have published on the expression and function of NCOA3 in some rarer, hormone-insensitive cancers and offer some incremental evidence of their larger impact on disease biology. NCOA3 expression has been found to be amplified in gastric cancer, but some discrepancy exists between the two studies with one citing amplification in 7% of cases [77] and the other finding amplification in 35% of cases [78]. In the former, overexpression of NCOA3 at the protein level was detected in 40% of the cases. Both groups found that all individuals with gene amplification also had overexpression but not the inverse. Likewise, both groups correlated NCOA3 overexpression with increased proliferation, invasion, metastasis, and worse prognosis. The later study by Shi et al. found that NCOA3 KD xenograft implantation had reduced proliferation, invasion, and tumorigenicity *in vivo*. Tumors from NCOA3 KD cells showed an increased population of cells in S-phase but also elevated apoptosis. Further, these changes were linked to modulations in ErbB and Wnt/β-catenin signaling. NCOA3 is also overexpressed in esophageal cancer in 46-64% of cases [79-81]. Overexpression was nearly universally detected in higher grade tumors and correlated with increased gene copy number and chemo- and radiotherapy resistance. He and colleagues also found that overexpression of NCOA3 correlated with increased distant lymph node metastasis but not to local node metastasis [79], a similar trend also observed in papillary thyroid cancer [82].

Nuclear Receptor Coactivator 4 (NCOA4)

Hormone Sensitive Cancers

While there is a lack of conclusive studies on NCOA4 in breast cancer, the role of NCOA4 in prostate cancer has been well-investigated. Interest in NCOA4 in prostate cancer began with the identification of NCOA4 as a specific coactivator of AR [83]. The association of NCOA4 with prostate cancer was further strengthened by a repository of GWAS studies showing mainly that two SNPs in chromosome 10q11 were significantly associated with risk of developing prostate cancer. The first and strongest of these was the rs10993994 SNP which is located upstream of, and in close proximity to the NCOA4 promoter region while the second SNP rs10761581 is located within the NCOA4 gene [84-87]. Interestingly, the rs10993994 was shown to cause an upregulation of NCOA4 transcript expression which was further correlated with an increase in anchorage dependent growth in prostate cancer cell lines [85]. Several follow up studies showed that NCOA4 may be involved in the development of castration resistance through agonist activities to classical antiandrogens thereby resulting in stimulation of AR transcriptional activity

[88, 89]. When the interaction of NCOA4 with AR is disrupted by mutations in NCOA4, the agonist effects of antiandrogens are significantly subdued [89]. Similarly, NCOA4 has been proposed as a mediator of the hormone refractory effects of overexpression of PSA in prostate cancer cell lines through coactivation of AR transcriptional programing [90]. Interestingly, further investigation into the function of NCOA4 in the setting of prostate cancer has suggested that full length NCOA4α may in fact impede prostate cancer proliferation in cell lines and tumor growth in xenograft models [91], whereas NCOA4β, a 35kDa splice variant, augments both AR dependent transcriptional activity and prostate cancer proliferation in cell lines [92]. Importantly, this group observed that NCOA4β expression increased with prostate cancer progression [92] whereas expression of NCOA4α was decreased with progression [91, 93] lending further support to the tumor promoting function of NCOA4β in prostate cancer [92]. In addition to its role as a coactivator of AR, NCOA4 has also been reported to function as a coactivator of aryl hydrocarbon receptor as well as PPARgamma [94, 95]. Interestingly, this interaction is dependent on the presence of ligand and squelched by co-expression of androgen receptor, but, at least in the case of aryl hydrocarbon receptor, is independent of the nuclear receptor binding domains of NCOA4 [94, 95]. Irrespective of these findings, the functional significance of aryl hydrocarbon receptor and PPARgamma coactivator activity in prostate cancer remains difficult to determine [94]. Cumulatively, these results indicate that NCOA4 may play a role as a coactivator of AR or other nuclear receptors in the setting of prostate cancer. However, the intricacies of the function of NCOA4 in prostate cancer cells, including the role of isoforms and splice variants and the function of various coactivator activities, largely remain to be parsed out by more detailed studies.

Hormone Insensitive Cancers

In hormone insensitive cancers, the actions of NCOA4 in cancer pathobiology do not seem to broadly involve the wild-type transcript. Many studies have indicated, however, that physical modification of chromosome 10 subjugates the NCOA4 and RET genes to a breakage-repair event resulting in a fusion product. Ghandhi et al. conducted chromosomal rearrangement analysis and observed that NCOA4 and RET genes are separated by nearly 8 Mb, but interphase coiling brings them within 0.87 µM of each other, closer than any other RET neighboring genes [96]. Despite the proximity of NCOA4 and RET, this fusion event is rare and has been detected in only a handful of patients with different malignancies such as thyroid cancers [97-101], NSCLC [98, 99, 102, 103], CRC [98, 104-106], and salivary carcinoma [107]. NCOA4-RET fusion seems to be more prominent in papillary thyroid carcinoma (PTC) with around 10% of patients harboring a RET fusion and about 90% of those fusions occur with NCOA4 [96-100]. These studies also demonstrate that the pathological consequences of NCOA4-RET fusion are likely attributed to the constitutive activation of the RET tyrosine kinase domain and subsequent

MAPK signaling. Interestingly, patients having this particular mutation were found to have no other mutations strongly implicated in their respective cancers. The conclusions made by each of these studies are that the minor groups of patients with tumors harboring NCOA4-RET fusions are likely to benefit from treatment with multi-tyrosine kinase inhibitors instead of other first-line therapies and personalized approaches to cancer intervention may likely require identifying this mutation at diagnosis.

Peroxisome Proliferator-activated Receptor-gamma Coactivator 1-alpha (PGC-1α, PPARGC1A)

PGC-1α is classically considered a coactivator of the peroxisome proliferator-activated receptor gamma transcription factor which together are considered to be a master regulator of mitochondrial biogenesis and function. Given the metabolic demands of malignant cells, it is not surprising that a significant body of literature has demonstrated the involvement of PGC-1α in the progression of breast and ovarian cancer. In a seminal paper on the subject, LeBleu et al. demonstrated that circulating tumor cells increased oxidative phosphorylation and elevated levels of PGC-1α expression compared to primary tumors in a 4T1 orthotopic tumor model and that knockdown of PGC-1α reduced the ability of 4T1 cells to metastasize particularly to lungs [108]. This was shown to occur through the production of energy necessary for metastasis rather than through the direct modulation of genes involved in the invasion-metastasis cascade [108]. Additional studies have supported this role of PGC-1α and demonstrated that expression of miRNA-485-3p and miRNA-485-5p directly target PGC-1α and that their expression in breast cancer cell lines suppresses the ability of cells to migrate *in vitro* and to metastasize *in vivo* [109]. A similar study showed that suppression of PGC-1α expression by miRNA-217 augmented the proliferative capacity of breast cancer cells; however, this is not exclusive of the less invasive phenotype observed with PGC-1α depletion in other studies [110]. Interestingly, additional studies on the role of PGC-1α identified cell lines with a propensity for metastasis to specific sites that is an organ tropism. Analysis of these lines identified that cell lines which preferentially metastasize to the lungs, and, to a lesser extent, the bones express higher levels of PGC-1α than those that do not metastasize or those that metastasize to the liver [111]. Additionally, forced expression of PGC-1α augmented the ability of breast cancer cells to metastasize in orthotopic and tail-vein injection models. Further, when these cells were treated with phenformin, an inhibitor of complex I of the electron transport chain, RNAseq analysis showed a decrease in the number of differentially regulated genes in PGC-1α overexpressing cells. This finding led to the hypothesis that expression of PGC-1α resulted in a plasticity of cellular metabolism that allowed for rapid cellular stress response [111]. Knockdown of PGC-1α resulted in increased cell death in response to acute and chronic metformin treatment suggesting a resistance to metabolic drugs

[111]. While this finding is interesting, it lacks immediate applicability. However, another study independently demonstrated that ovarian cancer stem-like cells showed greater expression of PGC-1α and that this expression drove the production of reactive oxygen species resulting in resistance to treatment with platinum and taxol drugs [112]. Specifically, overexpression of PGC-1α in non-stem-like cells reproduced the drug resistance associated with stem-like cell populations in this study [112]. Critically, high plasma levels of PGC-1α is an independent marker of poor prognosis in a cohort of 267 breast cancer patients [113]. In sum, the expression of PGC-1α in the setting of hormone sensitive cancer portends a metabolic plasticity that can act to drive both tumor metastasis as well as drug resistance.

While other nuclear-receptor coactivators were of significant interest in cancer due to their interactions with steroid hormone receptors, PGC-1α is most known for its role in cellular metabolism and thus has received significant attention in traditionally steroid hormone-independent cancer. Sustained activation of PGC-1α is prevalent in HCC and contributes to upregulation of peroxide-producing metabolic reactions and accumulation of DNA-damaging free radicals [114]. Investigation into the function of PGC-1α in HCC suggested that mitochondrial biogenesis is enhanced and may, therefore, fuel a cycle of perpetual cellular insult. Work by Li and colleagues demonstrated that SIRT1 KD abrogates mitochondrial biogenesis-mediated invasion and metastasis of HCC orthotopic tumors [115]. Further, PGC-1α is responsible for mediating the expression of gluconeogenic genes in liver cells while activated YAP, responsible for hepatic repair mechanisms and common to hepatic malignancies, suppresses these transcriptional activities as shown by Hu et al. [116]. In fact, Hu and team demonstrated that suppression of PGC-1α by HIF in renal cells inhibits mitochondrial biogenesis and acts to shift renal cells away from aerobic respiration; however, PGC-1α expression in renal cell carcinoma attenuates tumor growth and sensitizes cancer cells to cytotoxic drugs. A similar observation was made in which PGC-1α expression increased the efficacy of chemotherapeutic intervention in melanoma cells [117-119]. Counterintuitively, PGC-1α expression is inversely correlated with melanogenesis in melanoma cells, and hyperpigmentation has been shown to enhance resistance to chemo- and radio-therapies [120]. Colonocytes, on the other hand, have been shown to utilize PGC-1α expression to regulate ROS accumulation and promote physiologic apoptosis in a Bax-mediated manner [121] while PGC-1α overexpression in multiple myeloma facilitates ROS quenching and therapeutic resistance. Interestingly, pancreatic cancer stem cells demonstrate reduced MYC expression and upregulated PGC-1α which contribute to their rigid dependence on oxidative phosphorylation. Thus, the reversal of their expression has been suggested as a potential therapeutic strategy to eliminate the cancer cell population responsible for PC aggressiveness [122]. The sum of these findings suggests that the variable expression of PGC-1α in cancers makes its therapeutic targeting tissue- and malignancy-dependent.

Four-and-a-Half Lim Domains Protein 2 (FHL2)

FHL2 is classically a membrane associated protein, but has also been reported to function as a transcriptional coactivator or corepressor of several receptors associated with cancer. In breast cancer, FHL2 overexpression was shown to increase the expression of p21 both in a cell cycle dependent manner and in response to treatment with doxorubicin [123]. FHL2-mediated p21 regulation occurred through the activation of AP-1 transcriptional complex. This upregulation of p21 prevented cell cycle progression through G2/M phase. Cell cycle arrest induction by FHL2 is consistent with an additional study that showed FHL2 is an interaction partner of BRCA1, although this study did not specifically explore the role of FHL2 with respect to DNA repair mechanisms [124]. Despite the role of FHL2 in cell cycle arrest, which suggests a role of FHL2 as a tumor suppressor, its expression in a non-malignant cell line derived from breast tissue showed that FHL2 expression increased the ability of cells to grow in adhesion independent conditions [123]. These findings indicate that FHL2 may have the potential to transform cells or alternatively play a role in tumor initiating cells. In addition to the role of FHL2 in cell cycle arrest in breast cancer cells, studies have suggested other roles for FHL2. With respect to antihormone resistance, Fan et al. showed that long-term culture of breast cancer cells in the absence of estrogens results in sensitization to the apoptosis induced by physiologic concentrations of estrogen [125]. Interestingly, this was shown to be mediated by c-SRC, and inhibition of c-SRC with concomitant estrogen treatment causes cell proliferation and ultimately the emergence of a population of cells that is proliferative in the presence of both tamoxifen and estrogen independent of estrogen receptor [125]. Analysis of this cell population showed that estrogen and tamoxifen treatment resulted in upregulation of several genes including an enrichment of genes associated with the cytoplasmic membrane including FHL2 [126]. The authors claim that this model is reminiscent of antiestrogen resistance in humans, and thus FHL2 is implicated as a putative mediator of this resistance phenotype; however; more detailed studies of the role of FHL2 in this setting are required before firm conclusions can be drawn [126]. Similarly, another study investigated FHL2 in terms of cytoskeletal dynamics. In this study, the partial knockdown of FHL2 reduced breast cancer cell invasion. The study went on to show that expression of full length, but not N- or C-terminal domains of MabP1, inhibited the pro-invasive activity of FHL2 [127]. Ultimately, the pro-invasive activity of FHL2 was tied to its localization at focal adhesions [127]. This finding is similar to those associated with other nuclear receptor coactivators, such as NCOA3, indicating that transcriptional activity may only be a part of the functional role of a nuclear receptor coactivator in physiology and disease.

FHL2 has also been implicated in several facets of prostate cancer pathology. Notably, FHL2 was identified as a transcriptional coactivator of AR in prostate cancer cells and its interaction with RHO GTPase allows the transmission of signals from cell surface to the nucleus [128, 129]. In light of this, a subsequent study showed that nuclear expression of FHL2 was significantly associated with relapse in a prostate cancer cohort [130]. Interestingly, it was later shown that the induction of FHL2 is highly androgen dependent in luminal prostate cells and thus makes up a feed-forward loop with androgen receptor signaling [131]. The expression of FHL2 in prostate cancer cells was also shown to potentiate ETS-1-mediated increased prostate cancer cell invasion [132]. Perhaps more importantly in prostate cancer progression, the expression of AR splice variants is essential to overcome castrate conditions. McGrath et al. demonstrated that in castrate conditions, FHL2 is retained in the cytoplasm by interaction with filamin. When filamin is cleaved by calpain, FHL2 and a filamin cleavage product localize to the nucleus and are capable of co-activating AR splice variants including AR-V7 in a ligand-independent manner [133]. Importantly, the FHL2-filamin complex was only found in castrate resistant prostate cancer [133]. These findings point to FHL2 as a key mediator of castration resistance and a promising therapeutic target in the final stages of prostate cancer progression.

Transcriptional activation of the nuclear coactivator four-and-a-half LIM protein 2 (FLH2) demonstrates promiscuous nuclear binding partners and has been linked to an array of cellular responses. However, the extent of FLH2 expression and function in hormone-insensitive cancers has been understudied, with the exception of CRC. Its expression in CRC cells has been linked to Sp1 [134] and KLF8 expression resulting in increased tumorigenesis, invasion, and metastasis [135]. Further, others have used reciprocal co-immunoprecipitation assays and shown that it interacts with Snail1 in CRC [136], Smad2/3/4 to activate p21 [137], FOXK1 [138], and β-catenin [139, 140]. In CRC, FHL2 mediates upregulation of several genes such as vimentin, Snail, MMP2, MMP9, and E-cadherin [138, 141], suggesting that FHL2 is likely important for EMT in cancer cells. The FHL2 KD in PC cells resulted in decreased proliferation and survival, abrogation of radiotherapy resistance, and induction of MEK/ERK signaling and expression of cyclin D1, cyclin E, cyclin A, and cyclin B1 [142]. MEK inhibition in the same PC cells yielded similar results, suggesting that MEK activation may likely contribute to the signaling cascade upstream of FHL2; however, this relationship was not confirmed.

Targeting Nuclear Receptor Coactivators in Cancer

Because of the critical roles that they play in the development and progression of cancer, nuclear receptor coactivators represent therapeutic targets in cancer. In reality, the targeting of the nuclear receptor coactivators is difficult due to a high degree of conservation of domains involved in the interactions with nuclear receptors. Nonetheless, cardiac glycoside drugs were shown to cause decreased expression of NCOA1 and NCOA3 [143]. Of these, the cardiac glycoside bufalin was shown to be the most potent inhibitor of NCOA expression. Importantly, treatment of xenograft breast cancer model with nanoformulated bufalin inhibited tumor growth. Like bufalin, gambogic acid (GA) was also shown to cause decreased expression of NCOA3 in non-hodgkin lymphoma and chronic myelogenous leukemia cells. Further, GA treatment caused cell cycle arrest and increased apoptotic rate [144, 145]. However, these studies only associate the mechanism of GA with decreases in NCOA3 expression and thus the mechanism of action remains to be confirmed. Similarly, several miRNAs including miRNA

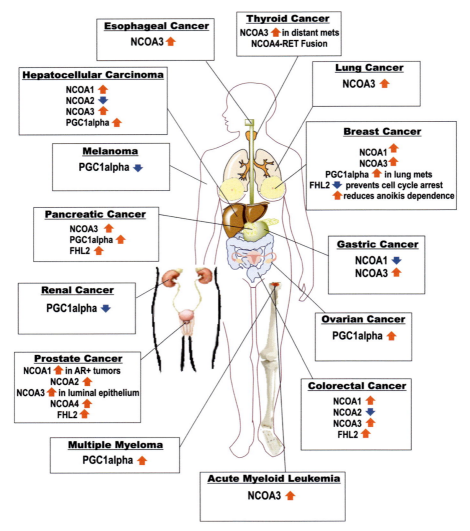

Fig. 1: Schematic of nuclear coactivator receptor expression in various cancers. Expression of coactivators unique to each cancer compared to respective normal tissues are listed along with the direction of change (increased expression = red up arrow; decreased expression = blue down arrow).

137, miRNA-217, miRNA-17 and -20b, miRNA-485-3p and miRNA-485-5p among others have all been shown to target nuclear receptor coactivators with some therapeutic potential. While these represent relatively specific targeting mechanism, it remains difficult to translate the use of miRNAs as therapeutic strategies into the clinical setting [109, 146, 147]. In contrast to cardiac glycosides which downregulate the expression of NCOA1 and NCOA3, gossypol was shown to directly interact with the NR-box of the NCOA3 and inhibit the downstream transcriptional activation of NCOA3. Additionally, treatment with gossypol decreases the cellular concentrations of both NCOA1 and NCOA3 [148]. In contrast to targeting the expression and interaction domains of NCOAs, it is also possible to target the signaling pathways leading to activation of NR coactivators through various PTMs. In a study by Oh et al., it was shown that phosphorylation of Y1357 was essential to the functioning of NCOA3. Importantly, the authors showed that the phosphorylation of this residue was mediated by the tyrosine kinase c-Abl and thus the activation of NCOA3 in terms of downstream transcriptional activation was abrogated to a great extent by treatment with imatinib [149]. Finally, a recent report used a combinatorial peptide phage display-based approach to identify peptides that are able to bind specific nuclear receptor coactivator interaction surfaces. This study went on to demonstrate that several of these peptides were able to inhibit the transcriptional activation downstream of nuclear receptor coactivators that interact with androgen receptor. This approach presents a major advantage in the sense that the targeting of specific coactivator domains would represent a means by which interactions with a nuclear receptor could be disturbed regardless of the nuclear receptor coactivator involved. However, this approach lacks specificity as targeting a highly conserved interaction domain would likely disrupt many pathways potentially involving many coactivator and nuclear receptors [150]. Additionally, the use of peptides in the clinical setting remains a difficult task.

Acknowledgement

Grant support: This work was supported in part by the grant from National Institute of Health (R01 CA78590, EDRN U01 CA111294, R21 AA 026428, R01 CA133774, R01 CA131944, SPORE P50 CA127297, T32 CA009476 and U54 TMEN CA163120).

References

1. Beischlag, T.V., S. Wang, D.W. Rose, J. Torchia, S. Reisz-Porszasz, K. Muhammad, et al. Recruitment of the NCoA/SRC-1/p160 family of transcriptional coactivators by the aryl hydrocarbon receptor/aryl hydrocarbon receptor nuclear translocator complex. Mol Cell Biol, 2002. 22(12): 4319-33.

2. Lindebro, M.C., L. Poellinger, M.L. Whitelaw. Protein-protein interaction via PAS domains: role of the PAS domain in positive and negative regulation of the bHLH/PAS dioxin receptor-Arnt transcription factor complex. Embo J, 1995. 14(14): 3528-39.
3. Litterst, C.M., S. Kliem, D. Marilley, E. Pfitzner. NCoA-1/SRC-1 is an essential coactivator of STAT5 that binds to the FDL motif in the alpha-helical region of the STAT5 transactivation domain. J Biol Chem, 2003. 278(46): 45340-51.
4. Huang, Z.J., I. Edery, M. Rosbash. PAS is a dimerization domain common to Drosophila period and several transcription factors. Nature, 1993. 364(6434): 259-62.
5. Razeto, A., V. Ramakrishnan, C.M. Litterst, K. Giller, C. Griesinger, T. Carlomagno, et al. Structure of the NCoA-1/SRC-1 PAS-B domain bound to the LXXLL motif of the STAT6 transactivation domain. J Mol Biol, 2004. 336(2): 319-29.
6. McInerney, E.M., D.W. Rose, S.E. Flynn, S. Westin, T.M. Mullen, A. Krones, et al. Determinants of coactivator LXXLL motif specificity in nuclear receptor transcriptional activation. Genes Dev, 1998. 12(21): 3357-68.
7. Lodrini, M., T. Munz, N. Coudevylle, C. Griesinger, S. Becker, E. Pfitzner. P160/SRC/NCoA coactivators form complexes via specific interaction of their PAS-B domain with the CID/AD1 domain. Nucleic Acids Res, 2008. 36(6): 1847-60.
8. Endler, A., L. Chen, F. Shibasaki. Coactivator recruitment of AhR/ARNT1. Int J Mol Sci, 2014. 15(6): 11100-10.
9. Onate, S.A., S.Y. Tsai, M.J. Tsai, B.W. O'Malley. Sequence and characterization of a coactivator for the steroid hormone receptor superfamily. Science, 1995. 270(5240): 1354-7.
10. Zhu, Y., C. Qi, C. Calandra, M.S. Rao, J.K. Reddy. Cloning and identification of mouse steroid receptor coactivator-1 (mSRC-1), as a coactivator of peroxisome proliferator-activated receptor gamma. Gene Expr, 1996. 6(3): 185-95.
11. Tong, Z., M. Li, W. Wang, P. Mo, L. Yu, K. Liu, et al. Steroid receptor coactivator 1 promotes human hepatocellular carcinoma progression by enhancing Wnt/beta-Catenin signaling. J Biol Chem, 2015. 290(30): 18596-608.
12. Qin, L., Y. Xu, Y. Xu, G. Ma, L. Liao, Y. Wu, et al. NCOA1 promotes angiogenesis in breast tumors by simultaneously enhancing both HIF1alpha- and AP-1-mediated VEGFa transcription. Oncotarget, 2015. 6(27): 23890-904.
13. Qin, L., Y.L. Wu, M.J. Toneff, D. Li, L. Liao, X. Gao, et al. NCOA1 directly targets M-CSF1 expression to promote breast cancer metastasis. Cancer Res, 2014. 74(13): 3477-88.
14. Qin, L., X. Chen, Y. Wu, Z. Feng, T. He, L. Wang, et al. Steroid receptor coactivator-1 upregulates integrin alpha(5) expression to promote breast cancer cell adhesion and migration. Cancer Res, 2011. 71(5): 1742-51.
15. Sieuwerts, A.M., M.B. Lyng, M.E. Meijer-van Gelder, V. de Weerd, F.C. Sweep, J.A. Foekens, et al. Evaluation of the ability of adjuvant tamoxifen-benefit gene signatures to predict outcome of hormone-naive estrogen receptor-positive breast cancer patients treated with tamoxifen in the advanced setting. Mol Oncol, 2014. 8(8): 1679-89.
16. Andres, S.A., K.E. Bickett, M.A. Alatoum, T.S. Kalbfleisch, G.N. Brock, and J.L. Wittliff. Interaction between smoking history and gene expression levels impacts survival of breast cancer patients. Breast Cancer Res Treat, 2015. 152(3): 545-56.

17. Luef, B., Luef, B., F. Handle, G. Kharaishvili, M. Hager, J. Rainer, G. Janetschek, et al. The AR/NCOA1 axis regulates prostate cancer migration by involvement of PRKD1. Endocr Relat Cancer, 2016. 23(6): 495-508.
18. Wang, J., J.X. Zou, X. Xue, D. Cai, Y. Zhang, Z. Duan, et al. ROR-gamma drives androgen receptor expression and represents a therapeutic target in castration-resistant prostate cancer. Nat Med, 2016. 22(5): 488-96.
19. Chen, J., A. Elfiky, M. Han, C. Chen, and M.W. Saif. The role of Src in colon cancer and its therapeutic implications. Clin Colorectal Cancer, 2014. 13(1): 5-13.
20. Jain, S., S. Pulikuri, Y. Zhu, C. Qi, Y.S. Kanwar, A.V. Yeldandi, et al. Differential expression of the peroxisome proliferator-activated receptor gamma (PPARgamma) and its coactivators steroid receptor coactivator-1 and PPAR-binding protein PBP in the brown fat, urinary bladder, colon, and breast of the mouse. Am J Pathol, 1998. 153(2): 349-54.
21. Meerson, A., H. Yehuda. Leptin and insulin up-regulate miR-4443 to suppress NCOA1 and TRAF4, and decrease the invasiveness of human colon cancer cells. BMC Cancer, 2016. 16(1): 882.
22. Frycz, B.A., D. Murawa, M. Borejsza-Wysocki, M. Wichtowski, A. Spychala, R. Marciniak, et al. mRNA expression of steroidogenic enzymes, steroid hormone receptors and their coregulators in gastric cancer. Oncol Lett, 2017. 13(5): 3369-78.
23. Ma, Y.S., T.M. Wu, Z.W. Lv, G.X. Lu, X.L. Cong, R.T. Xie, et al. High expression of miR-105-1 positively correlates with clinical prognosis of hepatocellular carcinoma by targeting oncogene NCOA1. Oncotarget, 2017. 8(7): 11896-905.
24. Lopez, G.N., C.W. Turck, F. Schaufele, M.R. Stallcup and P.J. Kushner. Growth factors signal to steroid receptors through mitogen-activated protein kinase regulation of p160 coactivator activity. J Biol Chem, 2001. 276(25): 22177-82.
25. Frigo, D.E., A. Basu, E.N. Nierth-Simpson, C.B. Weldon, C.M. Dugan, S. Elliott, et al. p38 mitogen-activated protein kinase stimulates estrogen-mediated transcription and proliferation through the phosphorylation and potentiation of the p160 coactivator glucocorticoid receptor-interacting protein 1. Mol Endocrinol, 2006. 20(5): 971-83.
26. Karmakar, S., E.A. Foster, J.K. Blackmore, C.L. Smith. Distinctive functions of p160 steroid receptor coactivators in proliferation of an estrogen-independent, tamoxifen-resistant breast cancer cell line. Endocr Relat Cancer, 2011. 18(1): 113-27.
27. Taylor, B.S., N. Schultz, H. Hieronymus, A. Gopalan, Y. Xiao, B.S. Carver, et al. Integrative genomic profiling of human prostate cancer. Cancer Cell, 2010. 18(1): 11-22.
28. Qin, J., H.J. Lee, S.P. Wu, S.C. Lin, R.B. Lanz, C.J. Creighton, et al. Androgen deprivation-induced NCoA2 promotes metastatic and castration-resistant prostate cancer. J Clin Invest, 2014. 124(11): 5013-26.
29. Dasgupta, S., N. Putluri, W. Long, B. Zhang, J. Wang, A.K. Kaushik, et al. Coactivator SRC-2-dependent metabolic reprogramming mediates prostate cancer survival and metastasis. J Clin Invest, 2015. 125(3): 1174-88.
30. Gregory, C.W., X. Fei, L.A. Ponguta, B. He, H.M. Bill, F.S. French, et al. Epidermal growth factor increases coactivation of the androgen receptor in recurrent prostate cancer. J Biol Chem, 2004. 279(8): 7119-30.
31. Yu, J., W.K. Wu, Q. Liang, N. Zhang, J. He, X. Li, et al. Disruption of NCOA2 by recurrent fusion with LACTB2 in colorectal cancer. Oncogene, 2016. 35(2): 187-95.

32. Suresh, S., D. Durakoglugil, X. Zhou, B. Zhu, S.A. Comerford, C. Xing, et al. SRC-2-mediated coactivation of anti-tumorigenic target genes suppresses MYC-induced liver cancer. PLoS Genet, 2017. 13(3): e1006650.
33. O'Donnell, K.A., V.W. Keng, B. York, E.L. Reineke, D. Seo, D. Fan, et al. A Sleeping Beauty mutagenesis screen reveals a tumor suppressor role for Ncoa2/Src-2 in liver cancer. Proc Natl Acad Sci USA, 2012. 109(21): E1377-86.
34. Anzick, S.L., J. Kononen, R.L. Walker, D.O. Azorsa, M.M. Tanner, X.Y. Guan, et al. AIB1, a steroid receptor coactivator amplified in breast and ovarian cancer. Science, 1997. 277(5328): 965-8.
35. Fereshteh, M.P., M.T. Tilli, S.E. Kim, J. Xu, B.W. O'Malley, A. Wellstein, et al. The nuclear receptor coactivator amplified in breast cancer-1 is required for Neu (ErbB2/HER2) activation, signaling, and mammary tumorigenesis in mice. Cancer Res, 2008. 68(10): 3697-706.
36. Font de Mora, J., M. Brown. AIB1 is a conduit for kinase-mediated growth factor signaling to the estrogen receptor. Mol Cell Biol, 2000. 20(14): 5041-7.
37. Oh, A., H.J. List, R. Reiter, A. Mani, Y. Zhang, E. Gehan, et al. The nuclear receptor coactivator AIB1 mediates insulin-like growth factor I-induced phenotypic changes in human breast cancer cells. Cancer Res, 2004. 64(22): 8299-308.
38. Lahusen, T., M. Fereshteh, A. Oh, A. Wellstein, and A.T. Riegel. Epidermal growth factor receptor tyrosine phosphorylation and signaling controlled by a nuclear receptor coactivator, amplified in breast cancer 1. Cancer Res, 2007. 67(15): 7256-65.
39. Ory, V., E. Tassi, L.R. Cavalli, G.M. Sharif, F. Saenz, T. Baker, et al. The nuclear coactivator amplified in breast cancer 1 maintains tumor-initiating cells during development of ductal carcinoma in situ. Oncogene, 2014. 33(23): 3033-42.
40. List, H.J., R. Reiter, B. Singh, A. Wellstein, A.T. Riegel. Expression of the nuclear coactivator AIB1 in normal and malignant breast tissue. Breast Cancer Res Treat, 2001. 68(1): 21-8.
41. Reiter, R., A. Wellstein, A.T. Riegel. An isoform of the coactivator AIB1 that increases hormone and growth factor sensitivity is overexpressed in breast cancer. J Biol Chem, 2001. 276(43): 39736-41.
42. Tilli, M.T., R. Reiter, A.S. Oh, R.T. Henke, K. McDonnell, G.I. Gallicano, et al. Overexpression of an N-terminally truncated isoform of the nuclear receptor coactivator amplified in breast cancer 1 leads to altered proliferation of mammary epithelial cells in transgenic mice. Mol Endocrinol, 2005. 19(3): 644-56.
43. Nakles, R.E., M.T. Shiffert, E.S. Diaz-Cruz, M.C. Cabrera, M. Alotaiby, A.M. Miermont, et al. Altered AIB1 or AIB1Delta3 expression impacts ERalpha effects on mammary gland stromal and epithelial content. Mol Endocrinol, 2011. 25(4): 549-63.
44. Reiter, R., A.S. Oh, A. Wellstein, A.T. Riegel. Impact of the nuclear receptor coactivator AIB1 isoform AIB1-Delta3 on estrogenic ligands with different intrinsic activity. Oncogene, 2004. 23(2): 403-9.
45. List, H.J., K.J. Lauritsen, R. Reiter, C. Powers, A. Wellstein, A.T. Riegel. Ribozyme targeting demonstrates that the nuclear receptor coactivator AIB1 is a rate-limiting factor for estrogen-dependent growth of human MCF-7 breast cancer cells. J Biol Chem, 2001. 276(26): 23763-68.
46. Wagner, M., M. Koslowski, C. Paret, M. Schmidt, O. Tureci, U. Sahin. NCOA3 is a selective co-activator of estrogen receptor alpha-mediated transactivation of PLAC1 in MCF-7 breast cancer cells. BMC Cancer, 2013. 13: 570.

47. Giamas, G., L. Castellano, Q. Feng, U. Knippschild, J. Jacob, R.S. Thomas, et al. CK1delta modulates the transcriptional activity of ERalpha via AIB1 in an estrogen-dependent manner and regulates ERalpha-AIB1 interactions. Nucleic Acids Res, 2009. 37(9): 3110-23.
48. Wang, L.H., X.Y. Yang, X. Zhang, P. An, H.J. Kim, J. Huang, et al. Disruption of estrogen receptor DNA-binding domain and related intramolecular communication restores tamoxifen sensitivity in resistant breast cancer. Cancer Cell, 2006. 10(6): 487-99.
49. Haugan Moi, L.L., M. Hauglid Flageng, S. Gandini, A. Guerrieri-Gonzaga, B. Bonanni, M. Lazzeroni, et al. Effect of low-dose tamoxifen on steroid receptor coactivator 3/amplified in breast cancer 1 in normal and malignant human breast tissue. Clin Cancer Res, 2010. 16(7): 2176-86.
50. You, D., H. Zhao, Y. Wang, Y. Jiao, M. Lu, and S. Yan. Acetylation enhances the promoting role of AIB1 in breast cancer cell proliferation. Mol Cells, 2016. 39(9): 663-8.
51. Burandt, E., G. Jens, F. Holst, F. Janicke, V. Muller, A. Quaas, et al. Prognostic relevance of AIB1 (NCoA3) amplification and overexpression in breast cancer. Breast Cancer Res Treat, 2013. 137(3): 745-53.
52. Burwinkel, B., M. Wirtenberger, R. Klaes, R.K. Schmutzler, E. Grzybowska, A. Forsti, et al. Association of NCOA3 polymorphisms with breast cancer risk. Clin Cancer Res, 2005. 11(6): 2169-74.
53. Hartmaier, R.J., S. Tchatchou, A.S. Richter, J. Wang, S.E. McGuire, T.C. Skaar, et al. Nuclear receptor coregulator SNP discovery and impact on breast cancer risk. BMC Cancer, 2009. 9: 438.
54. Gabrovska, P.N., R.A. Smith, G. O'Leary, L.M. Haupt, L.R. Griffiths. Investigation of the 1758G>C and 2880A>G variants within the NCOA3 gene in a breast cancer affected Australian population. Gene, 2011. 482(1-2): 68-72.
55. Chung, A.C., S. Zhou, L. Liao, J.C. Tien, N.M. Greenberg, J. Xu. Genetic ablation of the amplified-in-breast cancer 1 inhibits spontaneous prostate cancer progression in mice. Cancer Res, 2007. 67(12): 5965-75.
56. Yan, J., H. Erdem, R. Li, Y. Cai, G. Ayala, M. Ittmann, et al. Steroid receptor coactivator-3/AIB1 promotes cell migration and invasiveness through focal adhesion turnover and matrix metalloproteinase expression. Cancer Res, 2008. 68(13): 5460-8.
57. Zhou, G., Y. Hashimoto, I. Kwak, S.Y. Tsai, M.J. Tsai. Role of the steroid receptor coactivator SRC-3 in cell growth. Mol Cell Biol, 2003. 23(21): 7742-55.
58. Yan, J., C.T. Yu, M. Ozen, M. Ittmann, S.Y. Tsai, M.J. Tsai. Steroid receptor coactivator-3 and activator protein-1 coordinately regulate the transcription of components of the insulin-like growth factor/AKT signaling pathway. Cancer Res, 2006. 66(22): 11039-46.
59. Xie, D., J.S. Sham, W.F. Zeng, H.L. Lin, J. Bi, L.H. Che, et al. Correlation of AIB1 overexpression with advanced clinical stage of human colorectal carcinoma. Hum Pathol, 2005. 36(7): 777-83.
60. Mo, P., Q. Zhou, L. Guan, Y. Wang, W. Wang, M. Miao, et al. Amplified in breast cancer 1 promotes colorectal cancer progression through enhancing notch signaling. Oncogene, 2015. 34(30): 3935-45.
61. Li, Z., Z.Y. Fang, Y. Ding, W.T. Yao, Y. Yang, Z.Q. Zhu, et al. Amplifications of NCOA3 gene in colorectal cancers in a Chinese population. World J Gastroenterol, 2012. 18(8): 855-60.

62. Edvardsson, K., T. Nguyen-Vu, S.M. Kalasekar, F. Ponten, J.A. Gustafsson, C. Williams. Estrogen receptor beta expression induces changes in the microRNA pool in human colon cancer cells. Carcinogenesis, 2013. 34(7): 1431-41.
63. Li, Y., L. Li, M. Chen, X. Yu, Z. Gu, H. Qiu, et al. MAD2L2 inhibits colorectal cancer growth by promoting NCOA3 ubiquitination and degradation. Mol Oncol, 2018. 12(3): 391-405.
64. Al-Otaiby, M., E. Tassi, M.O. Schmidt, C.D. Chien, T. Baker, A.G. Salas, et al. Role of the nuclear receptor coactivator AIB1/SRC-3 in angiogenesis and wound healing. Am J Pathol, 2012. 180(4): 1474-84.
65. Gupta, K., T. Stefan, J. Ignatz-Hoover, S. Moreton, G. Parizher, Y. Saunthararajah, et al. GSK-3 Inhibition sensitizes acute myeloid leukemia cells to 1,25D-mediated differentiation. Cancer Res, 2016. 76(9): 2743-53.
66. Colo, G.P., R.R. Rosato, S. Grant, M.A. Costas. RAC3 down-regulation sensitizes human chronic myeloid leukemia cells to TRAIL-induced apoptosis. FEBS Lett, 2007. 581(26): 5075-81.
67. Song, J.M., M. Lu, F.F. Liu, X.J. Du, B.C. Xing. AIB1 as an independent prognostic marker in hepatocellular carcinoma after hepatic resection. J Gastrointest Surg, 2012. 16(2): 356-60.
68. Wang, Y., M.C. Wu, J.S. Sham, W. Zhang, W.Q. Wu and X.Y. Guan. Prognostic significance of c-myc and AIB1 amplification in hepatocellular carcinoma. A broad survey using high-throughput tissue microarray. Cancer, 2002. 95(11): 2346-52.
69. Xu, Y., Q. Chen, W. Li, X. Su, T. Chen, Y. Liu, et al. Overexpression of transcriptional coactivator AIB1 promotes hepatocellular carcinoma progression by enhancing cell proliferation and invasiveness. Oncogene, 2010. 29(23): 3386-97.
70. Cai, D., D.S. Shames, M.G. Raso, Y. Xie, Y.H. Kim, J.R. Pollack, et al. Steroid receptor coactivator-3 expression in lung cancer and its role in the regulation of cancer cell survival and proliferation. Cancer Res, 2010. 70(16): 6477-85.
71. He, L.R., H.Y. Zhao, B.K. Li, L.J. Zhang, M.Z. Liu, H.F. Kung, et al. Overexpression of AIB1 negatively affects survival of surgically resected non-small-cell lung cancer patients. Ann Oncol, 2010. 21(8): 1675-81.
72. Wang, H., D. Zhang, W. Wu, J. Zhang, D. Guo, Q. Wang, et al. Overexpression and gender-specific differences of SRC-3 (SRC-3/AIB1) immunoreactivity in human non-small cell lung cancer: an in vivo study. J Histochem Cytochem, 2010. 58(12): 1121-27.
73. Henke, R.T., B.R. Haddad, S.E. Kim, J.D. Rone, A. Mani, J.M. Jessup, et al. Overexpression of the nuclear receptor coactivator AIB1 (SRC-3) during progression of pancreatic adenocarcinoma. Clin Cancer Res, 2004. 10(18 Pt 1): 6134-42.
74. Ghadimi, B.M., E. Schrock, R.L. Walker, D. Wangsa, A. Jauho, P.S. Meltzer, et al. Specific chromosomal aberrations and amplification of the AIB1 nuclear receptor coactivator gene in pancreatic carcinomas. Am J Pathol, 1999. 154(2): 525-36.
75. Guo, S., J. Xu, R. Xue, Y. Liu, H. Yu. Overexpression of AIB1 correlates inversely with E-cadherin expression in pancreatic adenocarcinoma and may promote lymph node metastasis. Int J Clin Oncol, 2014. 19(2): 319-24.
76. Kumar, S., S. Das, S. Rachagani, S. Kaur, S. Joshi, S.L. Johansson, et al. NCOA3-mediated upregulation of mucin expression via transcriptional and post-

translational changes during the development of pancreatic cancer. Oncogene, 2015. 34(37): 4879-89.
77. Sakakura, C., A. Hagiwara, R. Yasuoka, Y. Fujita, M. Nakanishi, K. Masuda, et al. Amplification and over-expression of the AIB1 nuclear receptor co-activator gene in primary gastric cancers. Int J Cancer, 2000. 89(3): 217-23.
78. Shi, J., W. Liu, F. Sui, R. Lu, Q. He, Q. Yang, et al. Frequent amplification of AIB1, a critical oncogene modulating major signaling pathways, is associated with poor survival in gastric cancer. Oncotarget, 2015. 6(16): 14344-59.
79. He, L.R., M.Z. Liu, B.K. Li, H.L. Rao, H.X. Deng, X.Y. Guan, et al. Overexpression of AIB1 predicts resistance to chemoradiotherapy and poor prognosis in patients with primary esophageal squamous cell carcinoma. Cancer Sci, 2009. 100(9): 1591-6.
80. Li, L., P. Wei, M.H. Zhang, W. Zhang, Y. Ma, X. Fang, et al. Roles of the AIB1 protein in the proliferation and transformation of human esophageal squamous cell carcinoma. Genet Mol Res, 2015. 14(3): 10376-83.
81. Xu, F.P., D. Xie, J.M. Wen, H.X. Wu, Y.D. Liu, J. Bi, et al. SRC-3/AIB1 protein and gene amplification levels in human esophageal squamous cell carcinomas. Cancer Lett, 2007. 245(1-2): 69-74.
82. Liu, M.Y., H.P. Guo, C.Q. Hong, H.W. Peng, X.H. Yang, H. Zhang. Up-regulation of nuclear receptor coactivator amplified in breast cancer-1 in papillary thyroid carcinoma correlates with lymph node metastasis. Clin Transl Oncol, 2013. 15(11): 947-52.
83. Yeh, S., C. Chang. Cloning and characterization of a specific coactivator, ARA70, for the androgen receptor in human prostate cells. Proc Natl Acad Sci USA, 1996. 93(11): 5517-21.
84. Chang, B.L., S.D. Cramer, F. Wiklund, S.D. Isaacs, V.L. Stevens, J. Sun, et al. Fine mapping association study and functional analysis implicate a SNP in MSMB at 10q11 as a causal variant for prostate cancer risk. Hum Mol Genet, 2009. 18(7): 1368-75.
85. Pomerantz, M.M., Y. Shrestha, R.J. Flavin, M.M. Regan, K.L. Penney, L.A. Mucci, et al. Analysis of the 10q11 cancer risk locus implicates MSMB and NCOA4 in human prostate tumorigenesis. PLoS Genet, 2010. 6(11): e1001204.
86. Kwon, E.M., S.K. Holt, R. Fu, S. Kolb, G. Williams, J.L. Stanford, et al. Androgen metabolism and JAK/STAT pathway genes and prostate cancer risk. Cancer Epidemiol, 2012. 36(4): 347-53.
87. Grisanzio, C., L. Werner, D. Takeda, B.C. Awoyemi, M.M. Pomerantz, H. Yamada, et al. Genetic and functional analyses implicate the NUDT11, HNF1B, and SLC22A3 genes in prostate cancer pathogenesis. Proc Natl Acad Sci USA, 2012. 109(28): 11252-57.
88. Miyamoto, H., S. Yeh, G. Wilding, and C. Chang. Promotion of agonist activity of antiandrogens by the androgen receptor coactivator, ARA70, in human prostate cancer DU145 cells. Proc Natl Acad Sci USA, 1998. 95(13): 7379-84.
89. Rahman, M.M., H. Miyamoto, H. Takatera, S. Yeh, S. Altuwaijri, C. Chang. Reducing the agonist activity of antiandrogens by a dominant-negative androgen receptor coregulator ARA70 in prostate cancer cells. J Biol Chem, 2003. 278(22): 19619-26.
90. Niu, Y., S. Yeh, H. Miyamoto, G. Li, S. Altuwaijri, J. Yuan, et al. Tissue prostate-specific antigen facilitates refractory prostate tumor progression via enhancing ARA70-regulated androgen receptor transactivation. Cancer Res, 2008. 68(17): 7110-19.

91. Ligr, M., Y. Li, X. Zou, G. Daniels, J. Melamed, Y. Peng, et al. Tumor suppressor function of androgen receptor coactivator ARA70alpha in prostate cancer. Am J Pathol, 2010. 176(4): 1891-900.
92. Peng, Y., C.X. Li, F. Chen, Z. Wang, M. Ligr, J. Melamed, et al. Stimulation of prostate cancer cellular proliferation and invasion by the androgen receptor co-activator ARA70. Am J Pathol, 2008. 172(1): 225-35.
93. Li, P., X. Yu, K. Ge, J. Melamed, R.G. Roeder, Z. Wang. Heterogeneous expression and functions of androgen receptor co-factors in primary prostate cancer. Am J Pathol, 2002. 161(4): 1467-74.
94. Kollara, A., T.J. Brown. Functional interaction of nuclear receptor coactivator 4 with aryl hydrocarbon receptor. Biochem Biophys Res Commun, 2006. 346(2): 526-34.
95. Heinlein, C.A., H.J. Ting, S. Yeh, C. Chang. Identification of ARA70 as a ligand-enhanced coactivator for the peroxisome proliferator-activated receptor gamma. J Biol Chem, 1999. 274(23): 16147-52.
96. Gandhi, M., M. Medvedovic, J.R. Stringer, Y.E. Nikiforov. Interphase chromosome folding determines spatial proximity of genes participating in carcinogenic RET/PTC rearrangements. Oncogene, 2006. 25(16): 2360-66.
97. Bellelli, R., D. Vitagliano, G. Federico, P. Marotta, A. Tamburrino, P. Salerno, et al. Oncogene-induced senescence and its evasion in a mouse model of thyroid neoplasia. Mol Cell Endocrinol, 2018. 460: 24-35.
98. Li, G.G., R. Somwar, J. Joseph, R.S. Smith, T. Hayashi, L. Martin, et al. Antitumor activity of RXDX-105 in multiple cancer types with RET rearrangements or mutations. Clin Cancer Res, 2017. 23(12): 2981-90.
99. Okamoto, K., K. Kodama, K. Takase, N.H. Sugi, Y. Yamamoto, M. Iwata, et al. Antitumor activities of the targeted multi-tyrosine kinase inhibitor lenvatinib (E7080) against RET gene fusion-driven tumor models. Cancer Lett, 2013. 340(1): 97-103.
100. Richardson, D.S., T.S. Gujral, S. Peng, S.L. Asa, L.M. Mulligan. Transcript level modulates the inherent oncogenicity of RET/PTC oncoproteins. Cancer Res, 2009. 69(11): 4861-69.
101. Swierniak, M., A. Pfeifer, T. Stokowy, D. Rusinek, M. Chekan, D. Lange, et al. Somatic mutation profiling of follicular thyroid cancer by next generation sequencing. Mol Cell Endocrinol, 2016. 433: 130-7.
102. Dacic, S., A. Luvison, V. Evdokimova, L. Kelly, J.M. Siegfried, L.C. Villaruz, et al. RET rearrangements in lung adenocarcinoma and radiation. J Thorac Oncol, 2014. 9(1): 118-20.
103. Wang, R., H. Hu, Y. Pan, Y. Li, T. Ye, C. Li, et al. RET fusions define a unique molecular and clinicopathologic subtype of non-small-cell lung cancer. J Clin Oncol, 2012. 30(35): 4352-9.
104. Hechtman, J.F., A. Zehir, R. Yaeger, L. Wang, S. Middha, T. Zheng, et al. Identification of targetable kinase alterations in patients with colorectal carcinoma that are preferentially associated with wild-type RAS/RAF. Mol Cancer Res, 2016. 14(3): 296-301.
105. Le Rolle, A.F., S.J. Klempner, C.R. Garrett, T. Seery, E.M. Sanford, S. Balasubramanian, et al. Identification and characterization of RET fusions in advanced colorectal cancer. Oncotarget, 2015. 6(30): 28929-37.
106. Song, H.N., C. Lee, S.T. Kim, S.Y. Kim, N.K. Kim, J. Jang, et al. Molecular characterization of colorectal cancer patients and concomitant patient-derived tumor cell establishment. Oncotarget, 2016. 7(15): 19610-19.

107. Wang, K., J.S. Russell, J.D. McDermott, J.A. Elvin, D. Khaira, A. Johnson, et al. Profiling of 149 salivary duct carcinomas, carcinoma ex pleomorphic adenomas, and adenocarcinomas, not otherwise specified reveals actionable genomic alterations. Clin Cancer Res, 2016. 22(24): 6061-68.
108. LeBleu, V.S., J.T. O'Connell, K.N. Gonzalez Herrera, H. Wikman, K. Pantel, M.C. Haigis, et al. PGC-1alpha mediates mitochondrial biogenesis and oxidative phosphorylation in cancer cells to promote metastasis. Nat Cell Biol, 2014. 16(10): 992-1003, 1-15.
109. Lou, C., M. Xiao, S. Cheng, X. Lu, S. Jia, Y. Ren, et al. MiR-485-3p and miR-485-5p suppress breast cancer cell metastasis by inhibiting PGC-1alpha expression. Cell Death Dis, 2016. 7: e2159.
110. Zhang, S., X. Liu, J. Liu, H. Guo, H. Xu, and G. Zhang. PGC-1 alpha interacts with microRNA-217 to functionally regulate breast cancer cell proliferation. Biomed Pharmacother, 2017. 85: 541-8.
111. Andrzejewski, S., E. Klimcakova, R.M. Johnson, S. Tabaries, M.G. Annis, S. McGuirk, et al. PGC-1alpha promotes breast cancer metastasis and confers bioenergetic flexibility against metabolic drugs. Cell Metab, 2017. 26(5): 778-87 e5.
112. Kim, B., J.W. Jung, J. Jung, Y. Han, D.H. Suh, H.S. Kim, et al. PGC1alpha induced by reactive oxygen species contributes to chemoresistance of ovarian cancer cells. Oncotarget, 2017. 8(36): 60299-311.
113. Cai, F.F., C. Xu, X. Pan, L. Cai, X.Y. Lin, S. Chen, et al. Prognostic value of plasma levels of HIF-1a and PGC-1a in breast cancer. Oncotarget, 2016. 7(47): 77793-806.
114. Misra, P., J.K. Reddy. Peroxisome proliferator-activated receptor-alpha activation and excess energy burning in hepatocarcinogenesis. Biochimie, 2014. 98: 63-74.
115. Li, Y., S. Xu, J. Li, L. Zheng, M. Feng, X. Wang, et al. SIRT1 facilitates hepatocellular carcinoma metastasis by promoting PGC-1alpha-mediated mitochondrial biogenesis. Oncotarget, 2016. 7(20): 29255-74.
116. Hu, Y., D.J. Shin, H. Pan, Z. Lin, J.M. Dreyfuss, F.D. Camargo, et al. YAP suppresses gluconeogenic gene expression through PGC1alpha. Hepatology, 2017. 66(6): 2029-41.
117. Luo, C., J.H. Lim, Y. Lee, S.R. Granter, A. Thomas, F. Vazquez, et al. A PGC1alpha-mediated transcriptional axis suppresses melanoma metastasis. Nature, 2016. 537(7620): 422-6.
118. Torrens-Mas, M., D. Gonzalez-Hedstrom, M. Abrisqueta, P. Roca, J. Oliver, and J. Sastre-Serra. PGC-1alpha in melanoma: a key factor for antioxidant response and mitochondrial function. J Cell Biochem, 2017. 118(12): 4404-13.
119. Vazquez, F., J.H. Lim, H. Chim, K. Bhalla, G. Girnun, K. Pierce, et al. PGC1alpha expression defines a subset of human melanoma tumors with increased mitochondrial capacity and resistance to oxidative stress. Cancer Cell, 2013. 23(3): 287-301.
120. Grabacka, M., J. Wieczorek, D. Michalczyk-Wetula, M. Malinowski, N. Wolan, K. Wojcik, et al. Peroxisome Peroxisome proliferator-activated receptor alpha (PPARalpha) contributes to control of melanogenesis in B16 F10 melanoma cells. Arch Dermatol Res, 2017. 309(3): 141-57.
121. D'Errico, I., G. Lo Sasso, L. Salvatore, S. Murzilli, N. Martelli, M. Cristofaro, et al. Bax is necessary for PGC1alpha pro-apoptotic effect in colorectal cancer cells. Cell Cycle, 2011. 10(17): 2937-45.

122. Sancho, P., E. Burgos-Ramos, A. Tavera, T. Bou Kheir, P. Jagust, M. Schoenhals, et al. MYC/PGC-1alpha balance determines the metabolic phenotype and plasticity of pancreatic cancer stem cells. Cell Metab, 2015. 22(4): 590-605.
123. Martin, B.T., K. Kleiber, V. Wixler, M. Raab, B. Zimmer, M. Kaufmann, et al. FHL2 regulates cell cycle-dependent and doxorubicin-induced p21Cip1/Waf1 expression in breast cancer cells. Cell Cycle, 2007. 6(14): 1779-88.
124. Yan, J., J. Zhu, H. Zhong, Q. Lu, C. Huang, and Q. Ye. BRCA1 interacts with FHL2 and enhances FHL2 transactivation function. FEBS Lett, 2003. 553(1-2): 183-9.
125. Fan, P., F.A. Agboke, R.E. McDaniel, E.E. Sweeney, X. Zou, K. Creswell, et al. Inhibition of c-Src blocks oestrogen-induced apoptosis and restores oestrogen-stimulated growth in long-term oestrogen-deprived breast cancer cells. Eur J Cancer, 2014. 50(2): 457-68.
126. Fan, P., H.E. Cunliffe, O.L. Griffith, F.A. Agboke, P. Ramos, J.W. Gray, et al. Identification of gene regulation patterns underlying both oestrogen- and tamoxifen-stimulated cell growth through global gene expression profiling in breast cancer cells. Eur J Cancer, 2014. 50(16): 2877-86.
127. Boateng, L.R., D. Bennin, S. De Oliveira, A. Huttenlocher. Mammalian actin-binding protein-1/Hip-55 interacts with FHL2 and negatively regulates cell invasion. J Biol Chem, 2016. 291(27): 13987-98.
128. Muller, J.M., U. Isele, E. Metzger, A. Rempel, M. Moser, A. Pscherer, et al. FHL2, a novel tissue-specific coactivator of the androgen receptor. EMBO J, 2000. 19(3): 359-69.
129. Muller, J.M., E. Metzger, H. Greschik, A.K. Bosserhoff, L. Mercep, R. Buettner, et al. The transcriptional coactivator FHL2 transmits Rho signals from the cell membrane into the nucleus. EMBO J, 2002. 21(4): 736-48.
130. Kahl, P., L. Gullotti, L.C. Heukamp, S. Wolf, N. Friedrichs, R. Vorreuther, et al. Androgen receptor coactivators lysine-specific histone demethylase 1 and four and a half LIM domain protein 2 predict risk of prostate cancer recurrence. Cancer Res, 2006. 66(23): 11341-7.
131. Heemers, H.V., K.M. Regan, S.M. Dehm, and D.J. Tindall. Androgen induction of the androgen receptor coactivator four-and-a-half LIM domain protein-2: evidence for a role for serum response factor in prostate cancer. Cancer Res, 2007. 67(21): 10592-9.
132. Shaikhibrahim, Z., B. Langer, A. Lindstrot, A. Florin, A. Bosserhoff, R. Buettner, et al. Ets-1 is implicated in the regulation of androgen co-regulator FHL2 and reveals specificity for migration, but not invasion, of PC3 prostate cancer cells. Oncol Rep, 2011. 25(4): 1125-9.
133. McGrath, M.J., L.C. Binge, A. Sriratana, H. Wang, P.A. Robinson, D. Pook, et al. Regulation of the transcriptional coactivator FHL2 licenses activation of the androgen receptor in castrate-resistant prostate cancer. Cancer Res, 2013. 73(16): 5066-79.
134. Guo, Z., W. Zhang, G. Xia, L. Niu, Y. Zhang, X. Wang, et al. Sp1 upregulates the four-and-half lim 2 (FHL2) expression in gastrointestinal cancers through transcription regulation. Mol Carcinog, 2010. 49(9): 826-36.
135. Yan, Q., W. Zhang, Y. Wu, M. Wu, M. Zhang, X. Shi, et al. KLF8 promotes tumorigenesis, invasion and metastasis of colorectal cancer cells by transcriptional activation of FHL2. Oncotarget, 2015. 6(28): 25402-17.
136. Zhang, W., J. Wang, B. Zou, C. Sardet, J. Li, C.S. Lam, et al. Four and a half LIM protein 2 (FHL2) negatively regulates the transcription of E-cadherin through interaction with Snail1. Eur J Cancer, 2011. 47(1): 121-30.

137. Ding, L., Z. Wang, J. Yan, X. Yang, A. Liu, W. Qiu, et al. Human four-and-a-half LIM family members suppress tumor cell growth through a TGF-beta-like signaling pathway. J Clin Invest, 2009. 119(2): 349-61.
138. Wu, M., J. Wang, W. Tang, X. Zhan, Y. Li, Y. Peng, et al. FOXK1 interaction with FHL2 promotes proliferation, invasion and metastasis in colorectal cancer. Oncogenesis, 2016. 5(11): e271.
139. Wei, Y., C.A. Renard, C. Labalette, Y. Wu, L. Levy, C. Neuveut, et al. Identification of the LIM protein FHL2 as a coactivator of beta-catenin. J Biol Chem, 2003. 278(7): 5188-94.
140. Zhang, W., B. Jiang, Z. Guo, C. Sardet, B. Zou, C.S. Lam, et al. Four-and-a-half LIM protein 2 promotes invasive potential and epithelial-mesenchymal transition in colon cancer. Carcinogenesis, 2010. 31(7): 1220-9.
141. Amann, T., Y. Egle, A.K. Bosserhoff, C. Hellerbrand. FHL2 suppresses growth and differentiation of the colon cancer cell line HT-29. Oncol Rep, 2010. 23(6): 1669-74.
142. Zienert, E., I. Eke, D. Aust, N. Cordes. LIM-only protein FHL2 critically determines survival and radioresistance of pancreatic cancer cells. Cancer Lett, 2015. 364(1): 17-24.
143. Wang, Y., D.M. Lonard, Y. Yu, D.C. Chow, T.G. Palzkill, J. Wang, et al. Bufalin is a potent small-molecule inhibitor of the steroid receptor coactivators SRC-3 and SRC-1. Cancer Res, 2014. 74(5): 1506-17.
144. Zhao, Z., X. Zhang, L. Wen, S. Yi, J. Hu, J. Ruan, et al. Steroid receptor coactivator-3 is a pivotal target of gambogic acid in B-cell Non-Hodgkin lymphoma and an inducer of histone H3 deacetylation. Eur J Pharmacol, 2016. 789: 46-59.
145. Li, R., Y. Chen, L.L. Zeng, W.X. Shu, F. Zhao, L. Wen, et al. Gambogic acid induces G0/G1 arrest and apoptosis involving inhibition of SRC-3 and inactivation of Akt pathway in K562 leukemia cells. Toxicology, 2009. 262(2): 98-105.
146. Nilsson, E.M., K.B. Laursen, J. Whitchurch, A. McWilliam, N. Odum, J.L. Persson, et al. MiR137 is an androgen regulated repressor of an extended network of transcriptional coregulators. Oncotarget, 2015. 6(34): 35710-25.
147. Ao, X., P. Nie, B. Wu, W. Xu, T. Zhang, S. Wang, et al. Decreased expression of microRNA-17 and microRNA-20b promotes breast cancer resistance to taxol therapy by upregulation of NCOA3. Cell Death Dis, 2016. 7(11): e2463.
148. Wang, Y., D.M. Lonard, Y. Yu, D.C. Chow, T.G. Palzkill, B.W. O'Malley. Small molecule inhibition of the steroid receptor coactivators, SRC-3 and SRC-1. Mol Endocrinol, 2011. 25(12): 2041-53.
149. Oh, A.S., J.T. Lahusen, C.D. Chien, M.P. Fereshteh, X. Zhang, S. Dakshanamurthy, et al. Tyrosine phosphorylation of the nuclear receptor coactivator AIB1/SRC-3 is enhanced by Abl kinase and is required for its activity in cancer cells. Mol Cell Biol, 2008. 28(21): 6580-93.
150. Chang, C.Y., J. Abdo, T. Hartney, D.P. McDonnell. Development of peptide antagonists for the androgen receptor using combinatorial peptide phage display. Mol Endocrinol, 2005. 19(10): 2478-90.

CHAPTER 5

Liquid Biopsies for Pancreatic Cancer: A Step Towards Early Detection

Joseph Carmicheal, Rahat Jahan, Koelina Ganguly, Ashu Shah and Sukhwinder Kaur*

Department of Biochemistry and Molecular Biology, University of Nebraska Medical Center, Omaha, NE 68198, USA

Introduction

Pancreatic Cancer (PC) has a sole curative option of surgical resection. Unfortunately, the overwhelming majority of PC cases are diagnosed at later stages which are often difficult to resect and further, can be quite resistant to chemo- and radiotherapies. Limited and inefficient options are available to detect cancer in early stages at a time that is amenable to intervention modalities, namely excision. Tissue biopsies are inherently extremely invasive and high-risk procedures. The risk is due to the retroperitoneal location of the pancreatic head (where most PC originates) and the close proximity to the abdominal aorta, celiac trunk, and inferior vena cava. Contrarily, the non-invasive aspect of imaging via CT or MRI is quite alluring, yet these methods cannot detect early-stage disease thus providing no added benefit to clinically useful early detection.

Recently, blood-based liquid biopsies have become a highly investigated means by which PC may be detected. By using blood as the biopsy specimen, the risk to the patient is drastically reduced. Further, when the biopsy is not limited to a single solid tissue sample, tumor heterogeneity can be taken into account as well as various physiological responses to the tumor can be measured, thereby taking into account the overall gestalt of the disease. Myriad of molecules and particles have been studied in the attempt to find a PC specific liquid biopsy based biomarker including various glycoproteins [1] and miRNA [2]. While these modalities show great promise, no current biomarker is sufficient as of yet for accurate diagnosis. Novel emerging technologies and markers are required if this disease is to be detected at a curable stage. While not exhaustive, this chapter aims to elucidate some of the

---

*Corresponding author: skaur@unmc.edu

most promising molecules and methods that are currently being investigated including circulating cell-free tumor DNA (ctDNA), metabolomics, and exosomes.

Of note, the liquid biopsy studies discussed herein focus primarily on pancreatic cancer. While this insidious disease is an ideal representative for the necessity of early detection, prognostication, and accurate prediction of therapeutic efficiency, these techniques are not limited to this malignancy. The reader should be aware that the implications of the research present within this chapter can (and have been) be extended to myriad of other cancers.

Circulating Cell-free Tumor DNA (ctDNA)

Circulating cell-free DNA (cfDNA) are degraded DNA fragments released into the blood plasma. cfDNA can be used to describe various forms of DNA which are freely circulating in the bloodstream, such as circulating tumor DNA (ctDNA) and cell-free fetal DNA (cffDNA) [3]. Circulating tumor DNA (ctDNA) originates specifically from tumor cells and potentially harbors tumor-associated genetic mutations that inherently provide a marker for disease progression and/or therapeutic response [4]. Many cell death mechanisms including apoptosis, necrosis, oncosis, and phagocytosis release cfDNA/ctDNA into the bloodstream. These molecules can be present in their free form circulating by themselves or in the intravesicular compartment of various cell-derived vesicles [4]. They can also be bound to a DNA binding protein (specific or nonspecific) and the multi-nucleosome complex during vascular transport [4, 5].

Mandel and Metais reported, for the first time, the existence of cfDNA and their predictive role in tumor metastasis [6-8]. The half-life of ctDNA is between 15 minutes to hours, which can be cleared from the bloodstream by the kidney, liver, and spleen [9]. Amplicon-based studies have elucidated an important disparity showing that ctDNA is highly fragmented (size <100 bp), while the size of cfDNA is >400 bp [10]. Moreover, mice with bigger tumors shed more fragmented ctDNA as compared to those with smaller tumors, suggesting that not only fragment quantity but fragment size impacts tumor progression and notably may be utilized as a biomarker [5, 10]. The majority of time ctDNA is isolated from serum/plasma; however, its presence and purification are also carried out from urine, cerebrospinal fluid (CSF), and saliva [11], thus offering an array of biofluids from which ctDNA may be extracted.

Isolation and Analysis of ctDNA/Detection Methods

Due to limited yield and short half-life in circulation, a multitude of efforts are being made to improve the isolation and characterization technologies for ctDNA. To date, strategies used in ctDNA detection and analysis can be classified into two groups: site-specific (targeted) and genome-wide (untargeted), followed by subsequent administration of various applications (Fig. 1). The following section details the recent updates on targeted and untargeted approaches to isolate and characterize ctDNA.

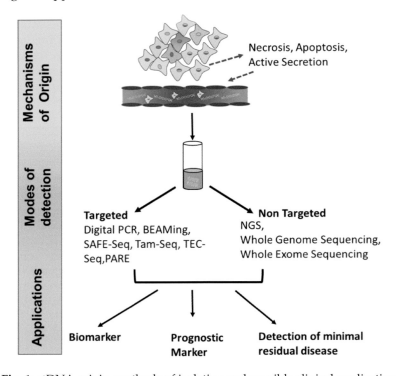

Fig. 1: ctDNA origin, methods of isolation and possible clinical applications.

Targeted Approach

The targeted approach focuses on identifying a specific somatic mutation, recurrent hot spot mutations, or short insertions/deletions that have been previously well documented and are associated with cancer progression [7]. Considering the scarcity of ctDNA, ultrasensitive detection methods must be used to detect mutations in ctDNA in the bloodstream. Polymerase Chain Reaction (PCR) is a widely used and conventional approach to identify known mutations in ctDNA. More recently, sensitive and specific digital PCR methods like digital droplet PCR (ddPCR) and BEAMing (beads, emulsions, amplification, and magnetics) are being increasingly utilized for the characterization and mutational profiling of circulating cfDNA. These are emulsion-based PCR methods that require a minimal percentage of mutated DNA (0.01-0.001% of total ctDNA) to be present in samples for accurate detection [12, 13].

In ddPCR, samples are prepared with fluorescently tagged primers, which are subsequently emulsified with the sample and the necessary reagents into thousands of individual droplets. After performing conventional PCR, the concentration of the DNA of interest can be measured by the number of fluorescent droplets. The benefit of ddPCR is twofold: it can detect target mutations at a much lower frequency and with greater precision than conventional qPCR methods [12]. Droplet digital PCR confers a level of specificity required for detection of ultralow mutation concentrations that may be present in ctDNA isolated from patient serum.

BEAMing is another form of sensitive and specific digital PCR, also based on the emulsification of the PCR mixture. This method, however, requires the added step of attaching streptavidin-coated magnetic beads to each strand of DNA prior to the PCR analysis. Fluorescently tagged complementary primers are next added which can bind with the DNA strands of interest attached to the beads, thus allowing for facile and specific isolation of our DNA targets of interest from the sample. These beads are analyzed using flow cytometry and the presence of DNA is validated by Sanger sequencing. These new advanced digital PCR methods can simultaneously track the number of normal and mutant DNA molecules in a given sample, thus providing another metric of quantification and blood-based characterization [13, 14].

Other targeted approaches are being developed based on Deep sequencing/Next Generation Sequencing (NGS) using site-specific primers to amplify a particular genomic region. Some examples of these regions include "Safe-SeqS", Tagged Amplicon Sequencing (TAm-Seq), AmpliSeq and Personalized Analysis of Rearranged Ends (PARE). Though they require further standardization and experimental study, these are some of the advanced technologies that will prove invaluable in the identification of specific hotspot mutations found in ctDNA [15, 16].

Untargeted Genome-wide Approach

Untargeted approaches are based on a comprehensive analysis of the tumor genome and do not require prior knowledge of specific mutations. These methods focus on elucidating novel mutations, insertions, deletions and chromosomal rearrangements at genome levels. NGS/Deep Sequencing of the entire genome (or exome) enables the identification of chromosomal aberrations, focal amplifications and gene rearrangements [17]. *De novo* identification of tumor-specific mutations is further combined with targeted approaches for further validation of newly identified mutation. Cost, coupled with longer analysis times, are cons of these untargeted processes, yet the comprehensive understanding of the complete ctDNA genome provides a significant means by which novel ctDNA biomarkers may be discovered.

Clinical Significance of ctDNA in Pancreatic Cancer

Numerous studies have demonstrated that elevated levels of ctDNA are found in PC. Fragmentation and concentration of tumor-derived ctDNA are positively correlated with tumor weight, progression, metastasis and shorter patient survival [10]. ctDNA in the plasma of PC patients has also been shown to correlate with primary tumor size, number of metastases, and the overall survival of the patients [18]. Moreover, a combination of KRAS mutations in ctDNA coupled with protein biomarker CA19.9 increased the detection sensitivity for PC at early stages [19]. Identification of a KRAS mutation in PC ctDNA is the central focus of multiple studies; however, other mutations in ctDNA for the detection of PC is gaining considerable attention. Table 1 summarizes the recent significant ctDNA findings in PC.

Overall, mutations in ctDNA not only mirror the mutations of the solid tumors from whence they originate, but also show significant potential as diagnostic and prognostic noninvasive markers. While many studies focus on the presence of KRAS mutations in ctDNA as a marker for PC, a combinatorial panel of p16/CDKN2A, TP53, SMAD4, and KRAS, has shown great potential as a biomarker panel for ctDNA profiling in PC patients. Profiling of ctDNA with a combination of known mutations can be used for diagnostic and prognostic purposes, or even to help elucidate the therapeutic efficacy of a treatment.

Challenges and Future Perspective

The importance of ctDNA as a biomarker for disease progression and therapy response is unequivocal. More efforts are being made to reduce pre-analytical, analytical, and biological variability. Due to the propensity for the minimal shedding of ctDNA from cells into the bloodstream, isolation of ctDNA poses a significant technical challenge. Serum/plasma concentration and size inconsistencies persist due to technical limitations and inherent variation in the platforms used for analysis. In the pre-analytical stage, consistency of ctDNA is heavily dependent on the type of collection tubes for blood/plasma, centrifugation duration/speed, collection timing, and extraction methods. Along with this, a caveat of isolation is that it should be performed within two hours of blood collection because of the short half-life of ctDNA. Further, the isolation of ctDNA needs to be performed with fresh serum/plasma as blood cells may lyse and release DNA that may dilute ctDNA. Also, blood should be collected in an EDTA containing tube instead of heparin because heparin can inhibit PCR [27].

Routinely used methods such as affinity column chromatography, magnetic bead or polymer isolation, and phenol-chloroform based plasma extraction methods can give rise to different sizes of ctDNA, which confers another layer of complexity and experimental artifact. Also, there is copious disparity regarding ctDNA concentration in various studies in association with healthy and disease states [9]. To reduce experimental variations in the analysis, utilizing both next-generation sequencing and digital PCR will be a required approach. To reduce the biological variation, information on stage and types of the tumor should also be considered while collecting and analyzing data. Though NGS is currently expensive and time-consuming, the field has seen an extensive cost reduction in recent years, and further

Table 1: Studies of circulating tumor DNA (ctDNA) in PC

Patient group	Specimen	Method	Target candidate	Findings	Reference
PC vs chronic pancreatitis (CP) and healthy	Plasma	Mutant allele-specific amplification method	Codon 12 of the KRAS gene	Detection of K-ras mutation was found in plasma DNA of PC patients that resembles mutation in solid tumors, while no mutation was found in CP and healthy subjects.	[20]
PC vs CP vs healthy individuals	Serum	PCR	KRAS mutation	K-ras gene mutations at codon 12 were detected in PC patients (14/20) while no mutation was found in CP and healthy individuals.	[21]
PC vs CP	Plasma	Restriction fragment length polymorphism-PCR and single-strand conformation polymorphism techniques	KRAS mutation	Presence of Kras mutation positively correlated with tumor stage, metastasis and shorter survival time.	[18]
PC vs CP	Serum	PCR and allele-specific amplification	Mutations at codon 12 of KRAS2	KRAS2 mutations were found in 47% PC patients and 13% of CP cases. The combination of KRAS2 and carbohydrate antigen 19.9 showed sensitivity and specificity of 98% and 77% respectively for the diagnosis of pancreatic cancer respectively.	[22]
PC	Plasma	Digital PCR	KRAS	26% of patients at all stages harbor KRAS mutation in ctDNA and it is correlated with overall survival.	[23]

(Contd.)

Table 1: (Contd.)

Patient group	Specimen	Method	Target candidate	Findings	Reference
Metastatic PC	Plasma	Exome sequencing and Droplet Digital PCR	KRAS, KDR, EGFR, ERBB2 exon17 and ERBB2 exon27	Mutations at BRCA2, KDR, EGFR, ERBB2 exon17 and ERBB2 exon27 in ctDNA were frequently mutated in metastatic PC. ERBB2 exon17 mutation ($p = 0.035$, HR = 1.61) was associated with overall survival of metastatic PC patients. Presence of ctDNA was positively correlated with treatment as assessed by CT imaging.	[24]
Metastatic PC, Intraductal papillary mucinous neoplasm (IPMN), healthy controls, patients with resected serous cystadenomas (SCAs) and borderline IPMN	Blood	Droplet Digital PCR	KRAS, Guanine nucleotide-binding protein (G-protein)-stimulating α-subunit (GNAS)	GNAS mutations in cfDNA from patients with IPMN, but not in patients with serous cystadenoma or controls. The KRAS codon 12 mutations (namely G12D and G12V mutation) were found 41.7% PC cases while not detectable in controls, SCA, and IPMN.	[25]
PC vs CP and healthy individuals	Serum	Digital PCR	G12V, G12D, and G12R in codon 12 of the K-ras gene	Rate of Kras mutation in ctDNA was 62.6% and was negatively correlated with patient survival.	[26]

KDR = Kinase Insert Domain Receptor; EGFR = Epidermal Growth Factor Receptor; ERBB2 = Erb-B2 Receptor Tyrosine Kinase 2; PC = Pancreatic Cancer; CP = Chronic Pancreatitis advances in technology are being pursued to make NGS economically feasible for disease detection.

New techniques must be established to improve the consistency, robustness, reproducibility, and accuracy of ctDNA analysis. Further research is required to understand what causes higher levels of ctDNA release in cancer patients when compared to healthy controls. Additionally, studies should be focused on identifying the threshold level of mutations in ctDNA that are required for disease initiation and progression. Moreover, an expert consensus must be formed on the concentration/number of ctDNA, which could thereby guide cancer patient care routinely. The most critical question remains unanswered: whether mutations in ctDNA adequately reflect solid tumor mutations in PC patients? This enhanced understanding is required to facilitate the eventual use of noninvasive ctDNA as a surrogate for tumor tissue samples. Recently, the U.S. Food and Drug Administration (FDA) has approved the first liquid biopsy test for the detection of exon 19 deletions or exon 21 (L858R) substitution mutations in metastatic non-small cell lung cancer patients to inform the clinical decision to treat with Erlotinib (a small molecule EGFR inhibitor) or not [28].

With the advent of personalized medicine and the continued evolution of cancer therapeutics, patients have new hope for disease detection or even cure. ctDNA will prove an invaluable component in deciphering the genomic milieu of a tumor via noninvasive liquid biopsy. It will continue to be at the forefront of this revolution in medicine due to its potential uses in early diagnosis, personalized/targeted therapies, therapeutic efficacy, and prognostication.

Metabolomics

Metabolomics can be defined as a robust approach to globally quantitate the measurable pool of small molecular weight intermediates of biochemical pathways, *i.e.* metabolites at the cell, tissue or organismal level. It exists at the bottom of the "omics" cascade and closely reflects variations in the phenotypic state of a biological system resulting from endogenous and/or exogenous influences [29]. Considering the dynamicity of living systems, genomics, transcriptomics, and proteomics often fail to mirror the actual clinical phenotype resulting from genetic/epigenetic mutations or adequately represent the specifically altered functionality of post-translationally modified enzymes. Metabolic fluxes in an organism can stem from either inherited genetic aberrations or subtle mutations in the genome. These fluxes can also result from machinery modifications acquired by the system over time due to environmental factors, gut microflora, antigen exposure, diseased state, and therapeutic interventions. Metabolomics is an emerging field of investigation with burgeoning interest amongst cancer researchers. Cancer pathogenesis is associated with hereditary predispositions, somatic mutations, and environmental exposures. Thus, combining a metabolic profile with mutational analysis may more comprehensively manifest the pathological grade, prognostic state or therapeutic response of pancreatic cancer and other malignancies.

Altered Metabolism in Cancer and Scope for Biomarker Discovery

One of the major hallmarks of neoplastic onset and progression is altered cellular metabolism. Tumor cells are highly proliferative cells with an elevated energy requirement and anabolic demand. Therefore, cancer cells have a higher glycolytic index than normal cells, referred to as the Warburg effect. Anaerobic glycolysis, the most prevalent form of glucose catabolism under the hypoxic tumor scenario, results in the accumulation of lactic acid in the tumor microenvironment. This has been implicated in various pro-tumorigenic properties including invasion of extracellular matrix components by tumor cells, enhanced angiogenesis, and immune cell functional exhaustion. In addition, tumor cells undergo high anaplerosis in order to produce the required building blocks necessary for dividing cells. This is manifested by an enhanced fatty acid, nucleotide and amino acid synthesis that can be produced from pyruvate or Kreb's cycle and Pentose Phosphate Pathway intermediates [30]. Moreover, cancer cells shunt the available nutrients from pathways in a distinct manner compared to normal cells. Thus, understanding the metabolic flux specific for pancreatic tumor cells may bolster the identification of metabolites as novel biomarkers in tissue specimens or body fluids.

Lipids being an integral part of malignant processes, cancer lipidomics has emerged as a promising arm of metabolomics-based biomarker studies. A wide range of glycerophospholipids and sphingolipids are essential for membrane biosynthesis in proliferating cancer cells, their eventual metastasis, and signal transduction properties. Thus, serum levels of phosphatidylcholines and their derivatives have been found to be upregulated in different malignancies including liver, breast, brain, and prostate [31, 32]. Besides their expression level, the active derivatives of some sphingolipids are well implicated in cancer cell growth and migration. For example, Sphingosine-1 phosphate (S1P), an active lysolipid formed by Sphingosine Kinase Type 1 enzyme, was found to induce breast cancer cell MCF-7 growth, transformation, and angiogenesis [33]. Interestingly, S1P was observed to be shed from the tumor cells and was found at elevated levels in ovarian cancer patient's serum [34]. These findings elucidate the universality of metabolomic profiling that is not inherently specific to pancreatic cancer but has broad application potential.

In addition to glucose and glutamine, tumor cells use non-essential amino acids like alanine and glycine as the major alternative energy sources to suffice for their elevated energy demand. These were found to be associated with hypoxic solid tumors and thus, studying their altered expression profile has facilitated biomarker discovery [35]. Elevated levels of alanine in serum can be an indication of enhanced cellular utilization of glucose or glutamine as energy sources by the tumor cells because pyruvate formed from glycolysis or glutaminolysis can undergo transamination to alanine under hypoxic conditions. Alanine levels are high in hypoxic liver and brain tumors [35, 36]. On the other hand, glycine levels decrease upon hypoxia-inducible factor-1 (HIF1) signaling [36].

Rapidly proliferating tumor cells undergo ROS mediated spontaneous DNA damage and aberrant chemical modifications like DNA methylation and histone acetylation/deacetylation leading to the formation of a pool of modified bases. Owing to DNA repair mechanisms and enhanced cellular turnover, those altered bases are generally excised and expelled from the system. Elevated levels of such modified nucleosides in the urine are evolving as important biomarkers in different cancers including liver, breast, colorectal cancer, hepatocellular carcinoma and leukemia [37-39]. Sarcosine is produced by transfer of an N-methyl group on glycine during DNA methylation and its level in the urine is shown to increase during prostate cancer progression and metastasis [40]. This finding indicates that one of the potential metabolites that can be used as a cancer biomarker are modified nucleosides.

If one considers hotspot mutations, seminal epigenetic or pathological modifications as the disease progresses, most malignancies undergo comparable metabolic alterations leading to the identification of global metabolite signature that is significantly deviated from normal physiologic metabolism. However, identification of metabolic biomarkers for specific cancers needs an in-depth understanding of the disease etiology along with awareness of the crosstalk-talk among different cellular components in the tumor microenvironment. In that context, metabolomics studies on pancreatic cancer seem challenging owing to the dynamic coupling between the tumor and stromal cells and the complex molecular interplay associated with the disease.

Methodology

One of the emerging areas in metabolomics research involves the discovery of novel biomarkers for early-stage PC detection and markers to differentiate metastatic from non-metastatic PC cases. Within the metabolomics field there are broadly two approaches: targeted and untargeted approach. Targeted metabolomics aims at detecting the level of a specific predefined set of metabolites while untargeted metabolomics has a global application in identifying and measuring as many metabolites as possible present in a biological sample [41]. The greatest advantage of a metabolite based biomarker is the scope of examining a large set of samples for an array of molecules. Another benefit of this approach is the possibility of using low sample volume, i.e. 0.5 ml of bio-fluids are said to be sufficient for metabolomics analysis. Another advantage of metabolic biomarkers in patients is obtaining tissue biopsies and body fluids which are part of the normal cancer therapy regimen and as such no special procedures would be required. The general scheme for using metabolomics in the cancer biomarker discovery field is as follows:

Sample Acquisition

Myriad forms of patient samples are used in biomarker research including urine, saliva, bile & pancreatic juice, plasma, serum, feces, and tissue biopsies [42]. The benefit of metabolite based biomarker studies is its overall non-invasive approach. Bio-fluids from PC patients may be compared with that from normal individuals. This comparison is superior to cancer tissue biopsies where the comparison is drawn with adjacent healthy tissue as the representation of "normal" which may often be misleading.

Sample Processing and Analysis

Generally, Gas Chromatography-Mass Spectrometry (GC-MS), Liquid Chromatography-Mass Spectrometry (LC-MS), Capillary Electrophorese-Mass Spectrometry (CE-MS), Fourier Transform Ion Cyclotron Resonance-Mass Spectrometry (FTICR-MS), Nuclear Magnetic Resonance (NMR) and High Resolution Magic Angle Spinning Nuclear Magnetic Spectroscopy (HR-MAS-NMR) are used for metabolic profiling from biological specimens. These techniques have their own advantages/disadvantages but data acquired by each separate modality complement one another [42, 43]. Also, due to the technical difficulties, hydrophilic and hydrophobic metabolites have not yet been able to be analyzed using a single technique. Thus, for obtaining a robust and global metabolite profile of the patient and in order to discern the disease versus normal states, a combination of various analytical techniques is used [43]. Recent advances in the field have enabled metabolites to be visualized spatially within biological samples using imaging mass spectrometry [44].

Pattern Recognition and Validation

The metabolite data acquired is then subjected to a series of bioinformatics and biostatistical platforms in order to normalize the data, remove outliers, and map them to common cancer-associated pathways. This is done in order to glean a possible mechanistic insight or identify an underlying pattern of the analyzed metabolites. Principal component analysis is a commonly used statistical approach to classify the samples based on multiple dimensions. This is followed by the use of various supervised data analysis approaches like partial least squares-discriminate analysis (PLS-DA) and artificial neural networks (ANN). These methods are used in conjunction with histopathological scores, and other "omics" models along with pathological outcome [45, 46]. For verification analysis of metabolic shunts prevalent in the system across a certain period of time, radiolabeled metabolites such as C^{13} labeled glucose are widely used [47]. Once the metabolites are identified as potential diagnostic or prognostic markers, validation studies are conducted in independent cohorts with a multi-center approach.

Applications of Metabolomics Studies in Pancreatic Cancer

One of the significant reason for poor prognosis of pancreatic cancer is the unavailability of highly sensitive and specific early detection markers. Since the pancreas is a vital metabolic organ of the body, identification and classification of its physiologic versus pathological metabolic signatures seem a logical pursuit for PC biomarker discovery. Since an altered metabolome significantly reflects the changing phenotypic status of an individual over time, metabolomics has emerged as a promising readout for differentiating high-risk benign groups including type 2 diabetes mellitus and chronic pancreatitis from pancreatic cancer cases. Table 2 enlists the recent significant studies that have identified crucial metabolites for PC detection.

Challenges and Future Perspective

The greatest concern in metabolomics-based biomarker studies is the collection of samples. Along with considering age, gender, and geographical area, factors like dietary plan, fasting state, physical activity, ethnic origin, etc. must be taken into account. This is required in order to design a metabolomics pipeline with adequate experimental controls. When considering pancreatic cancer, candidate selections for metabolomic studies need further precision because the pancreas is an organ crucial for normal physiological metabolism. As such, the alterations in the metabolic profile associated with pancreatic cancer closely resemble other metabolic disorders including type 2 diabetes, cachexia, oxidative stress, etc.

Exosomes

In recent years, the academic medical community has begun to realize the potential of exosomes to serve as noninvasive liquid biopsy based biomarkers for the early detection of pancreatic cancer. Exosomes are 50-200 nm extracellular vesicles that are present in a variety of biofluids including serum, urine, ascitic fluid, and saliva [60, 61] making them an easily accessible target for noninvasive liquid biopsies (Fig. 2). In addition to their ubiquitous nature, exosomal production is largely increased in the setting of cancer [62]. This provides a valuable and useful increase in the exosome population of interest that may be utilized as a means of detection.

Inherent to exosomal creation is the intraexosomal content representative of the cell of origin. Exosomes arise from the inward pinching of the late endosomal membrane and the formation of a multivesicular body (MVB). This MVB fuses with the cellular plasma membrane and the formed vesicles are released into the extracellular space, which are now termed exosomes [63]. Due to the formation of a phospholipid bilayer membrane, exosomes are resistant to degradation in the extracellular environment and the bloodstream, thus conferring some level of protection to the carried contents. This biogenesis naturally provides an elegant and natural means of sampling

Table 2: Key recent reports of metabolomics analysis on PC patients

Classification	Specimen	Methods	Altered metabolites in PC cases	Reference
Healthy vs PC	Plasma	NMR LC/MS, HPLC-ESI-MS/MS	Altered lipid metabolome; decrease in membrane phosphatidylinositols	[48]
	Serum	NMR	Elevated choline, taurine, glucose, triglycerides	[49]
	Serum	GC-QMS	Elevated lactate, thioglycolate, 7- hydroxyoctanoate, asparagine, aconitate, homogentisate, N-acetyl tyrosine, reduced glycine, urea, glycerate, laurate, myristate, palmitate, and stearate.	[50]
	Serum	NMR	Elevated isoleucine, leucine, creatinine, triglyceride; reduced hydroxybutyrate, lactate, trimethylamine-N-oxide	[51]
Healthy vs chronic pancreatitis vs early stage PC	Plasma	GC-TOF-MS, RP-LC/ESI-MS, HILIC-LC/ESI-MS	Elevated N-methylalanine, lysine, glutamine, phenylalanine, arachidonic acid, taurocholic acid, PCh, lysoPCh, PE and glycine variants; reduced glutamine, inosine, tryptamine, phenylalanine, hydrocinnamate	[52]
Hepatobiliary disease vs PC	Serum	NMR	Altered levels of PI, PG	[53]
Healthy vs chronic pancreatitis vs PC (diagnostic model)	Serum	GC-QMS	Elevated arabinose, ribulose; reduced valine, histidine, tyrosine, urate, methionine, 1,5-Anhydro-D-glucitol, valine, etc.	[53]
Chronic pancreatitis vs PC	Plasma, Serum	GC-MS; LC-Tandem/MS	Differential levels of complex lipids and fatty acids (Sphingomyelin, PC, Sphinganine-1-P, Ceramide, Sphingomyelin), Pyruvate, Histidine, Proline	[54]
PP vs PA; Long term PC survivors	Tissue	HR-MAS	Higher level of myo-inositol and glycerol in PP; elevated levels of glucose, lactate, ascorbate, ethanolamine, taurine; higher ethanolamine associated with poor survival	[55]

(Contd.)

Healthy vs tumor areas	Tissue	UPLC-TOF-MS	Elevated taurine, reduced succinate, malate, uridine, glutathione, UDP-N-Acetyl-D-Glucosamine, NAD, UMP, AMP	[56]
Healthy vs different cancers	Saliva	CE-TOF-MS	Differential levels of 57 metabolites including polyamines, carnitine, amino acids, purine, amino-alcohols	[57]
Healthy vs PC	Urine	NMR	Elevated acetoacetate, leucine, glucose, 2-phenylacetamide, acetylated compounds, reduced citrate, creatinine, glycine, hippurate, 3-hydroxyisoverate, trigonelline.	[58]
			Significant alterations in urinary metabolite	[59]

PE: Phosphatidyl Ethanolamine, PCh: Phosphatidyl Choline, PI: Phosphatidyl Inositols; PG: Phosphatidyl Glycerols, PP: Pancreatic Parenchyma, PA: Pancreatic Adenocarcinoma

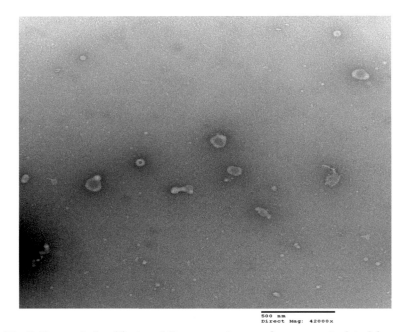

Fig. 2: Transmission Electron Microscopy image of exosomes isolated from pancreatic cystic fluid with characteristic "cup-shape" morphology

the intracellular environment of tumors that are difficult to access or are inaccessible by tissue biopsy, such as pancreatic cancer.

Exosomes have been shown to have a multitude of functional properties exerting influence on many necessary cellular mechanisms, as well as those found in pancreatic cancer. Two of the most important of these functional capabilities are their role in intercellular communication [64] and the facilitation of the induction/suppression of the immune response [65]. Costa-Silva et al. studied the ability of pancreatic cancer exosomes to form a pre-metastatic niche in the liver. Their group discovered that exosomes, originating from PC cells, cause Kupfer cells to secrete Transforming Growth Factor-β (TGF-β), thereby forming a fibrous microenvironment in the liver that is conducive to the propagation of metastases [66]. Additionally, tumor-derived exosomes have also been implicated in angiogenesis within the local tumor microenvironment [67] which has direct implications on tumor access to oxygen and nutrients. These particles have also been shown to modify the glucose uptake of cancer cells, thereby having a direct impact on metabolism and overall cell survival [68].

Potential Exosomal Biomarkers in Pancreatic Cancer

Various components found on the surface and within exosomes have been proposed as novel diagnostic markers for the early detection of pancreatic cancer. These include miRNA profiling, dsDNA mutation detection, surface lipidomics/proteomics, and metabolomics. The amenability of exosomes to a multitude of analytic techniques is precisely what makes them such a valuable diagnostic target. Table 3 lists various exosomal diagnostic markers and methods of isolation recently investigated.

MicroRNA (miRNA)

Since one of the primary functions of exosomes is intercellular communication, the presence of miRNA is expected. The miRNA profile of exosomes has great potential as a possible PC biomarker and as such, has been an active area of research in recent years. miRNAs are noncoding RNAs, 19-25 nucleic acids in length, that modulate the expression of specific mRNA(s) by binding to the 3' UTR end. In conjunction, it has recently been shown that exosomes transport the RISC complex which is necessary for the processing of miRNA into the functional unit required to alter mRNA expression [78].

Along with the functional and communication aspects of miRNA in exosomes, their differential expression levels, or more simply their presence/absence, can be measured for the purposes of cancer detection. Many exosomal miRNA profiles have been found to be highly sensitive and specific to pancreatic cancer. One study combined miRNA expression with computer learning, and was able to correctly characterize exosomes as arising from healthy or cancer cell lines without error [69]. Further, miRNAs have been shown to have predictive validity for patient prognosis and chemotherapeutic response [70]. These findings display the capabilities that exosomes have moved beyond diagnostics and into the clinical realm of patient outcome and therapeutic efficacy.

Double Stranded DNA (dsDNA)

Along with miRNA, dsDNA has also been found in exosomes. Due to dsDNA ability to yield information on specific genetic mutations within the tumor cell of origin, identification of dsDNA in exosome has created much excitement as a target for cancer detection. This amazing potential was confirmed in a study done by Yang et al. This group was able to detect mutations in KRAS and p53 via exosomes purified from patient serum [72], thus elucidating the initiating driver mutations for that patient. Another group was able to use exosomal dsDNA to realize the genomic mutations that lead to chemotherapy resistance [79]. Yet another use of exosomes could be used as a sampling of genetic data to be utilized in the personalization of medicine for certain patients.

Surface Lipidomics and Proteomics

Surface analysis of exosomes has become an important aspect of biomarker investigation. ELISA has been used extensively to attempt ultrasensitive detection of specific exosomal surface markers. More recently, surface lipidomics and proteomics, i.e. the exosomal "surfaceome", has shown great potential for discovering highly specific pancreatic cancer biomarkers. In an

Table 3: Exosomal biomarkers in pancreatic cancer

Exosome origin	Molecule of interest	Methods	Findings	Reference
PC patient serum, PC cell lines	miRNA	Ultracentrifugation, next-generation small RNA sequencing, qRT-PCR	miR-196a and miR-1246 are potential indicators of PC	[69]
Patient serum	miRNA	Ultracentrifugation, miRNA microarray, qRT-PCR	miR-23b-3p is increased in PC patient serum	[70]
In silico analysis	miRNA	Exosomal genes database and GeneCards metadata tools	26 different miRNAs as possible PC targets	[71]
PC patient serum, Healthy patient serum	dsDNA	Ultracentrifugation, Digital PCR	KRAS and TP53 mutations are detectable in exosomal DNA	[72]
Mouse serum	Lipid profile	Ultracentrifugation, ELISA, KPC Murine model	Elevated exosomal Phosphatidylserine in tumor baring KPC murine models	[73]
PC cell lines, PC patient serum	Protein (Surface)	Ultracentrifugation, biotinylation, LC-MS/MS	CLDN4, EPCAM, CD151, LGALS3BP, HIST2H2BE, HIST2H2BF could be PC specific	[74, 75]
PC cell lines, PC patient serum	Protein (Surface)	Ultracentrifugation, LC-MS/MS	Glypican1 is a sensitive and specific marker for PC	[75]
PC cell lines, KPCY murine serum	mRNA	ExoTENPO-based exosome capture	Machine learning and mRNA profile are able to differentiate PC from healthy and PanIN	[76]
PC patient serum before and after chemo	Protein (Total)	Ultracentrifugation, iTRAQ method	Exosomes carried 700-800 proteins per sample with many associated with metastasis and treatment resistance	[77]

elegant study, Castillo et al. used mass spectrometry to assess the protein composition of the exosomal membrane surface [74]. They were able to identify nine PC specific proteins that were then specifically targeted via capture antibodies, to increase the yield of exosomes originating from tumor cells. Another exosomal surface glycoprotein, glypican 1, was shown to be 100% sensitive and specific for pancreatic cancer [75]. Also, recently it was discovered that the sphingomyelin content of the exosomal membrane could be used for early diagnosis of pancreatic cancer (and breast cancer) prior to any outward signs of disease or metastasis [73].

Methodology

Various engineering advances have been utilized in the effort to rapidly and inexpensively purify exosomes from biofluids. A promising application is in the use of micro/nanofluidics for purification of exosomes predicated on the physical properties of density and size, compared to the rest of the serum milieu [80]. This technology, if refined, could prove invaluable for translation to the clinic. It has been shown that combining sound with fluidics, in a technique called acoustofluidics, can further improve the yield and purity of exosome populations from patient serum [81].

Affinity-based chromatography isolation via heparinized beads has also been attempted. While this method works well with a pure exosome population, the efficacy drops considerably for isolating exosomes directly from biofluids. Another bead-based application is flow cytometry. However, these beads have antibodies against exosome specific markers attached to them rather than a heparin coating. This has been used to assess the relative quantity of exosomes within a sample as well as to check for the presence of specific surface markers like CD63 and CD81. Ultracentrifugation, however, is the primary means of purification of exosomes from biofluids but considerable drawbacks persist as will be further elucidated.

Challenges and Future Perspective

While exosomes have vast potential, they have some shortcomings that must be overcome in order to go from bench to bedside and enter into everyday use in clinic. The primary issue revolves around exosome purification from biofluids. The accepted standard practice involves ultracentrifugation and density gradient fractionation. This technique yields highly pure exosome samples but is exceedingly time-consuming and labor intensive. Future investigation and refinement of the aforementioned purification methodologies must be undertaken to a point that they become standard practice, thereby decreasing the opportunity cost associated with exosomal collection and making them more amenable to analysis. Another hindrance for exosome use as biomarkers is the fact that many different cell types contribute to the exosomal population found in biofluids. Congruently, at early pancreatic cancer stages, the majority of the pancreatic exosomes originate from healthy epithelial cells and not tumor cells.

These obstacles provide readily apparent avenues for the future direction of pancreatic cancer exosomal research. They point out the necessity for continued surface proteomic refinement and the continued pursuit for a pancreatic cancer-specific marker. This could subsequently be utilized as the mechanism for exosomal isolation and facilitate the inspection of the intravesicular contents of exosomes specifically originating from cancer cells. Finally, novel sources of exosomes, and their component profiles, namely, cystic fluid, saliva, and urine, must be investigated further. The advantages of these sources are different yet equally important. Saliva and urine are easily collected with virtually no risk to the patient and can be reassessed at any time to monitor tumor progression, patient reaction to therapy, or for prognostic value. Cystic fluid may be an invaluable source of exosomes as they will only originate from the pancreas and not be diluted in the bloodstream. Further, cystic fluid exosomes could offer a means of detecting those at greatest risk for developing this deadly disease, before they even get it.

Conclusion

Pancreatic cancer is a disease with the single curative option of complete resection. Lack and ambiguity of early symptoms related to the disease makes it impossible to be detected after surgical intervention. Assessment of cell-free DNA, metabolomes and exosome profiles in patient serum are emerging technologies with great potential as circulating biomarkers for the early detection of PC. These methods offer a noninvasive and sensitive means of cancer discovery, thus conferring little risk to the patient. However, to actualize the potential of these methods as detection modalities, the overall accuracy of the assays must be improved to make them clinically relevant. Continued investigation and research into these biomarker techniques, and combinations thereof, will improve our ability to detect and treat pancreatic cancer.

References

1. Llop, E., P. EG, A. Duran, S. Barrabes, A. Massaguer, M. Jose Ferri, et al. Glycoprotein biomarkers for the detection of pancreatic ductal adenocarcinoma. World J Gastroent, 2018. 24(24): 2537-54.
2. Guo, S., A. Fesler, H. Wang, J. Ju. MicroRNA based prognostic biomarkers in pancreatic cancer. Biomark Res, 2018. 6: 18.
3. Chang, Y., B. Tolani, X. Nie, X. Zhi, M. Hu, B. He. Review of the clinical applications and technological advances of circulating tumor DNA in cancer monitoring. Ther Clin Risk Manag, 2017. 13: 1363-74.
4. Thierry, A.R., S. El Messaoudi, P.B. Gahan, P. Anker, M. Stroun. Origins, structures, and functions of circulating DNA in oncology. Cancer Metastasis Rev, 2016. 35(3): 347-76.

5. Thierry, A.R., F. Mouliere, C. Gongora, J. Ollier, B. Robert, M. Ychou, et al. Origin and quantification of circulating DNA in mice with human colorectal cancer xenografts. Nucleic Acids Res, 2010. 38(18): 6159-75.
6. Heitzer, E., S. Perakis, J.B. Geigl, M.R. Speicher. The potential of liquid biopsies for the early detection of cancer. NPJ Precis Oncol, 2017. 1(1): 36.
7. Perakis, S., M.R. Speicher. Emerging concepts in liquid biopsies. BMC Med, 2017. 15(1): 75.
8. Mandel, P., P. Metais. Les acidesnucleiques du plasma sanguin chez l'homme. C R Seances Soc Biol Fil, 1948. 142(3-4): 241-3.
9. Han, X., J. Wang, Y. Sun. Circulating tumor DNA as biomarkers for cancer detection. Genomics Proteomics Bioinformatics, 2017. 15(2): 59-72.
10. Mouliere, F., B. Robert, E. ArnauPeyrotte, M. Del Rio, M. Ychou, F. Molina, et al. High fragmentation characterizes tumour-derived circulating DNA. PLoS One, 2011. 6(9): e23418.
11. Bardelli, A., K. Pantel. Liquid biopsies, what we do not know (yet). Cancer Cell, 2017. 31(2): 172-9.
12. Perkins, G., H. Lu, F. Garlan, V. Taly. Droplet-based digital PCR: application in cancer research. Adv Clin Chem, 2017. 79: 43-91.
13. Diehl, F., M. Li, Y. He, K.W. Kinzler, B. Vogelstein, D. Dressman. BEAMing: single-molecule PCR on microparticles in water-in-oil emulsions. Nat Methods, 2006. 3(7): 551-9.
14. Diehl, F., M. Li, D. Dressman, Y. He, D. Shen, S. Szabo, et al. Detection and quantification of mutations in the plasma of patients with colorectal tumors. Proc Natl Acad Sci USA, 2005. 102(45): 16368-73.
15. Kinde, I., J. Wu, N. Papadopoulos, K.W. Kinzler, B. Vogelstein. Detection and quantification of rare mutations with massively parallel sequencing. Proc Natl Acad Sci USA, 2011. 108(23): 9530-5.
16. Vendrell, J.A., F.T. Mau-Them, B. Beganton, S. Godreuil, P. Coopman, J. Solassol. Circulating cell free tumor DNA detection as a routine tool for lung cancer patient management. Int J Mol Sci, 2017. 18(2): 264.
17. Wan, J.C.M., C. Massie, J. Garcia-Corbacho, F. Mouliere, J.D. Brenton, C. Caldas, et al. Liquid biopsies come of age: towards implementation of circulating tumour DNA. Nat Rev Cancer, 2017. 17(4): 223-38.
18. Castells, A., P. Puig, J. Mora, J. Boadas, L. Boix, E. Urgell, et al. K-ras mutations in DNA extracted from the plasma of patients with pancreatic carcinoma: diagnostic utility and prognostic significance. J Clin Oncol, 1999. 17(2): 578-84.
19. Cohen, J.D., A.A. Javed, C. Thoburn, F. Wong, J. Tie, P. Gibbs, et al. Combined circulating tumor DNA and protein biomarker-based liquid biopsy for the earlier detection of pancreatic cancers. Proc Natl Acad Sci USA, 2017. 114(38): 10202-07.
20. Yamada, T., S. Nakamori, H. Ohzato, S. Oshima, T. Aoki, N. Higaki, et al. Detection of K-ras gene mutations in plasma DNA of patients with pancreatic adenocarcinoma: correlation with clinicopathological features. Clin Cancer Res, 1998. 4(6): 1527-32.
21. Theodor, L., E. Melzer, M. Sologov, G. Idelman, E. Friedman, S. Bar-Meir, et al. Detection of pancreatic carcinoma: diagnostic value of K-ras mutations in circulating DNA from serum. Dig Dis Sci, 1999. 44(10): 2014-9.
22. Maire, F., S. Micard, P. Hammel, H. Voitot, P. Lévy, P.H. Cugnenc, et al. Differential diagnosis between chronic pancreatitis and pancreatic cancer: value of the detection of KRAS2 mutations in circulating DNA. Br J Cancer, 2002. 87(5): 551-4.

23. Earl, J., S. Garcia-Nieto, J.C. Martinez-Avila, J. Montans, A. Sanjuanbenito, M. Rodríguez-Garrote, et al. Circulating tumor cells (CTC) and kras mutant circulating free DNA (cfDNA) detection in peripheral blood as biomarkers in patients diagnosed with exocrine pancreatic cancer. BMC Cancer, 2015. 15: 797.
24. Cheng, H., Liu, J. Jiang, G. Luo, Y. Lu, K. Jin, et al. Analysis of ctDNA to predict prognosis and monitor treatment responses in metastatic pancreatic cancer patients. Int J Cancer, 2017. 140(10): 2344-50.
25. Berger, A.W., D. Schwerdel, I.G. Costa, T. Hackert, O. Strobel, S. Lam, et al. Detection of hot-spot mutations in circulating cell-free DNA from patients with intraductal papillary mucinous neoplasms of the pancreas. Gastroenterology, 2016. 151(2): 267-70.
26. Kinugasa, H., K. Nouso, K. Miyahara, Y. Morimoto, C. Dohi, K. Tsutsumi, et al. Detection of K-ras gene mutation by liquid biopsy in patients with pancreatic cancer. Cancer, 2015. 121(13): 2271-80.
27. Volik, S., M. Alcaide, R.D. Morin, C. Collins. Cell-free DNA (cfDNA): clinical significance and utility in cancer shaped by emerging technologies. Mol Cancer Res, 2016. 14(10): 898-908.
28. Kwapisz, D. The first liquid biopsy test approved. Is it a new era of mutation testing for non-small cell lung cancer? Ann Transl Med, 2017. 5(3): 46.
29. Fiehn, O. Metabolomics – the link between genotypes and phenotypes, in functional genomics. 2002, Springer. 155-71.
30. Kim, J.W., C.V. Dang. Cancer's molecular sweet tooth and the Warburg effect. Cancer Res, 2006. 66(18): 8927-30.
31. Glunde, K., C. Jie, Z.M. Bhujwalla. Molecular causes of the aberrant choline phospholipid metabolism in breast cancer. Cancer Res, 2004. 64(12): 4270-6.
32. Hilvo, M., C. Denkert, L. Lehtinen, B. Muller, S. Brockmoller, T. Seppanen-Laakso, et al. Novel theranostic opportunities offered by characterization of altered membrane lipid metabolism in breast cancer progression. Cancer Res, 2011. 71(9): 3236-45.
33. Sarkar, S., M. Maceyka, N.C. Hait, S.W. Paugh, H. Sankala, S. Milstien, et al. Sphingosine kinase 1 is required for migration, proliferation and survival of MCF-7 human breast cancer cells. FEBS Lett, 2005. 579(24): 5313-7.
34. Knapp, P., L. Bodnar, A. Blachnio-Zabielska, M. Swiderska, A. Chabowski. Plasma and ovarian tissue sphingolipids profiling in patients with advanced ovarian cancer. Gynecol Oncol, 2017. 147(1): 139-44.
35. Griffiths, J.R., M. Stubbs. Opportunities for studying cancer by metabolomics: preliminary observations on tumors deficient in hypoxia-inducible factor 1. Adv Enzyme Regul, 2003. 43: 67-76.
36. Ben-Yoseph, O., R.S. Badar-Goffer, P.G. Morris, H.S. Bachelard. Glycerol 3-phosphate and lactate as indicators of the cerebral cytoplasmic redox state in severe and mild hypoxia respectively: a 13C- and 31P-n.m.r. study. Biochem J, 1993. 291(Pt 3): 915-9.
37. Zambonin, C.G., A. Aresta, F. Palmisano, G. Specchia, V. Liso. Liquid chromatographic determination of urinary 5-methyl-2'-deoxycytidine and pseudouridine as potential biological markers for leukaemia. J Pharm Biomed Anal, 1999. 21(5): 1045-51.
38. Struck, W., M. Waszczuk-Jankowska, R. Kaliszan, M.J. Markuszewski. The state-of-the-art determination of urinary nucleosides using chromatographic techniques "hyphenated" with advanced bioinformatic methods. Anal Bioanal Chem, 2011. 401(7): 2039-50.

39. Woo, H.M., K.M. Kim, M.H. Choi, B.H. Jung, J. Lee, G. Kong, et al. Mass spectrometry based metabolomic approaches in urinary biomarker study of women's cancers. Clin Chim Acta, 2009. 400(1-2): 63-9.
40. Sreekumar, A., L.M. Poisson, T.M. Rajendiran, A.P. Khan, Q. Cao, J. Yu, et al. Metabolomic profiles delineate potential role for sarcosine in prostate cancer progression. Nature, 2009. 457(7231): 910-4.
41. Patti, G.J., O. Yanes, G. Siuzdak. Innovation: metabolomics – the apogee of the omics trilogy. Nat Rev Mol Cell Biol, 2012. 13(4): 263-9.
42. Armitage, E.G., M. Ciborowski. Applications of metabolomics in cancer studies. Adv Exp Med Biol, 2017. 965: 209-34.
43. Armitage, E.G., C. Barbas. Metabolomics in cancer biomarker discovery: current trends and future perspectives. J Pharm Biomed Anal, 2014. 87: 1-11.
44. Koizumi, S., S. Yamamoto, T. Hayasaka, Y. Konishi, M. Yamaguchi-Okada, N. Goto-Inoue, et al. Imaging mass spectrometry revealed the production of lyso-phosphatidylcholine in the injured ischemic rat brain. Neuroscience, 2010. 168(1): 219-25.
45. Katajamaa, M., M. Oresic. Data processing for mass spectrometry-based metabolomics. J Chromatogr A, 2007. 1158(1-2): 318-28.
46. Sugimoto, M., M. Kawakami, M. Robert, T. Soga, M. Tomita. Bioinformatics tools for mass spectroscopy-based metabolomic data processing and analysis. Curr Bioinform, 2012. 7(1): 96-108.
47. Lane, A.N., T.W. Fan, R.M. Higashi, J. Tan, M. Bousamra, D.M. Miller. Prospects for clinical cancer metabolomics using stable isotope tracers. Exp Mol Pathol, 2009. 86(3): 165-73.
48. Beger, R.D., L.K. Schnackenberg, R.D. Holland, D. Li, Y. Dragan, et al. Metabonomic models of human pancreatic cancer using 1D proton NMR spectra of lipids in plasma. Metabolomics, 2006. 2(3): 125-34.
49. Tesiram, Y.A., M. Lerner, C. Stewart, C. Njoku, D.J. Brackett, et al. Utility of nuclear magnetic resonance spectroscopy for pancreatic cancer studies. Pancreas, 2012. 41(3): 474-80.
50. Nishiumi, S., M. Shinohara, A. Ikeda, T. Yoshie, N. Hatano, S. Kakuyama, et al. Serum metabolomics as a novel diagnostic approach for pancreatic cancer. Metabolomics, 2010. 6(4): 518-28.
51. OuYang, D., J. Xu, H. Huang, Z. Chen. Metabolomic profiling of serum from human pancreatic cancer patients using 1 H NMR spectroscopy and principal component analysis. Appl Biochem Biotech, 2011. 165(1): 148-54.
52. Urayama, S., W. Zou, K. Brooks, V. Tolstikov. Comprehensive mass spectrometry based metabolic profiling of blood plasma reveals potent discriminatory classifiers of pancreatic cancer. RCM 2010. 24(5): 613-20.
53. Kobayashi, T., et al. A novel serum metabolomics-based diagnostic approach to pancreatic cancer. Cancer Epidemiol. Biomarkers Prev, 2013. 22(4): 571-9
54. Mayerle, J., H. Kalthoff, R. Reszka, B. Kamlage, E. Peter, B. Schniewind, et al. Metabolic biomarker signature to differentiate pancreatic ductal adenocarcinoma from chronic pancreatitis. Gut, 2017: gutjnl-2016-312432.
55. Battini, S., F. Faitot, A. Imperiale, A.E. Cicek, C. Heimburger, G. Averous, et al. Metabolomics approaches in pancreatic adenocarcinoma: tumor metabolism profiling predicts clinical outcome of patients. BMC Medicine, 2017. 15(1): 56.
56. Kaur, P., K. Sheikh, A. Kirilyuk, K. Kirilyuk, R. Singh, H. Ressom, et al. Metabolomic profiling for biomarker discovery in pancreatic cancer. Int. J. Mass Spectrom., 2012. 310: 44-51.

57. Sugimoto, M., D.T. Wong, A. Hirayama, T. Soga, M. Tomita, et al. Capillary electrophoresis mass spectrometry-based saliva metabolomics identified oral, breast and pancreatic cancer-specific profiles. Metabolomics, 2010. 6(1): 78-95.
58. Napoli, C., N. Sperandio, R.T. Lawlor, A. Scarpa, H. Molinari, M. Assfalg M, et al. Urine metabolic signature of pancreatic ductal adenocarcinoma by (1)h nuclear magnetic resonance: identification, mapping, and evolution. J Proteome Res, 2012. 11(2): 1274-83.
59. Davis, V.W., D. Eurich, O.F. Bathe, M.B. Sawyer, et al. Pancreatic ductal adenocarcinoma is associated with a distinct urinary metabolomic signature. Annals of Surgical Oncol, 2013. 20(3): 415-23.
60. Butz, H., R. Nofech-Mozes, Q. Ding, H.W.Z. Khella, P.M. Szabo, M. Jewett, et al. Exosomal microRNAs are diagnostic biomarkers and can mediate cell-cell communication in renal cell carcinoma. Eur Urol Focus, 2016. 2(2): 210-18.
61. Sun, Y., S. Liu, Z. Qiao, Z. Shang, Z. Xia, X. Niu, et al. Systematic comparison of exosomal proteomes from human saliva and serum for the detection of lung cancer. Anal Chim Acta, 2017. 982: 84-95.
62. Tamkovich, S.N., N.V. Yunusova, A.K. Somov, G.V. Kakurina, E.S. Kolegova, E.A. Tugurova, et al. Comparative sub-population analysis of exosomes from blood plasma of cancer patients. Biomed Khim, 2018. 64(1): 110-14.
63. Kowal, J., M. Tkach, C. Thery. Biogenesis and secretion of exosomes. Curr Opin Cell Biol, 2014. 29: 116-25.
64. Rackov, G., N. Garcia-Romero, S. Esteban-Rubio, J. Carrion-Navarro, C. Belda-Iniesta, A. Ayuso-Sacido. Vesicle-mediated control of cell function: the role of extracellular matrix and microenvironment. Front Physiol, 2018. 9: 651.
65. Maybruck, B.T., L.W. Pfannenstiel, M. Diaz-Montero, B.R. Gastman. Tumor-derived exosomes induce CD8(+) T cell suppressors. J Immunother Cancer, 2017. 5(1): 65.
66. Costa-Silva, B., N.M. Aiello, A.J. Ocean, S. Singh, H. Zhang, B.K. Thakur et al. Pancreatic cancer exosomes initiate pre-metastatic niche formation in the liver. Nat Cell Biol, 2015. 17(6): 816-26.
67. Ludwig, N., S.S. Yerneni, B.M. Razzo, T.L. Whiteside. Exosomes from HNSCC promote angiogenesis through reprogramming of endothelial cells. Mol Cancer Res, 2018. 16(11): 1798-808.
68. Zhao, H., L. Yang, J. Baddour, A. Achreja, V. Bernard, T. Moss, et al. Tumor microenvironment derived exosomes pleiotropically modulate cancer cell metabolism. Elife, 2016. 5: e10250.
69. Xu, Y.F., B.N. Hannafon, Y.D. Zhao, R.G. Postier, W.Q. Ding. Plasma exosome miR-196a and miR-1246 are potential indicators of localized pancreatic cancer. Oncotarget, 2017. 8(44): 77028-40.
70. Chen, D., X. Wu, M. Xia, F. Wu, J. Ding, Y. Jiao, et al. Upregulated exosomic miR23b3p plays regulatory roles in the progression of pancreatic cancer. Oncol Rep, 2017. 38(4): 2182-88.
71. Makler, A., R. Narayanan. Mining exosomal genes for pancreatic cancer targets. Cancer Genomics Proteomics, 2017. 14(3): 161-72.
72. Yang, S., S.P. Che, P. Kurywchak, J.L. Tavormina, L.B. Gansmo, P. Correa de Sampaio, et al. Detection of mutant KRAS and TP53 DNA in circulating exosomes from healthy individuals and patients with pancreatic cancer. Cancer Biol Ther, 2017. 18(3): 158-65.
73. Sharma, R., X. Huang, R.A. Brekken, A.J. Schroit. Detection of phosphatidylserine-positive exosomes for the diagnosis of early-stage malignancies. Br J Cancer, 2017. 117(4): 545-52.

74. Castillo, J., V. Bernard, F.A. San Lucas, K. Allenson, M. Capello, D.U. Kim, et al. Surfaceome profiling enables isolation of cancer-specific exosomal cargo in liquid biopsies from pancreatic cancer patients. Ann Oncol, 2018. 29(1): 223-9.
75. Melo, S.A., L.B. Luecke, C. Kahlert, A.F. Fernandez, S.T. Gammon, J. Kaye, et al. Glypican-1 identifies cancer exosomes and detects early pancreatic cancer. Nature, 2015. 523(7559): 177-82.
76. Ko, J., N. Bhagwat, S.S. Yee, N. Ortiz, A. Sahmoud, T. Black, et al. Combining machine learning and nanofluidic technology to diagnose pancreatic cancer using exosomes. ACS Nano, 2017. 11(11): 11182-93.
77. An, M., I. Lohse, Z. Tan, J. Zhu, J. Wu, H. Kurapati, et al. Quantitative proteomic analysis of serum exosomes from patients with locally advanced pancreatic cancer undergoing chemoradiotherapy. J Proteome Res, 2017. 16(4): 1763-72.
78. Gibbings, D.J., C. Ciaudo, M. Erhardt, O. Voinnet. Multivesicular bodies associate with components of miRNA effector complexes and modulate miRNA activity. Nat Cell Biol, 2009. 11(9): 1143-9.
79. Thakur, B.K., H. Zhang, A. Becker, I. Matei, Y. Huang, B. Costa-Silva, et al. Double-stranded DNA in exosomes: a novel biomarker in cancer detection. Cell Res, 2014. 24(6): 766-9.
80. Zhang, H., D. Freitas, H.S. Kim, K. Fabijanic, Z. Li, H. Chen, et al. Identification of distinct nanoparticles and subsets of extracellular vesicles by asymmetric flow field-flow fractionation. Nat Cell Biol, 2018. 20(3): 332-43.
81. Wu, M., Y. Ouyang, Z. Wang, R. Zhang, P.H. Huang, C. Chen, et al. Isolation of exosomes from whole blood by integrating acoustics and microfluidics. Proc Natl Acad Sci USA, 2017. 114(40): 10584-89.

CHAPTER 6

Targeting Subgroup-specific Cancer Epitopes for Effective Treatment of Pediatric Medulloblastoma

Sidharth Mahapatra[1,2,3]* and Naveenkumar Perumal[1]

[1] Department of Biochemistry and Molecular Biology, University of Nebraska Medical Center (UNMC), Omaha, NE, 68198-5870, USA
[2] Department of Pediatrics, University of Nebraska Medical Center (UNMC), Omaha, NE, 68198-5870, USA
[3] Buffett Cancer Center, University of Nebraska Medical Center (UNMC), Omaha, NE, 68198-5870, USA

Introduction

Perhaps one of the well-suited malignancies for the development of targeted cancer therapeutics is medulloblastoma, the most common malignant brain tumor of childhood. Despite its low incidence, i.e. approximately 500 cases per year, five-year event-free survival can be less than 50% (non-SHH/WNT medulloblastoma) [1, 2]. Even with standard treatment options, i.e. maximal surgical resection, whole neuroaxis radiation, and chemotherapy, patients with a relatively benign course still suffer significant treatment-related neurocognitive and endocrinologic abnormalities [3, 4]. Moreover, not only do the younger patients have a more insidious course, but their age poses considerable limitations to post-surgical radiation and chemotherapy [5, 6]. As a result, the added risk of mortality to the sickest patients has become a palpable challenge. The need for better modalities of treatment now is dire.

Traditional risk stratification of medulloblastoma has divided patients into two basic categories, i.e. average-risk and high-risk groups, based on age, extent of post-operative residual disease, and metastases at diagnosis (Table 1). Older children (diagnosed after three years of age) with near total tumor resection (<1.5 cm^2 of post-operative residual disease by MRI) and no metastases at diagnosis are considered to have the best prognosis, being placed in the average-risk group with a five-year event-free survival of approximately 80% [7-9]. These patients are treated with a combination of

---

*Corresponding author: sidharth.mahapatra@unmc.edu

chemo- and radiation therapy status post maximal surgical resection of tumor [5, 9-12]. However, they suffer appreciable post-treatment complications, especially endocrinologic abnormalities, like growth delay, thyroid dysfunction, hyperprolactinemia, and gonadal damage, likely secondary to direct craniospinal irradiation during critical years of growth [3]. Notably, the introduction of chemotherapy as an adjunct allowed a reduction in total radiation dosing from 36.0 Gy to 23.4 Gy without any detriment to cure rates [11, 12]. Using an eight-cycle regimen of either vincristine/lomustine/cisplatin or vincristine/cyclophosphamide/cisplatin following reduced dose craniospinal irradiation of 23.4 Gy (including 55 Gy boost to tumor bed), five-year event-free survival remained >80% in both arms with no additional toxicity [12]. Unfortunately, children younger than eight years of age demonstrated neurocognitive delays even with a reduction to 23.4 Gy [4, 12]. As a result, a further reduction to 18.0 Gy is currently being studied in a phase III trial in average-risk patients younger than eight years of age by the Children's Oncology Group (ACNS0331).

Table 1: Risk stratification of patients diagnosed with medulloblastoma

	Average risk*	High risk
Age at diagnosis	≥ 3 years old	< 3 years old
Extent of post-operative disease (by MRI)	≤ 1.5 cm	> 1.5 cm
Presence of extraneural metastasis	No	Yes
Five-year event free survival	~ 80%	~ 60%

*To qualify as average-risk, all three criteria must be met; on the other hand, presence of one or more high-risk criteria may qualify a patient into the high-risk group.

Patients in the high-risk group expectedly fare worse due to age-related restrictions on available treatment modalities. For example, aside from complications associated with surgery at such a young age, infants are particularly vulnerable to the neurotoxic effects of radiation. The employment of maximal surgical resection followed by chemotherapy without radiation therapy or with delayed or reduced dosage has led to tumor recurrence as early as nine months post-treatment [7, 13]. Using radiation alone has resulted in <40% survival at five years [14]. Addition of standard chemotherapy post-irradiation has brought about some improvement in survival. Current standard of therapy for high-risk patients older than three years is craniospinal irradiation with 36 to 39 Gy (55 Gy boost to tumor bed), followed by cisplatin/cyclophosphamide/vincristine, which has resulted in five-year overall survival of 60-65% [6, 8, 15, 16]. An on-going phase III clinical trial by the Children's Oncology Group is currently testing different chemo-radiotherapy regimens in children with newly-diagnosed high-risk medulloblastoma (NCT00392327).

The perplexing issues of: 1) suboptimal cancer therapy for high-risk patients leading to higher mortality and 2) treatment-related long-term morbidity in average-risk patients serve as the impetus for discovery of improved modalities of treatment, without age-based restrictions or post-treatment toxic effects. The discovery of clustering cytogenetic events in distinct development-specific signal transduction pathways combined with genome-wide high-throughput analytic techniques has facilitated the classification of medulloblastoma now into four subgroups, i.e. WNT (Wingless), SHH (Sonic Hedgehog), non-SHH/WNT groups 3 and 4, each with unique transcriptomes, proteomes, histologic signatures, and clinical outcomes [2, 17]. Incorporation of unique clinical features and patient outcomes has further refined the subgrouping of medulloblastoma patients into as many as 12 subtypes [18, 19]. Aside from generating a more accurate prognostic picture for patients, subgrouping has triggered the discovery of promising targeted therapeutics. With an evolution in mechanistic insights of medulloblastoma pathophysiology, efforts are currently underway to target and disrupt critical steps in signal transduction with the ultimate goal of abrogating phenotypic transformation (Table 2).

Wingless (WNT) Medulloblastoma

The WNT pathway plays a critical role in the proliferation and terminal differentiation of neural progenitor cells [7, 20]. More specifically, signaling through the canonical (WNT-β-catenin) pathway leads to regulation of gene transcription. In healthy cells, the canonical pathway remains inactive due to the absence of WNT, which results in low cytoplasmic concentrations of β-catenin [7]. A multi-protein complex comprised of axin, adenomatous polyposis coli (APC), casein kinase 1α (CSK1α), and glycogen synthase kinase 3 (GSK3) binds and phosphorylates β-catenin, thereby promoting polyubiquitination and complete proteolysis [7, 21]. The pathway is activated through the binding of WNT to the Frizzled (FZD) receptor protein on the neuronal cell surface. FZD then phosphorylates Disheveled (DSH), which in turn inactivates the multimeric protein complex responsible for β-catenin degradation [7]. The rising β-catenin levels shuttle to the nucleus to transcribe proto-oncogenes, such as *cyclinD1* (*CCND1*) and *MYC* [7, 22].

Targeting WNT/β-catenin Signaling

WNT medulloblastoma ensues aberrant WNT/β-catenin pathway activation via oncogenic mutations within the proto-oncogene *CTNNB1*, in an area encoding β-catenin [23, 24]. As a result, β-catenin degradation is abrogated, leading to constitutive activation of the WNT signaling pathway [22, 25, 26]. More specifically, the impediment to glycogen synthase kinase (GSK)-3β-dependent phosphorylation and subsequent degradation of β-catenin results in elevated nuclear β-catenin, which, in turn, interacts with T-cell factor/lymphoid enhancer factor-1 (TCF/LEF-1) transcription factors and increases the expression of cell proliferation and survival genes such as *c-Myc* and

Table 2: Targeted therapeutics for the four subgroups of medulloblastoma

Subgroup	Histology	Frequency	Prognosis	Therapies	Evidence
WNT (Wingless)	Classic	~10%	Good	OSU03012	*In vitro and in vivo*
				Ginkgetin	*In vitro*
				Curcumin	*In vitro*
				Norcantharidin	*In vitro and in vivo*
SHH (Sonic Hedgehog)	Desmoplastic/Nodular	~30%	Intermediate	Cyclopamine	*In vitro and in vivo*
				PF-5274857	*In vitro and in vivo*
				Saridegib	Phase I
				Sodinegib	Phase II
				Vismodegib	Phase I/II
				HhAntag	*In vitro and in vivo*
				Itraconazole	*In vitro and in vivo*
				Arsenic trioxide	*In vitro and in vivo*
				GANT61	*In vitro*
				Pyrvinium	*In vitro and in vivo*
				I-BET151	*In vitro and in vivo*
Non-SHH/WNT Group 3	Large cell/ Anaplastic	~30%	Poor	JQ1	*In vitro and in vivo*
				Alsterpaullone	*In vitro and in vivo*
				Pemetrexed	*In vitro and in vivo*
				MK-1775	Phase I/II
				Aurora A/B inhib	*In vitro and in vivo*
Non-SHH/WNT Group 4	Classic	~30%	Poor	BI 2536	*In vitro*
				Palbociclib	*In vitro and in vivo*
				HDAC inhibitors	*In vitro*
				Hu5F9-G4	*In vitro and in vivo*
				CART B7-H3	*In vitro and in vivo*

cyclin D1 [27, 28]. Coordinating efforts towards deciphering a molecule or compound that targets the molecular underpinning of this subgroup could yield an effective treatment strategy for WNT medulloblastoma. A few promising candidates have thus been generated.

OSU03012, a celecoxib derivative, specifically targets WNT/β-catenin signaling by inhibiting GSK-3β^{Ser9} phosphorylation and inducing β-catenin phosphorylation and subsequent degradation, effectively reducing *c-Myc* and *cyclin D1* transcription. OSU03012 was designed for specific inhibition of PDK1 activity and has demonstrated cytotoxic, cell cycle arrest (S-Phase) and apoptotic effects in D283 and D324 medulloblastoma cell lines by reducing the phosphorylation of PDK1^{Ser241}, Akt^{Thr308}, and Akt^{Ser473}. Oral supplements of OSU03012 to D283 xenograft athymic mice showed reduced tumor volume with decreased levels of β-catenin, c-MYC and cyclin D1 [29]. The natural bioflavonoid Ginkgetin isolated from *Cephalotaxus fortunei* var. *alpina* potently inhibits WNT/β-catenin signaling by downregulating its target proteins, i.e. axin 2, cyclin D1 and survivin. By this mechanism, it was shown to induce G_2/M phase cell cycle arrest in DAOY and D283 cell lines [30]. With a similar mechanism, Curcumin (diferuloymethane) exhibits time- and dose-dependent cytotoxicity via cell cycle arrest in DAOY cells [31]. Norcantharidin, a demethylated analog of cantharidin which is a terpenoid secreted by blister beetles, possesses less toxicity and the ability to cross the blood brain barrier compared to its methylated counterpart. It has demonstrated antitumorogenic properties in medulloblastoma given its ability to inhibit protein phosphatase 2A (PP2A), which plays a critical role in regulating β-catenin stability [32]. By further inhibiting nuclear β-catenin translocation, Norcantharidin induces G_2 cell cycle arrest and activates apoptosis both *in vitro* and *in vivo* [33].

Thus, drugs that specifically target WNT/β-catenin signaling may provide a preferred alternative approach to the current conventional cytotoxic therapies. However, WNT subgroup MB possesses the best prognostic profile amongst medulloblastomas with an approximate 90% five-year event-free survival [2, 34]. As a result, efforts are now underway to study the effects of potentially limiting cytotoxic therapies (NCT02066220, NCT01878617, and NCT02724579) [8, 35]. Hence, these targeted therapeutics have yet to find their way to clinical trials. That said, this subset of patients may benefit from further restrictions in cytotoxic therapies with the addition of these promising therapeutic options.

Sonic Hedgehog (SHH) Medulloblastoma

During embryonic growth, neural precursor cells generated in the rhombic lip of the dorsal hindbrain migrate along the surface of the cerebellum to form the external granule layer (EGL). The secretion of bone morphogenic proteins (BMPs), such as BMP6, BMP7, and GDF7, encourages further proliferation generating a rich pool of granule cell precursors (GCPs) in the developing EGL [20, 36]. As older cells exit the EGL and migrate through a layer of Purkinje cells, they encounter the Hedgehog pathway ligand, Sonic hedgehog (SHH), a highly conserved embryonic signaling system which binds to its receptor, Patched 1 (PTCH1), expressed on GCPs in the EGL [37, 38]. Downstream effectors include the GLI family of transcription factors (GLI1, GLI2, and GLI3), which activate transcription of genes, such as *cyclinD1* (*CCND1*) and *MYC*, thereby facilitating GCP proliferation and migration [7, 20, 36, 39]. After post-natal cerebellar development, this pathway goes dormant with the 12-pass transmembrane receptor, PTCH1, keeping the 7-pass transmembrane protein, Smoothened (SMO), in an inactivated state [40]. This, in turn, leads to the sequestration of downstream effectors of the SHH pathway by Suppressor-of-Fused (SUFU), effectively silencing gene expression [7]. Deregulated binding of SHH to PTCH1 releases and constitutively activates SMO which, in turn, inhibits SUFU, leading to release and nuclear translocation of GLI1-3; as a result, aberrant gene transcription is activated, facilitating phenotypic transformation into medulloblastoma [7].

Targeting Sonic Hedgehog Signaling

Among medulloblastoma subtypes, SHH medulloblastoma has been the most well studied and well defined subgroup. SHH signaling plays a crucial role in central nervous system development, including neuronal precursor proliferation, dorso-ventral patterning, specification of oligodendrocytes, and axonal growth control [41]. Aberrant activation of SHH pathway is not exclusive to medulloblastoma but has also been noted in basal cell carcinoma (BCC), pancreatic cancer, ovarian cancer, and colorectal cancer [42-44]. Constitutive activation of SHH signaling due to mutations in PTCH1/2, SUFU (suppressor-of-fused) and SMO (smoothened) leads to activation of GLI1 and GLI2 favoring transcription of pro-proliferative, pro-survival and pro-angiogenic genes leading to enhanced tumor growth, metastasis and therapeutic resistance [45]. Thus, targeting the key molecules of canonical SHH signaling pathway could yield better therapeutic options against SHH medulloblastoma.

Several natural and synthetic antagonists for SMO and GLI have been studied in medulloblastoma. The steroidal alkaloid cyclopamine and its derivatives have been the best studied due to their ability to selectively bind SMO and exert inhibitory action on the SHH pathway [46, 47]. Cyclopamine treatment, for example, inhibits proliferation and induces neuronal differentiation with subsequent loss of neuronal stem cell-like character both *in vitro* and in murine tumor allografts [48]. Cyclopamine derivatives with superior potency and bioavailability that were later developed include vismodegib (GDC-0449), sodinegib (LDE-225), saridegib (IPI-926), and PF-5274857. All have shown efficacy in tumor reduction and prolonged survival in a SHH-driven medulloblastoma mouse models [49-53]. Moreover, Phase I and II trials of vismodegib showed significant tumor regression in both pediatric and adult patients with refractory or relapsed SHH pathway-driven medulloblastoma [52, 54, 55]. Similarly, Novartis has recently completed a phase II trial on the safety and efficacy of oral sodinegib in patients with Hh-pathway activated, relapsed medulloblastoma (NCT01708174) [56]. Given the success of these inhibitors in both pre-clinical and clinical trials, SMO-inhibitors may mark the first successful signaling pathway-specific therapeutic option for patients with SHH subgroup MB.

Despite these successes, multiple other SHH-signaling antagonists are being currently studied. HhAntag-691 is a benzimidazole derivative with the ability to cross the blood brain barrier and suppresses SHH signaling by binding to and inhibiting SMO. HhAntag-691 was shown in PTCH1$^{+/-}$p53$^{-/-}$ mice to exert dose-dependent downregulation of *GLI1* and *GLI2* genes, thereby significantly reducing tumor volume and prolonging tumor-free survival [57]. Itraconazole, a systemic antifungal molecule, was shown to bind to a different pocket on SMO and effectively inhibiting SHH signaling, resulting in decreased expression of *GLI1* and favoring tumor regression in medulloblastoma allograft mice [58]. Arsenic trioxide (ATO) treatment was shown to inhibit *in vivo* tumor growth in mice bearing SHH pathway-dependent medulloblastoma allografts through alteration in transcriptional activation of *GLI2* and decreasing ciliary trafficking [59]. Moreover, combining itraconazole and ATO decreased tumor growth and prolonged survival of mice with all functional SMO resistant mutants [60]. GANT61, a GLI inhibitor, was shown to promote mitochondrial-mediated apoptosis in DAOY medulloblastoma cells [61]. Pyrvinium, a Casein Kinase-1α (CK1α) agonist, acts as a potent SHH inhibitor by destabilizing the GLI proteins and exhibits antitumor action in PTCH$^{+/-}$ derived medulloblastoma allografts [62]. Recent study on bromodomain and extra terminal domain (BET) protein inhibitor (I-BET151) showed modulation in SHH signaling downstream of SMO. I-BET151 potentiated antitumor activity *in vitro* and *in vivo* against SHH-dependent medulloblastoma [63].

These numerous studies have resulted in the largest advances in the development of targeted therapeutics for patients with SHH subgroup medulloblastoma, many attaining clinical trials status. Emerging studies could focus on multimodal therapies targeting different effector proteins in the SHH signaling pathway, that may help effectively limit cytotoxic therapies and improve both the morbidity and mortality of patients with SHH subgroup medulloblastoma.

Groups 3 and 4 (non-SHH/WNT) Medulloblastoma

The majority of cases (~60%) of sporadic medulloblastoma falls into this final group, which has been further divided into Groups 3 and 4 based on gene expression data, methylation profiling and somatic copy number aberration analyses [2, 34, 64-66]. The two groups share a key cytogenetic aberration, i.e. isochromosome 17q (i17q), which contributes to their high aggressive nature (overall five-year survival: group 3 <50%, group 4 50-75%) with high rates of metastasis (30-40%) and recurrence [2, 8, 34, 67-71]. Due to not only key similarities in cytogenetic features, but also a high congruence in transcriptomes and proteomes of these subgroups, the World Health Organization currently categorizes groups 3 and 4 medulloblastoma provisionally into the non-SHH/WNT subgroup [72, 73]. Distinguishing features for Group 3 are NPR3 staining, MYC amplification (10-20%) and aberrations in NOTCH and TGFb signaling, and that for Group 4 are KCNA1 staining, CDK6 and MYCN amplifications (5-10%) and inactivating mutations of histone demethylase KDM6A (10%), affecting epigenetic regulation of these genes, the functional significance of which remains unknown [8, 65, 67, 74, 75].

Promising Targets for Non-SHH/WNT Medulloblastoma

While noteworthy advancements have been made in the generation of targeted therapeutics for WNT and SHH medulloblastoma subgroups, facilitating the translation of bench successes to the bedside, the same does not hold true for non-SHH/WNT medulloblastoma. Groups 3 and 4 medulloblastoma together constitute the most frequently-occurring subgroups of medulloblastoma (cumulative overall incidence 55-60%) [2, 34, 64, 76]. More specifically, Group 3 MB occurs exclusively in children, with highest incidence in the infant age group, i.e. under three years of age. A high prevalence of metastases at diagnosis (40%) coupled with the large cell/anaplastic histology gives this subgroup a particularly dismal prognosis (survival<50%) [2, 34, 65, 76]. Group 4 MB, on the other hand, is not exclusive to children, presenting with classic histologic features, and possessing an intermediate prognosis [2, 34, 65, 76]. Other high-risk features of these tumors include cytogenetic aberrations, including isochromosome 17q (i17q), MYC amplification and ERBB2 overexpression [2, 34, 67-70, 77]. Furthermore, the absence of a clear trigger, signal transduction pathway, or pathophysiologic mechanism underlying phenotypic transformation has made it difficult to identify mechanism-specific therapeutic options that may be explored to abrogate tumor growth. The dearth of good pre-clinical *in vivo* models has further impeded progress in the generation of targeted therapeutics for groups 3 and 4 MB. That said, pathways for further study have been identified and have shown some success *in vitro*.

MYC Inhibition

MYC overexpression has been linked with poor prognosis, specifically in Group 3 MB which has the highest incidence of MYC overexpression [2, 34, 70, 78, 79]. The overall incidence of MYC amplification in Group 3 MB is approximately 10-17% [65]. Studies that combined whole genome sequencing with transcription profiling shed light on the role of epigenetic silencing in the development of Group 3 and 4 MB; they revealed the importance of transcriptional silencing of histones [80]. A clustering of mutations influencing the methylation of H3K27 (histone H3 lysine 27) and H3K4 (histone 3 lysine 4) was noted and linked with tumor stemness and invasiveness [75, 76]. MYC is a strong activator of EZH2, an H3K27 methyl-transferase, overexpressed in non-SHH/WNT MB [75, 81]. Both proteins promote stem-like states and antagonize the process of neural cell differentiation, leading to elevations in stem cell markers, Oct4, Sox2, and Nanog [82, 83]. These markers have, in turn, been associated with aggressiveness of MB [83]. Thus, a link between high MYC expression and tumor aggressiveness has been elucidated from these studies. In turn, MYC deprivation in orthotopic xenograft models of non-SHH/WNT MB has demonstrated tumor cell senescence and apoptosis [84]. As a result, MYC targeting has become a preferred model of study for the generation of targeted therapeutics in non-SHH/WNT MB.

The bromodomain and extraterminal bromodomain (BET) protein inhibitor, JQ1, which inhibits BRD4 by competitive inhibition of the acetyl-lysine recognition motif, has been shown to reduce cell viability due to arrest at G1 phase followed by an increase in tumor cell apoptosis in an MYC-amplified MB model. JQ1 suppressed MYC expression and inhibited MYC-associated targets [85-87]. Similarly, other BRD4 inhibitors are under current investigation [88]. A cyclin-dependent kinase inhibitor, alsterpaullone (ALP), was shown recently to reduce cell proliferation *in vitro* and improve mortality in an *in vivo* mouse model of Group 3 MB via the downregulation of MYC expression [89]. Additionally, the folate synthesis inhibitor, pemetrexed, and the nucleoside analog, gemcitabine, demonstrated a synergistic effect in reducing neurosphere proliferation *in vitro*, inhibiting tumor cell proliferation *in vivo*, and increasing the survival of mice bearing MYC-overexpressing tumors [90, 91].

Kinase Inhibitors

Integrated genomic analysis approaches starting with gene expression profiling and subsequent gene enrichment analysis identified twenty-nine cell cycle-related kinases that were dysregulated in medulloblastoma. Within this subset, six genes (*Aurora kinase A, Aurora kinase B, CDK2, PLK, TTK,* and *WEE1*) were identified that are both overexpressed in medulloblastoma and can suppress tumor cell proliferation [92]. These six targets were further studied using specific inhibitors to explore effects on tumor cell growth.

The WEE 1 tyrosine kinase regulates entry into mitosis by arresting cells with DNA damage at the G2 phase. Since tumor cells lack normal DNA repair mechanisms, they rely heavily on this kinase to serve as a DNA damage checkpoint [93]. In conjunction with cytotoxic agents, inhibition of WEE 1 kinase (via the pyrazolopyrimidine derivative MK-1775) has been adopted for the study of adult solid tumors in Phase I/II trials (NCT01748825, NCT02095132, NCT01357161) [94]. In DAOY and UW228 medulloblastoma cell lines, MK-1775 at nanomolar concentrations inhibited colony formation. Furthermore, in nude mice injected with DAOY cells subcutaneously, oral treatment with MK-1775 led to tumor regression. When tested in conjunction with cisplatin, MK-1775 accelerated apoptosis and inhibited repair of cisplatin-induced DNA damage [92].

Aurora kinase A and B maintain genomic stability through their involvement in chromosome separation during mitosis [95]. Overexpression has been shown to lead to oncogenesis in multiple types of cancers, aside from medulloblastoma [96]. Aurora kinase A (AURKA) inhibition was shown to induce apoptosis in DAOY cells and to lead to a time-dependent G2/M phase arrest [97]. Moreover, AURKA inhibition reduced the IC$_{50}$ of etoposide and cisplatin concurrently enhancing their cytotoxic effects *in vitro* [97]. Aurora kinase B (Aurora B) co-expression has been shown specifically in the high MYC-expressing group 3 tumors. Inhibition of Aurora B in MYC overexpressing cells not only reduced tumor cell proliferation but also sensitized them to apoptosis [98]. In MB xenograft models, Aurora B inhibition impaired cerebellar tumor growth and augmented survival [98].

Polo-like kinase 1 (PLK1) facilitates mitotic entry and exit. Overexpression leads to aneuploidy, chromosomal instability, and neoplastic transformation [99]. Likewise, a small molecule inhibitor of PLK1 (BI 2536) suppressed colony and tumor sphere formation, inhibited cell growth, and increased apoptosis *in vitro*; moreover, PLK1 inhibition enhanced radiosensitivity of high-MYC expressing medulloblastoma cell lines [100]. Finally, the CDK4/6 inhibitor, palbociclib, was shown to decrease tumor cell proliferation, to increase apoptosis, and to improve survival in an orthotopic mouse model of MYC-overexpressing Group 3 MB [101].

HDAC Inhibitors

Histone deacetylases (HDAC) are a superfamily of zinc-dependent enzymes that have varying biological roles; they are subclassified into four groups, i.e. I, IIa, IIb, and IV [102]. HDAC inhibitors have been used successfully *in vitro* and *in vivo* in the treatment of a wide variety of malignancies. In fact, Vorinostat was the first HDAC inhibitor to receive FDA approval for the treatment of cutaneous T-cell lymphoma [103]. It has since been shown to improve median overall survival in recurrent glioblastoma multiforme and advanced non-small cell lung cancer [104, 105]. However, this drug is a non-specific HDAC inhibitor of all classes of HDACs. Given the varying important biological functions of HDACs, global inhibition can be associated with important side effects. Thus, an impetus exists for selective targeting of HDACs. Chromatin immunoprecipitation assays have confirmed an MYC binding motif in the HDAC2 promoter region; hence, MYC-amplified tumors would serve as good targets for selective HDAC2 inhibitors [106]. In fact, depletion of HDAC2 in medulloblastoma has been shown to be cytotoxic [106]. The class IIa HDACs, HDAC5 and 9, have been shown to be upregulated in patients with aggressive high-risk medulloblastoma [107]. In turn, silencing HDAC5 and 9 expressions via siRNA led to a reduction in medulloblastoma tumor cell growth *in vitro* [107].

Immunotherapy

Malignant cells have the capacity to evade immunologic surveillance through the expression of unique cell surface markers making them immune resistant. One of these markers that is highly expressed in a variety of tumor types, CD47, activates the signal regulatory protein alpha (SIRPα) on myeloid cells, which in turn, protects the malignant cells from phagocytosis by macrophages [108]. Using the anti-CD47 antibody, Hu5F9-G4, authors demonstrated not only elevated tumor cell phagocytosis and growth inhibition but also an inhibition of neuroaxis spread in xenograft models of high MYC-expressing medulloblastoma [109]. Similarly, B7-H3 (CD276), which is an immune checkpoint member of the B7 family, plays a role in the functional inhibition of T-cells and is overexpressed in a variety of solid tumors, often correlated with poor prognosis [110]. A variety of clinical-translational advances have been made in B7-H3 targeting for cancer therapy, including blocking antibodies, bispecific antibodies, small molecule inhibitors, and chimeric antigen receptor T-cell (CAR T-cell) therapy [110, 111]. In an MYC-amplified medulloblastoma xenograft model, B7-H3 CAR T-cell infusion into the posterior fossa led to dramatic reduction in tumor burden and prolonged survival [111]. Finally, an emerging and provocative new approach to high-risk medulloblastoma treatment is exploiting telomerase targeting. By incorporating a telomerase substrate, 6-thio-dG, into the telomeres of tumor cells, authors demonstrated a dose-dependent inhibition of MB cell growth *in vitro* with reduced tumor sphere formation and elevated apoptosis [112].

Future Perspectives

Although the mainstay of therapy for medulloblastoma remains unchanged, current focus has shifted to not only improving five-year event-free survival but also long-term treatment-related sequelae resulting from cytotoxic therapies instituted during critical periods of growth and development. With a more refined understanding of molecular pathways implicated in phenotypic transformation to medulloblastoma, multiple viable pre-clinical models have been generated for the SHH and WNT subgroups, thereby facilitating the development and testing of targeted anti-neoplastic therapies, some of which are now under clinical investigation.

Contrarily, the non-SHH/WNT subgroups 3 and 4, which carry the highest incidence and worst prognosis, have few good pre-clinical models. Thus, good conceptual therapeutic tactics have been restricted to *in vitro* successes and xenograft mouse models. With continued refinement of our understanding of these two subgroups, the generation of the first pre-clinical [non-xenograft] mouse model of non-SHH/WNT subgroup medulloblastoma may be undertaken, allowing the fair assessment of targeted therapies, such as myc-inhibitors, kinase inhibitors, HDAC inhibitors, and immunotherapy, on tumor behavior in the context of non-SHH/WNT medulloblastoma.

Hereafter, the largest impact may be instituted on the subgroup with the direst need for such therapies given the poor overall survival and high-risk of the fragile patients afflicted with this malignancy.

References

1. Smoll, N.R., K.J. Drummond. The incidence of medulloblastomas and primitive neurectodermal tumours in adults and children. J Clin Neurosci, 2012. 19(11): 1541-4.
2. Taylor, M.D., P.A. Northcott, A. Korshunov, M. Remke, Y.J. Cho, S.C. Clifford, et al. Molecular subgroups of medulloblastoma: the current consensus. Acta Neuropathol, 2012. 123(4): 465-72.
3. Brown, I.H., T.J. Lee, O.B. Eden, J.A. Bullimore, D.C. Savage. Growth and endocrine function after treatment for medulloblastoma. Arch Dis Child, 1983. 58(9): 722-7.
4. Ris, M.D., R. Packer, J. Goldwein, D. Jones-Wallace, J.M. Boyett. Intellectual outcome after reduced-dose radiation therapy plus adjuvant chemotherapy for medulloblastoma: a Children's Cancer Group study. J Clin Oncol, 2001. 19(15): 3470-6.
5. Packer, R.J., J. Goldwein, H.S. Nicholson, L.G. Vezina, J.C. Allen, M.D. Ris, et al. Treatment of children with medulloblastomas with reduced-dose craniospinal radiation therapy and adjuvant chemotherapy: a Children's Cancer Group study. J Clin Oncol, 1999. 17(7): 2127-36.
6. Gajjar, A., M. Chintagumpala, D. Ashley, S. Kellie, L.E. Kun, T.E. Merchant, et al. Risk-adapted craniospinal radiotherapy followed by high-dose chemotherapy and stem-cell rescue in children with newly diagnosed medulloblastoma (St Jude Medulloblastoma-96): long-term results from a prospective, multicentre trial. Lancet Oncol, 2006. 7(10): 813-20.
7. Gilbertson, R.J. Medulloblastoma: signalling a change in treatment. Lancet Oncol, 2004. 5(4): 209-18.
8. Ramaswamy, V., M.D. Taylor. Medulloblastoma: from myth to molecular. J Clin Oncol, 2017. 35(21): 2355-63.
9. Taylor, R.E., C.C. Bailey, K. Robinson, C.L. Weston, D. Ellison, J. Ironside, et al. Results of a randomized study of preradiation chemotherapy versus radiotherapy alone for nonmetastatic medulloblastoma. The International Society of Paediatric Oncology/United Kingdom Children's Cancer Study Group PNET-3 Study. J Clin Oncol, 2003. 21(8): 1581-91.
10. Kortmann, R.D., J. Kuhl, B. Timmermann, U. Mittler, C. Urban, V. Budach, et al. Postoperative neoadjuvant chemotherapy before radiotherapy as compared to immediate radiotherapy followed by maintenance chemotherapy in the treatment of medulloblastoma in childhood: results of the German prospective randomized trial HIT '91. Int J Radiat Oncol Biol Phys, 2000. 46(2): 269-79.
11. Strother, D., D. Ashley, S.J. Kellie, A. Patel, D. Jones-Wallace, S. Thompson, et al. Feasibility of four consecutive high-dose chemotherapy cycles with stem-cell rescue for patients with newly diagnosed medulloblastoma or supratentorial primitive neuroectodermal tumor after craniospinal radiotherapy: results of a collaborative study. J Clin Oncol, 2001. 19(10): 2696-704.

12. Packer, R.J., A. Gajjar, G. Vezina, L. Rorke-Adams, P.C. Burger, P.L. Robertson, et al. Phase III study of craniospinal radiation therapy followed by adjuvant chemotherapy for newly diagnosed average-risk medulloblastoma. J Clin Oncol, 2006. 24(25): 4202-08.
13. Heideman, R.L. Overview of the treatment of infant central nervous system tumors: medulloblastoma as a model. J Pediatr Hematol Onocol, 2001. 23(5): 268-71.
14. Merchant, T.E., M.H. Wang, T. Haida, K.L. Lindsley, J. Finlay, I.J. Dunkel, et al. Medulloblastoma: long-term results for patients treated with definitive radiation therapy during the computed tomography era. Int J Radiat Oncol Biol Phys, 1996. 36(1): 29-35.
15. Lannering, B., S. Rutkowski, F. Doz, B. Pizer, G. Gustafsson, A. Navajas, et al. Hyperfractionated versus conventional radiotherapy followed by chemotherapy in standard-risk medulloblastoma: results from the randomized multicenter HIT-SIOP PNET 4 trial. J Clin Oncol, 2012. 30(26): 3187-93.
16. Tarbell, N.J., H. Friedman, W.R. Polkinghorn, T. Yock, T. Zhou, Z. Chen, et al. High-risk medulloblastoma: a pediatric oncology group randomized trial of chemotherapy before or after radiation therapy (POG 9031). J Clin Oncol, 2013. 31(23): 2936-41.
17. Gottardo, N.G., J.R. Hansford, J.P. McGlade, F. Alvaro, D.M. Ashley, S. Bailey, et al. Medulloblastoma down under 2013: a report from the third annual meeting of the International Medulloblastoma Working Group. Acta Neuropathol, 2014. 127(2): 189-201.
18. Bavle, A., D.W. Parsons. From one to many: further refinement of medulloblastoma subtypes offers promise for personalized therapy. Cancer Cell, 2017. 31(6): 727-9.
19. Cavalli, F.M.G., M. Remke, L. Rampasek, J. Peacock, D.J.H. Shih, B. Luu, et al. Intertumoral heterogeneity within medulloblastoma subgroups. Cancer Cell, 2017. 31(6): 737-54 e736.
20. Wechsler-Reya, R., M.P. Scott. The developmental biology of brain tumors. Annu Rev Neurosci, 2001. 24: 385-428.
21. Kang, D.E., S. Soriano, X. Xia, C.G. Eberhart, B. De Strooper, H. Zheng, et al. Presenilin couples the paired phosphorylation of beta-catenin independent of axin: implications for beta-catenin activation in tumorigenesis. Cell. 2002. 110(6): 751-62.
22. Rossi, A., V. Caracciolo, G. Russo, K. Reiss, A. Giordano. Medulloblastoma: from molecular pathology to therapy. Clin Cancer Res, 2008. 14(4): 971-6.
23. Fattet, S., C. Haberler, P. Legoix, P. Varlet, A. Lellouch-Tubiana, S. Lair, et al. Beta-catenin status in paediatric medulloblastomas: correlation of immunohistochemical expression with mutational status, genetic profiles, and clinical characteristics. J Pathol, 2009. 218(1): 86-94.
24. Rogers, H.A., S. Miller, J. Lowe, M.A. Brundler, B. Coyle, R.G. Grundy. An investigation of WNT pathway activation and association with survival in central nervous system primitive neuroectodermal tumours (CNS PNET). Br J Cancer, 2009. 100(8): 1292-1302.
25. Koch, A., A. Waha, J.C. Tonn, N. Sorensen, F. Berthold, M. Wolter, et al. Somatic mutations of WNT/wingless signaling pathway components in primitive neuroectodermal tumors. Int J Cancer, 2001. 93(3): 445-9.
26. Zurawel, R.H., S.A. Chiappa, C. Allen, C. Raffel. Sporadic medulloblastomas contain oncogenic beta-catenin mutations. Cancer Res, 1998. 58(5): 896-9.

27. Tetsu, O., F. McCormick. Beta-catenin regulates expression of cyclin D1 in colon carcinoma cells. Nature, 1999. 398(6726): 422-6.
28. He, T.C., A.B. Sparks, C. Rago, H. Hermeking, L. Zawel, L.T. da Costa, et al. Identification of c-MYC as a target of the APC pathway. Science, 1998. 281(5382): 1509-12.
29. Baryawno, N., B. Sveinbjornsson, S. Eksborg, C.S. Chen, P. Kogner, J.I. Johnsen. Small-molecule inhibitors of phosphatidylinositol 3-kinase/Akt signaling inhibit Wnt/beta-catenin pathway cross-talk and suppress medulloblastoma growth. Cancer Res, 2010. 70(1): 266-76.
30. Ye, Z.N., M.Y. Yu, L.M. Kong, W.H. Wang, Y.F. Yang, J.Q. Liu, et al. Biflavone ginkgetin, a novel Wnt inhibitor, suppresses the growth of medulloblastoma. Nat Prod Bioprospect, 2015. 5(2): 91-97.
31. He, M., Y. Li, L. Zhang, L. Li, Y. Shen, L. Lin, et al. Curcumin suppresses cell proliferation through inhibition of the Wnt/beta-catenin signaling pathway in medulloblastoma. Oncol Rep, 2014. 32(1): 173-80.
32. Yang, J., J. Wu, C. Tan, P.S. Klein. PP2A:B56epsilon is required for Wnt/beta-catenin signaling during embryonic development. Development (Cambridge, England), 2003. 130(23): 5569-78.
33. Cimmino, F., M.N. Scoppettuolo, M. Carotenuto, P. De Antonellis, V.D. Dato, G. De Vita, et al. Norcantharidin impairs medulloblastoma growth by inhibition of Wnt/beta-catenin signaling. J Neurooncol, 2012. 106(1): 59-70.
34. Kool, M., A. Korshunov, M. Remke, D.T. Jones, M. Schlanstein, P.A. Northcott, et al. Molecular subgroups of medulloblastoma: an international meta-analysis of transcriptome, genetic aberrations, and clinical data of WNT, SHH, Group 3, and Group 4 medulloblastomas. Acta Neuropathol, 2012. 123(4): 473-84.
35. Leary, S.E., J.M. Olson. The molecular classification of medulloblastoma: driving the next generation clinical trials. Curr Opin Pediatr, 2012. 24(1): 33-9.
36. Gilbertson, R.J., D.W. Ellison. The origins of medulloblastoma subtypes. Annu Rev Pathol, 2008. 3: 341-65.
37. Wechsler-Reya, R.J., M.P. Scott. Control of neuronal precursor proliferation in the cerebellum by Sonic Hedgehog. Neuron, 1999. 22(1): 103-14.
38. Raffel, C. Medulloblastoma: molecular genetics and animal models. Neoplasia (New York, NY), 2004. 6(4): 310-22.
39. Goodrich, L.V., M.P. Scott. Hedgehog and patched in neural development and disease. Neuron, 1998. 21(6): 1243-57.
40. Stone, D.M., M. Hynes, M. Armanini, T.A. Swanson, Q. Gu, R.L. Johnson, et al. The tumour-suppressor gene patched encodes a candidate receptor for Sonic hedgehog. Nature, 1996. 384(6605): 129-34.
41. Heussler, H.S., M. Suri. Sonic hedgehog. Mol Pathol, 2003. 56(3): 129-31.
42. Xie, J., C.M. Bartels, S.W. Barton, D. Gu. Targeting hedgehog signaling in cancer: research and clinical developments. Onco Targets Ther, 2013. 6: 1425-35.
43. Yauch, R.L., S.E. Gould, S.J. Scales, T. Tang, H. Tian, C.P. Ahn, et al. A paracrine requirement for hedgehog signalling in cancer. Nature, 2008. 455(7211): 406-10.
44. Varnat, F., A. Duquet, M. Malerba, M. Zbinden, C. Mas, P. Gervaz, et al. Human colon cancer epithelial cells harbour active HEDGEHOG-GLI signalling that is essential for tumour growth, recurrence, metastasis and stem cell survival and expansion. EMBO Mol Med, 2009. 1(6-7): 338-51.
45. Ng, J.M., T. Curran. The Hedgehog's tale: developing strategies for targeting cancer. Nat Rev Cancer, 2011. 11(7): 493-501.

46. Chen, J.K., J. Taipale, M.K. Cooper, P.A. Beachy. Inhibition of Hedgehog signaling by direct binding of cyclopamine to smoothened. Genes & Dev, 2002. 16(21): 2743-48.
47. Taipale, J., J.K. Chen, M.K. Cooper, B. Wang, R.K. Mann, L. Milenkovic, et al. Effects of oncogenic mutations in smoothened and patched can be reversed by cyclopamine. Nature, 2000. 406(6799): 1005-09.
48. Berman, D.M., S.S. Karhadkar, A.R. Hallahan, J.I. Pritchard, C.G. Eberhart, D.N. Watkins, et al. Medulloblastoma growth inhibition by hedgehog pathway blockade. Science, 2002. 297(5586): 1559-61.
49. Tremblay, M.R., A. Lescarbeau, M.J. Grogan, E. Tan, G. Lin, B.C. Austad, et al. Discovery of a potent and orally active hedgehog pathway antagonist (IPI-926). J Med Chem, 2009. 52(14): 4400-18.
50. Lee, M.J., B.A. Hatton, E.H. Villavicencio, P.C. Khanna, S.D. Friedman, S. Ditzler, et al. Hedgehog pathway inhibitor saridegib (IPI-926) increases lifespan in a mouse medulloblastoma model. PNAS, 2012. 109(20): 7859-64.
51. Buonamici, S., J. Williams, M. Morrissey, A. Wang, R. Guo, A. Vattay, et al. Interfering with resistance to smoothened antagonists by inhibition of the PI3K pathway in medulloblastoma. Sci Transl Med, 2010. 2(51): 51ra70.
52. Rudin, C.M., C.L. Hann, J. Laterra, R.L. Yauch, C.A. Callahan, L. Fu, et al. Treatment of medulloblastoma with hedgehog pathway inhibitor GDC-0449. NEJM, 2009. 361(12): 1173-78.
53. Rohner, A., M.E. Spilker, J.L. Lam, B. Pascual, D. Bartkowski, Q.J. Li, et al. Effective targeting of hedgehog signaling in a medulloblastoma model with PF-5274857, a potent and selective smoothened antagonist that penetrates the blood-brain barrier. Mol Cancer Ther, 2012. 11(1): 57-65.
54. Gajjar, A., C.F. Stewart, D.W. Ellison, S. Kaste, L.E. Kun, R.J. Packer, et al. Phase I study of vismodegib in children with recurrent or refractory medulloblastoma: a pediatric brain tumor consortium study. Clin Cancer Res, 2013. 19(22): 6305-12.
55. Robinson, G.W., B.A. Orr, G. Wu, S. Gururangan, T. Lin, I. Qaddoumi, et al. Vismodegib exerts targeted efficacy against recurrent sonic hedgehog-subgroup medulloblastoma: results from phase II pediatric brain tumor consortium studies PBTC-025B and PBTC-032. J Clin Oncol, 2015. 33(24): 2646-54.
56. Das, P. Where are we with targeted therapies for medulloblastoma? Brain Disord Ther, 2015. S2: 007.
57. Romer, J.T., H. Kimura, S. Magdaleno, K. Sasai, C. Fuller, H. Baines, et al. Suppression of the SHH pathway using a small molecule inhibitor eliminates medulloblastoma in Ptc1(+/-)p53(-/-) mice. Cancer Cell, 2004. 6(3): 229-40.
58. Kim, J., J.Y. Tang, R. Gong, J. Kim, J.J. Lee, K.V. Clemons, et al. Itraconazole, a commonly used antifungal that inhibits hedgehog pathway activity and cancer growth. Cancer Cell, 2010. 17(4): 388-99.
59. Kim, J., J.J. Lee, J. Kim, D. Gardner, P.A. Beachy. Arsenic antagonizes the hedgehog pathway by preventing ciliary accumulation and reducing stability of the Gli2 transcriptional effector. Proc Natl Acad Sci USA, 2010. 107(30): 13432-37.
60. Kim, J., B.T. Aftab, J.Y. Tang, D. Kim, A.H. Lee, M. Rezaee, et al. Itraconazole and arsenic trioxide inhibit hedgehog pathway activation and tumor growth associated with acquired resistance to smoothened antagonists. Cancer Cell, 2013. 23(1): 23-34.
61. Lin, Z., S. Li, H. Sheng, M. Cai, L.Y. Ma, L. Hu, et al. Suppression of GLI

sensitizes medulloblastoma cells to mitochondria-mediated apoptosis. J Cancer Res Clin Oncol, 2016. 142(12): 2469-78.
62. Li, B., D.L. Fei, C.A. Flaveny, N. Dahmane, V. Baubet, Z. Wang, et al. Pyrvinium attenuates hedgehog signaling downstream of smoothened. Cancer Res, 2014. 74(17): 4811-21.
63. Long, J., B. Li, J. Rodriguez-Blanco, C. Pastori, C.H. Volmar, C. Wahlestedt, et al. The BET bromodomain inhibitor I-BET151 acts downstream of smoothened protein to abrogate the growth of hedgehog protein-driven cancers. J Biol Chem, 2014. 289(51): 35494-502.
64. Northcott, P.A., A. Korshunov, H. Witt, T. Hielscher, C.G. Eberhart, S. Mack, et al. Medulloblastoma comprises four distinct molecular variants. J Clin Oncol, 2011. 29(11): 1408-14.
65. Northcott, P.A., D.J. Shih, J. Peacock, L. Garzia, A.S. Morrissy, T. Zichner, et al. Subgroup-specific structural variation across 1,000 medulloblastoma genomes. Nature, 2012. 488(7409): 49-56.
66. Hovestadt, V., D.T. Jones, S. Picelli, W. Wang, M. Kool, P.A. Northcott, et al. Decoding the regulatory landscape of medulloblastoma using DNA methylation sequencing. Nature, 2014. 510(7506): 537-41.
67. Ellison, D.W., J. Dalton, M. Kocak, S.L. Nicholson, C. Fraga, G. Neale, et al. Medulloblastoma: clinicopathological correlates of SHH, WNT, and non-SHH/WNT molecular subgroups. Acta Neuropathol, 2011. 121(3): 381-96.
68. Gilbertson, R., C. Wickramasinghe, R. Hernan, V. Balaji, D. Hunt, D. Jones-Wallace, et al. Clinical and molecular stratification of disease risk in medulloblastoma. Br J Cancer, 2001. 85(5): 705-12.
69. Min, H.S., J.Y. Lee, S.K. Kim, S.H. Park. Genetic grouping of medulloblastomas by representative markers in pathologic diagnosis. Transl Oncol, 2013. 6(3): 265-72.
70. Ryzhova, M.V., O.G. Zheludkova, E.V. Kumirova, L.V. Shishkina, T.N. Panina, S.K. Gorelyshev, et al. Characteristics of medulloblastoma in children under age of three years. Zhurnal voprosy neirokhirurgii imeni N N Burdenko, 2013. 77(1): 3-10; discussion 11.
71. Ramaswamy, V., M. Remke. E. Bouffet, S. Bailey, S.C. Clifford, F. Doz, et al. Risk stratification of childhood medulloblastoma in the molecular era: the current consensus. Acta Neuropathol, 2016. 131(6): 821-31.
72. Louis, D.N., A. Perry, G. Reifenberger, A. von Deimling, D. Figarella-Branger, W.K. Cavenee, et al. The 2016 world health organization classification of tumors of the central nervous system: a summary. Acta Neuropathol, 2016. 131(6): 803-20.
73. Rood, B. The proteo(epi)genomics of medulloblastoma. 4th Bienneial Conference on Pediatric Neuro-Oncology. New York, 2017.
74. Northcott, P.A., I. Buchhalter, A.S. Morrissy, V. Hovestadt, J. Weischenfeldt, T. Ehrenberger, et al. The whole-genome landscape of medulloblastoma subtypes, Nature, 2017. 547(7663): 311-7.
75. Robinson, G., M. Parker, T.A. Kranenburg, C. Lu, X. Chen, L. Ding, et al. Novel mutations target distinct subgroups of medulloblastoma. Nature, 2012. 488(7409): 43-48.
76. Roussel, M.F., G.W. Robinson. Role of MYC in medulloblastoma. Cold Spring Harb Perspect Med., 2013. 3(11).
77. Wasson, J.C., R.L. Saylors 3rd, P. Zeltzer, H.S. Friedman, S.H. Bigner, P.C. Burger, et al. Oncogene amplification in pediatric brain tumors. Cancer Res, 1990. 50(10): 2987-90.

78. Cho, Y.J., A. Tsherniak, P. Tamayo, S. Santagata, A. Ligon, H. Greulich, et al. Integrative genomic analysis of medulloblastoma identifies a molecular subgroup that drives poor clinical outcome. J Clin Oncol, 2011. 29(11): 1424-30.
79. Pfister, S., M. Remke, A. Benner, F. Mendrzyk, G. Toedt, J. Felsberg, et al. Outcome prediction in pediatric medulloblastoma based on DNA copy-number aberrations of chromosomes 6q and 17q and the MYC and MYCN loci. J Clin Oncol, 2009. 27(10): 1627-36.
80. Northcott, P.A., Y. Nakahara, X. Wu, L. Feuk, D.W. Ellison, S. Croul, et al. Multiple recurrent genetic events converge on control of histone lysine methylation in medulloblastoma. Nature Genetics, 2009. 41(4): 465-72.
81. Salvatori, B., I. Iosue, N. Djodji Damas, A. Mangiavacchi, S. Chiaretti, M. Messina, et al. Critical role of c-Myc in acute myeloid leukemia involving direct regulation of miR-26a and histone methyltransferase EZH2. Genes & Cancer, 2011. 2(5): 585-92.
82. Lin, C.H., C. Lin, H. Tanaka, M.L. Fero, R.N. Eisenman. Gene regulation and epigenetic remodeling in murine embryonic stem cells by c-Myc. PloS One, 2009. 4(11): e7839.
83. Kawauchi, D., G. Robinson, T. Uziel, P. Gibson, J. Rehg, C. Gao, et al. A mouse model of the most aggressive subgroup of human medulloblastoma. Cancer Cell, 2012. 21(2): 168-80.
84. Pei, Y., C.E. Moore, J. Wang, A.K. Tewari, A. Eroshkin, Y.J. Cho, et al. An animal model of MYC-driven medulloblastoma. Cancer Cell. 2012. 21(2): 155-67.
85. Bandopadhayay, P., G. Bergthold, B. Nguyen, S. Schubert, S. Gholamin, Y. Tang, et al. BET bromodomain inhibition of MYC-amplified medulloblastoma. Clin Cancer Res, 2014. 20(4): 912-25.
86. Henssen, A., T. Thor, A. Odersky, L. Heukamp, N. El-Hindy, A. Beckers, et al. BET bromodomain protein inhibition is a therapeutic option for medulloblastoma. Oncotarget, 2013. 4(11): 2080-95.
87. Delmore, J.E., G.C. Issa, M.E. Lemieux, P.B. Rahl, J. Shi, H.M. Jacobs, et al. BET bromodomain inhibition as a therapeutic strategy to target c-Myc. Cell, 2011. 146(6): 904-17.
88. McKeown, M.R., J.E. Bradner. Therapeutic strategies to inhibit MYC. Cold Spring Harb Perspect Med, 2014. 4(10): pii: a014266.
89. Faria, C.C., S. Agnihotri, S.C. Mack, B.J. Golbourn, R.J. Diaz, S. Olsen, et al. Identification of alsterpaullone as a novel small molecule inhibitor to target group 3 medulloblastoma. Oncotarget, 2015. 6(25): 21718-29.
90. Morfouace, M., A. Shelat, M. Jacus, B.B. Freeman 3rd, D. Turner, S. Robinson, et al. Pemetrexed and gemcitabine as combination therapy for the treatment of Group 3 medulloblastoma. Cancer Cell, 2014. 25(4): 516-29.
91. Chattopadhyay, S., R.G. Moran, I.D. Goldman. Pemetrexed: biochemical and cellular pharmacology, mechanisms, and clinical applications. Mol Cancer Ther., 2007. 6(2): 404-17.
92. Harris, P.S., S. Venkataraman, I. Alimova, D.K. Birks, I. Balakrishnan, B. Cristiano, et al. Integrated genomic analysis identifies the mitotic checkpoint kinase WEE1 as a novel therapeutic target in medulloblastoma. Molecular Cancer, 2014. 13: 72.
93. Leijen, S., J.H. Beijnen, J.H. Schellens. Abrogation of the G2 checkpoint by inhibition of Wee-1 kinase results in sensitization of p53-deficient tumor cells to DNA-damaging agents. Curr Clin Pharmacol, 2010. 5(3): 186-91.

94. Do, K., D. Wilsker, J. Ji, J. Zlott, T. Freshwater, R.J. Kinders, et al. Phase I study of single-agent AZD1775 (MK-1775), a Wee1 kinase inhibitor, in patients with refractory solid tumors. J Clin Oncol, 2015. 33(30): 3409-15.
95. Marumoto, T., D. Zhang, H. Saya. Aurora – a guardian of poles. Nat Rev Cancer, 2005. 5(1): 42-50.
96. Vader, G., S.M. Lens. The Aurora kinase family in cell division and cancer. Biochim Biophys Acta, 2008. 1786(1): 60-72.
97. El-Sheikh, A., R. Fan, D. Birks, A. Donson, N.K. Foreman, R. Vibhakar. Inhibition of Aurora kinase A enhances chemosensitivity of medulloblastoma cell lines. Pediatr Blood Cancer, 2010. 55(1): 35-41.
98. Diaz, R.J., B. Golbourn, C. Faria, D. Picard, D. Shih, D. Raynaud, et al. Mechanism of action and therapeutic efficacy of Aurora kinase B inhibition in MYC overexpressing medulloblastoma. Oncotarget, 2015. 6(5): 3359-74.
99. Strebhardt, K. Multifaceted polo-like kinases: drug targets and antitargets for cancer therapy. Nat Rev Drug Discov, 2010. 9(8): 643-60.
100. Harris, P.S., S. Venkataraman, I. Alimova, D.K. Birks, A.M. Donson, J. Knipstein, et al. Polo-like kinase 1 (PLK1) inhibition suppresses cell growth and enhances radiation sensitivity in medulloblastoma cells. BMC Cancer, 2012. 12: 80.
101. Hanaford, A.R., T.C. Archer, A. Price, U.D. Kahlert, J. Maciaczyk, G. Nikkhah, et al. DiSCoVERing innovative therapies for rare tumors: combining genetically accurate disease models with in silico analysis to identify novel therapeutic targets. Clin Cancer Res, 2016. 22(15): 3903-14.
102. Witt, O., H.E. Deubzer, T. Milde, I. Oehme. HDAC family: what are the cancer relevant targets? Cancer Lett, 2009. 277(1): 8-21.
103. Mann, B.S., J.R. Johnson, M.H. Cohen, R. Justice, R. Pazdur. FDA approval summary: vorinostat for treatment of advanced primary cutaneous T-cell lymphoma. The Oncologist, 2007. 12(10): 1247-52.
104. Bezecny, P. Histone deacetylase inhibitors in glioblastoma: pre-clinical and clinical experience. Medical Oncology (Northwood, London, England), 2014. 31(6): 985.
105. Ramalingam, S.S., M.L. Maitland, P. Frankel, A.E. Argiris, M. Koczywas, B. Gitlitz, et al. Carboplatin and Paclitaxel in combination with either vorinostat or placebo for first-line therapy of advanced non-small-cell lung cancer. J Clin Oncol, 2010. 28(1): 56-62.
106. Bhandari, D.R., K.W. Seo, J.W. Jung, H.S. Kim, S.R. Yang, K.S. Kang. The regulatory role of c-MYC on HDAC2 and PcG expression in human multipotent stem cells. J Cell Mol Med, 2011. 15(7): 1603-14.
107. Milde, T, I. Oehme, A. Korshunov, A. Kopp-Schneider, M. Remke, P. Northcott, et al. HDAC5 and HDAC9 in medulloblastoma: novel markers for risk stratification and role in tumor cell growth. Clin Cancer Res, 2010. 16(12): 3240-52.
108. Barclay, A.N., T.K. Van den Berg. The interaction between signal regulatory protein alpha (SIRPalpha) and CD47: structure, function, and therapeutic target. Annu Rev Immunol, 2014. 32: 25-50.
109. Gholamin, S., S.S. Mitra, A.H. Feroze, J. Liu, S.A. Kahn, M. Zhang, et al. Disrupting the CD47-SIRPalpha anti-phagocytic axis by a humanized anti-CD47 antibody is an efficacious treatment for malignant pediatric brain tumors. Sci Transl Med. 2017. 9(381).
110. Picarda, E., K.C. Ohaegbulam, X. Zang. Molecular pathways: targeting B7-H3 (CD276) for human cancer immunotherapy. Clin Cancer Res. 2016. 22(14): 3425-31.

111. Theruvath, J., S. Heitzeneder, R. Majzner, C. Moritz Graef, K. Cui, A. Nellan, et al. Checkpoint molecule B7-H3 is highly expressed on medulloblastoma and proves to be a promising candidate for CAR T-cell immunotherapy. Neuro-oncology, 2017. 19(Suppl_4): iv28-iv29.
112. Sengupta, S., M. Sobo, S. Kumar, I. Mender, J.W. Shay, R. Drissi. Targeting telomeres to treat therapy-resistant pediatric brain tumors. Neuro-oncology, 2016. 18(Suppl_3): iii26.

CHAPTER 7

Aberrant Methylation of UC Promoters in Human Pancreatic Ductal Carcinomas

Michiyo Higashi[1*] and Seiya Yokoyama[2]

[1] Department of Pathology, Kagoshima University Hospital, Center for the Research of Advanced Diagnosis and Therapy of Cancer, Graduate School of Medical and Dental Sciences, Kagoshima University, 8-35-1 Sakuragaoka, Kagoshima 890-8544, Japan

[2] Department of Pathology, Research Field in Medicine and Health Sciences, Medical and Dental Sciences Area, Research and Education Assembly, Kagoshima University. Center for the Research of Advanced Diagnosis and Therapy of Cancer, Graduate School of Medical and Dental Sciences, Kagoshima University, Japan

Introduction

Pancreatic ductal adenocarcinoma (PDAC) is still a lethal disease in spite of the improvement in diagnosis and treatment. The overall five-year survival rate for all patients with or without pancreatectomy after diagnosis is 13% in Japan [1]. On the other hand, patients with a successful resection of PDAC at an early stage (Stage IA: tumors located within pancreas; tumor size <2 cm, without metastasis) have a 46% five year survival rate [1, 2]. However, most patients with PDAC are diagnosed in the advanced stages because of the anatomical location of the pancreas, lack of specific symptoms, infiltration to the surrounding organs, or distant metastasis even from a small primary tumor less than 2 cm in diameter. Thus, a diagnostic technique for small pancreatic adenocarcinomas without symptoms is urgently needed.

Mucins (MUC) play crucial roles in carcinogenesis and tumor invasion in pancreatic neoplasms. MUC1 and MUC4 are large membrane-bound glycoproteins that are translated as single polypeptides. These mucins undergo intracellular autocatalytic proteolytic cleavage into two subunits that form stable non-covalent heterodimers that are transported to the cell surface. MUC1 contributes to oncogenesis by promoting the loss of epithelial cell polarity, promoting growth and survival pathways, activating

*Corresponding author: east@m2.kufm.kagoshima-u.ac.jp

receptor tyrosine kinase signaling pathways, and conferring resistance to the stress-induced cell death pathway [3, 4]. MUC4 plays an important role in cell proliferation and differentiation of epithelial cells by inducing specific phosphorylation of ErbB2 and enhancing expression of the cyclin dependent kinase inhibitor p27, which inhibits cell cycle progression [5, 6]. MUC2 is expressed in indolent pancreatobiliary neoplasms, but these tumors sometimes show invasive growth with MUC1 expression in invasive areas [7]. MUC5AC shows *de novo* high expression in many types of precancerous lesions of pancreatobiliary cancers and is an effective marker for early detection of the neoplasms [8].

MUC1 and MUC4 are often overexpressed in epithelial tumors and precursor lesions. Aberrant expression of MUC1 and MUC4 are associated with invasive proliferation of tumors and they are related to a poor outcome for patients in our immunohistochemical studies [9, 10]. The expression of *MUC1/MUC4* mRNA is regulated by epigenetic mechanisms such as DNA methylation in the promoter region in our biomolecular studies [11-13]. Combined evaluation of mucin expression may be effective for early detection, classification and evaluation of potential for malignancy of human neoplastic lesions. However, a robust assay for pancreatic or bile juice is required for early detection of pancreatic neoplasms with a poor clinical outcome.

Mucin Expression Profiles in Pancreatic Lesions

In normal pancreatic tissue, each MUC1 glycoform (MUC1/CORE, MUC1/DF3, MUC1/MY.1E12 and MUC1/HMFG-1) is expressed at cell apices in the centroacinar cells, intercalated ducts and intralobular ducts. MUC1/CORE and MUC1/DF3 are not expressed in interlobular ducts, whereas MUC1/MY.1E12 and MUC1/HMFG-1 are sometimes expressed in the interlobular ducts. They were not expressed in the main pancreatic ducts, acini or islets [7, 8, 10]. MUC2, MUC4 and MUC5AC are not expressed in normal pancreatic tissue [8, 14].

In PDAC, each MUC1 glycoform has a high expression level. Expression of MUC1 occurs in cell apical areas along the luminal side as well as in the luminal secretion of tubular structures, and at the basolateral membrane and in the cytoplasm, particularly in poorly differentiated areas. This expression pattern in PDACs differs from the pattern in the normal pancreas [8, 15, 16]. MUC1/DF3 high expression (more than 50% of cancer cells positive) was found to be significantly more frequent in pTNM stage IV than in stage III, and was associated with unfavorable overall survival in stage IV cases [17]. These data suggest that expression of MUC1/DF3 is associated with progression of PDAC. MUC2 is not expressed except mucinous (colloid) carcinoma or large extracellular stromal mucin pools containing suspended neoplastic cells [18]. MUC4 was highly expressed mainly in the cytoplasm [14]. The high MUC4 expression was an independent factor for a poor prognosis in PDAC [14]. MUC5AC also showed high expression [14, 19].

Pancreatic intraepithelial neoplasia (PanIN) is the most common precursor lesions of PDAC. Histological characteristics of PanIN are microscopic papillary or flat, noninvasive epithelial composed of columnar to cuboidal cells with variable mucin [18, 20]. Another precursor lesions of PDAC is intraductal papillary mucinous neoplasm (IPMN) which is intraductal neoplasms whose characteristics are macroscopic papillary epithelium composed of columnar cells with abundant mucin. The expression of MUC1 and MUC4 increases with increasing PanIN and/or IPMN grade [10, 21, 22].

Epigenetic Regulation of Mucin Genes

DNA methylation has been a focus in expression analysis of mucin genes. The epigenetic mechanisms of MUC1 were not examined for more than a decade after initial methylation analysis of the MUC1 coding region [23]. Yamada et al. revealed that MUC1 gene expression is regulated by DNA methylation and histone H3 lysine 9 (H3-K9) modification at the MUC1 promoter for the first time [13]. In approximately 3,000 bp of the MUC1 promoter region, there are 184 CpG sites. Analysis of all these sites using the MassARRAY compact system for base-specific cleavage of nucleic acids was performed in MUC1-positive and MUC1-negative breast, pancreas, and colon cancer cell lines. The methylation status of nine CpG sites near the transcriptional start site and histone H3-K9 modification in the vicinity of the start site were both related to MUC1 gene expression. Collectively, a series of epigenetic analyses revealed that DNA demethylation, histone H3-K9 demethylation, and histone H3-K9 acetylation in the 5′-flanking region of MUC1 might all be necessary for MUC1 gene expression. Downregulation of MUC1 expression by micro-RNA has been reported by several groups [24, 25].

Epigenetic analysis of the MUC2 gene promoter region has been widely studied [26-28]. CpG methylation near the MUC2 transcriptional start site (−338 to +158) plays a critical role in MUC2 gene expression [29, 30]. Several *in vitro* studies suggest that epigenetic mechanisms are tightly correlated with MUC2 expression in epithelial cells in various organs [30, 31]. Hypomethylation of MUC2 plays an important role in the high level of MUC2 expression in mucinous colorectal cancer taken from a human [32].

In the MUC3, methylation in the vicinity of the transcriptional start site plays the important role among the methylation status of 30 CpG sites, although histone H3-K4 and H3-K9 do not play a critical role in MUC3 regulation [33].

In the *MUC4* promoter, five CpG sites in the 5′ flanking region of MUC4 (−170 bp to −102 bp) which were associated with expression of MUC4 [12, 34] reported that five CpG sites (−121 bp to −81 bp) were correlated with expression of *MUC4* in these cells; however, two of these CpG sites (−93 bp and −81 bp) were unrelated to expression of MUC4 [12]. Aberrant MUC4 promoter hypomethylation may be involved in pancreatic carcinogenesis and malignant development of pancreatic ductal adenocarcinoma in clinical samples [35].

Pathological events such as carcinogenesis can be caused by a drastic alteration in established methylation patterns or dysregulation of chromatin remodeling processes, both of which can cause significant and consequential changes in gene expression. Thus, investigation of DNA methylation, chromatin modification and micro-RNA expression is important for diagnosis of carcinogenic risk and prediction of outcomes in patients with cancer [11, 13].

Methylation Specific Electrophoresis (MSE): Novel Method for DNA Methylation Analysis

Methylation of CpG sites in genomic DNA plays an important role in gene regulation and especially in gene silencing. Mechanisms of epigenetic regulation for expression of mucins are markers of malignancy potential and early detection of human neoplasms. Epigenetic changes in promoter regions appear to be the first step in expression of mucins. Thus, detection of promoter methylation status is important for early diagnosis of cancer, monitoring of tumor behavior, and evaluating the response of tumors to targeted therapy. However, conventional analytical methods for DNA methylation require a large amount of DNA and have low sensitivity. Several methods for analyzing DNA methylation have been developed, each of which ideally requires a large amount of high-quality DNA, such as that obtained from cultured cells or fresh tissue samples. However, DNA recovered from clinical samples such as pancreatic juice is often limited in quantity or is degraded [36]. Here, we report a modified version of the bisulfite-denaturing gradient gel electrophoresis (DGGE) using a nested PCR approach. We designated this method as methylation specific electrophoresis (MSE, international patent open: WO 2011/132798) [37]. In Fig. 1, in analysis of crude samples, several methods for analyzing DNA methylation have been shown. For example, there are samples that methylation pattern might be A (includes pattern) or B (does not include pattern). MSP and qMSP may indicate that "a region was 75% methylated" while massARRAY indicates that "all CpG sites were 75% methylated". These two patterns are very different, but current methods cannot distinguish between A and B. This is a pitfall for accurate diagnosis in clinical application of DNA methylation analysis. MSE has resolved this issue. Next generation sequencer (NGS) could also resolve this issue, but it is too expensive to use for daily work.

The other advantage of methylation specific electrophoresis is a few requirement of the amount of input DNA. The lower detection limit for distinguishing different methylation status is less than 0.1% and the detectable minimum amount of DNA is 20 pg, which can be obtained from only a few cells. It can be used for analysis of challenging samples such as human pancreatic juices, from which only a small amount of DNA can be extracted. The MSE method can provide a qualitative information of methylated sequence profile, sensitive and specific analysis of the DNA methylation pattern of almost any block of multiple CpG sites.

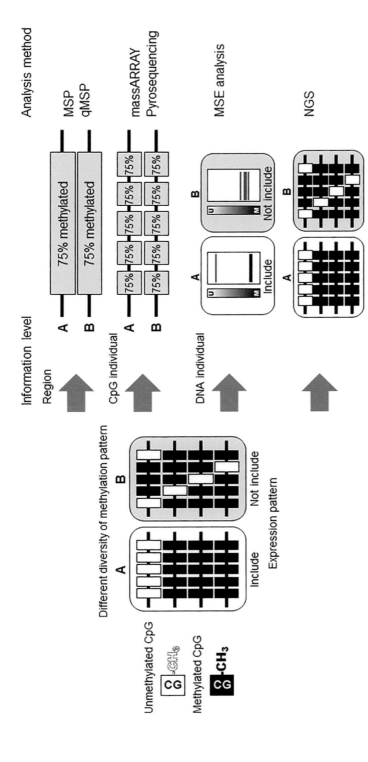

Fig. 1: Methylation analysis methods: Several methods for analyzing DNA methylation have been shown. For example, there are two different samples that methylation pattern might be A (include pattern) or B (not include pattern). MSP and qMSP may indicate that "a region was 75% methylated", while massARRAY indicates that "all CpG sites were 75% methylated". These methods cannot distinguish between A and B. This is a pitfall for accurate diagnosis in clinical application of DNA methylation analysis. MSE has been able to resolve this issue. Next generation sequencer (NGS) could also resolve this issue, but it is too expensive to use in daily work.

Distinction of Pancreatic Neoplastic Lesions Based on Aberrant Methylation of Three Mucins

MUC1, MUC2, and MUC4 are key mucins in pathological diagnosis of pancreatic neoplasms [10, 16, 38]. In analyses of pancreatic neoplastic and non-neoplastic tissues of PDAC samples, there is a strong relationship between the mRNA expression level and DNA methylation status for MUC1. This is similar to the results in pancreatic cancer cell lines in our previous study [13] and suggests that DNA methylation has a key role in MUC1 regulation in human pancreatic tissue. Thus, evaluation of the DNA methylation status of MUC1 can provide important information for diagnosis of human pancreatic neoplasms. MUC2 was not expressed in PDAC and/or non-neoplastic pancreas [15, 16]. Similarly, PDAC and non-neoplastic pancreas showed low expression level of MUC2 mRNA [39]. In MUC4, PDAC showed hypomethylation status of DNA than paired non-neoplastic area. However, significant correlation was not found between DNA methylation status and expression level of mRNA. This result suggests that other factors affect MUC4 expression. Although there was no relationship between the mRNA expression level and DNA methylation status for MUC2 and MUC4 in the tissue samples, the DNA methylation status of MUC2 and MUC4 could be applied in the analysis of pancreatic juice as follows.

Representative cases have shown the DNA methylation status using MSE of pancreatic juice and expression of mucins examined by immunohistochemical examination in paired pancreatic tissues from PDAC, intestinal-type IPMN and gastric-type IPMN (Fig. 2). Pancreatic juice from patients with PDAC showed unmethylated MUC1, methylated MUC2, and unmethylated MUC4 in MSE analysis, and paired pancreatic tissues were MUC1-positive, MUC2-negative and MUC4-positive (Fig. 2A). In the case of intestinal type IPMN (Fig. 2B) and gastric type IPMN (Fig. 2C), mucin expression and methylation status correlate with each other like PDAC (Fig. 2A). Thus, MSE analysis of DNA methylation status of MUC1, MUC2, and MUC4 in pancreatic juices may be useful for assessment of mucin expression levels and differential diagnosis of human neoplasms, i.e. PDAC, intestinal-type IPMN and gastric-type IPMN, with high specificity and sensitivity [39].

Relationships Among the Expression of MUC1 and MUC4, and Expression of DNA Methylation-related Enzymes

An analysis of the correlation between expression and hypoxic environment revealed that MUC1 and MUC4 expression was correlated with a hypoxic environment, as was expression of *CAIX*. This is similar to the results of our previous study in which we showed that a hypoxic environment upregulates MUC1 expression in a pancreatic cancer cell line, and hypoxia-inducible MUC1 contributes to hypoxia-driven angiogenesis through the activation of

Aberrant Methylation of UC Promoters in Human Pancreatic Ductal... 157

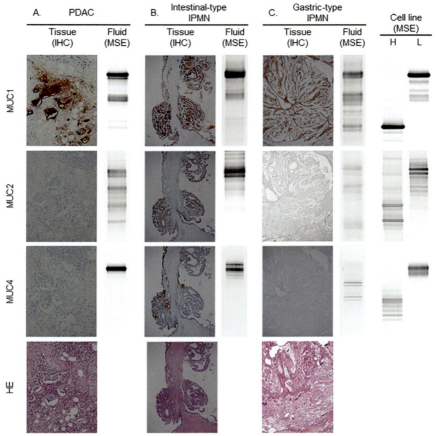

Fig. 2: Correlation between DNA methylation status in pancreatic juice and mucin expression. Representative cases have shown the DNA methylation status and expression of mucins in paired pancreatic tissues. A: Pancreatic juice from patients with PDAC showed unmethylated MUC1, methylated MUC2, and unmethylated MUC4 in MSE analysis, and paired pancreatic tissues were MUC1-positive, MUC2-negative and MUC4-positive. B: Pancreatic juice from patients with intestinal-type IPMN showed unmethylated MUC1, unmethylated MUC2, and unmethylated MUC4 in MSE analysis, and paired pancreatic tissues were MUC1-positive, MUC2-positive and MUC4-positive. C: Pancreatic juice from patients with Gastric-type IPMN showed unmethylated MUC1, methylated MUC2, and methylated MUC4 in MSE analysis, and paired pancreatic tissues were MUC1-positive, MUC2-negative and MUC4-negative.

proangiogenic factors in pancreatic cancer [39]. This is similar to the results that a hypoxic environment upregulates MUC1 expression in a pancreatic cancer cell line, and hypoxia-inducible MUC1 contributes to hypoxia-driven angiogenesis through the activation of proangiogenic factors in pancreatic cancer [40]. MUC4 expression showed a similar result in pancreatic tissue.

These results suggest that both hypoxia and methylation status play key roles in the regulation of expression of MUC1 and MUC4 [37, 39].

Members of the Ten-Eleven Translocation (TET) family and/or activation-induced deaminase (AID)/APOBEC family were demethylated by conversion from 5-methylcytosine (5mC) to 5-hydroxymethylcytosine (5hmC) and further oxidized products in mammalian genomes (i.e. active DNA) [41, 42]. Thus, we evaluated differences in expression of DNA methylation-related enzymes in pancreatic neoplastic regions and non-neoplastic regions; *MUC1* and *MUC4* mRNA expression correlate with DNA methylation-related enzymes or *MUC1* and *MUC4* methylation status [43]. In general, neoplastic regions expressed lower levels of DNA demethylation factors (*TET1, TET2*) and DNA methylation factors (*DNMT1*) than non-neoplastic regions (p<0.001 in all three factors). Conversely, the neoplastic regions expressed more MUC4 and hypoxia biomarker (*CAIX*) than the non-neoplastic regions (p < 0.001 in both factors). This result suggested that neoplastic regions have altered regulation of epigenetic status. A multiple regression analysis revealed significant correlations for non-neoplastic samples between promoter hypomethylation status and the expression of enzymes related to DNA methylation. However, neoplastic samples showed no correlation between promoter hypomethylation status and expression of enzymes related to DNA methylation. These results suggest that epigenetic regulation of *MUC1* and *MUC4* by these enzymes was ineffective or altered in neoplastic regions.

The relationship between MUC1 and MUC4 promoter methylation status and clinicopathological information of pancreatic lesions was also investigated. Those patients in early stage (IA, IB, and IIA: without lymph node nor distant metastasis) with hypomethylated MUC1 or MUC4 had a significantly decreased overall survival as compared to those with hypermethylated MUC1 or MUC4. When considered together, the methylation status of MUC1 and MUC4 was predictive of survival: patients with hypermethylation of both genes had significantly increased overall survival. Thus, aberrant methylation of MUC1 and MUC4 promoters are potential prognostic biomarkers for PDAC, and suggest further MSE analysis of human clinical samples to determine its utility for early diagnosis of pancreatic neoplasms and for stratifying patients with respect to modes of treatment. The translational research into mucin gene expression mechanisms, including epigenetics, may provide new tools for early and accurate detection of human neoplasms, not only pancreatic neoplasms but also other organs, because similar data is shown in lung neoplasms [44].

Acknowledgements

We thank Kei Matsuo for his assistance with clinical sampling and DNA methylation analysis. We thank Yukari Nishimura for her excellent technical assistance with immunohistochemistry.

This study was supported in part by JSPS KAKENHI Grant Numbers Scientific Research (C) 15K08466 to M. Higashi, Scientific Research (C) 15K11297, to T. Hamada and Young Scientists (B) 15K21247, and to S. Yokoyama from the Ministry of Education, Science, Sports, Culture and Technology, Japan, by the Kodama Memorial Foundation, Japan (S. Yokoyama and M. Higashi). The funders had no role in study design, data collection and analysis, decision to publish, or preparation of the manuscript.

References

1. Egawa, S., H. Toma, H. Ohigashi, T. Okusaka, A. Nakao, T. Hatori, et al. Japan Pancreatic Cancer Registry; 30th year anniversary: Japan Pancreas Society. Pancreas, 2012. 41: 985-92.
2. Isaji, S., Y. Kawarada, S. Uemoto. Classification of pancreatic cancer: comparison of Japanese and UICC classifications. Pancreas, 2004. 28: 231-4.
3. Hollingsworth, M.A., B.J. Swanson. Mucins in cancer: protection and control of the cell surface. Nat Rev Cancer, 2004. 4: 45-60.
4. Kaur, S., S. Kumar, N. Momi, A.R. Sasson, S.K. Batra. Mucins in pancreatic cancer and its microenvironment. Nat Rev Gastroenterol Hepatol, 2013. 10: 607-20.
5. Jepson, S., M. Komatsu, B. Haq, M.E. Arango, D. Huang, C.A. Carraway, et al. MUC4/sialomucin complex, the intramembrane ErbB2 ligand, induces specific phosphorylation of ErbB2 and enhances expression of p27(kip), but does not activate mitogen-activated kinase or protein kinaseB/Akt pathways. Oncogene, 2002. 21: 7524-32.
6. Rachagani, S., M.A. Macha, M.P. Ponnusamy, D. Haridas, S. Kaur, M. Jain, et al. MUC4 potentiates invasion and metastasis of pancreatic cancer cells through stabilization of fibroblast growth factor receptor 1. Carcinogenesis, 2012. 33: 1953-64.
7. Yonezawa, S., M. Higashi, N. Yamada, S. Yokoyama, M. Goto. Significance of mucin expression in pancreatobiliary neoplasms. J Hepatobiliary Pancreat Sci, 2010. 17: 108-24.
8. Horinouchi, M., K. Nagata, A. Nakamura, M. Goto, S. Takao, M. Sakamoto, et al. Expression of different glycoforms of membrane mucin (MUC1) and secretory mucin (MUC2, MUC5AC and MUC6) in pancreatic neoplasms. Acta Histochemica et Cytochemica, 2003. 36: 443-53.
9. Higashi, M., S. Yokoyama, T. Yamamoto, Y. Goto, I. Kitazono, T. Hiraki, et al. Mucin expression in endoscopic ultrasound-guided fine-needle aspiration specimens is a useful prognostic factor in pancreatic ductal adenocarcinoma. Pancreas, 2015. 44: 728-34.
10. Yonezawa, S., M. Goto, N. Yamada, M. Higashi, M. Nomoto. Expression profiles of MUC1, MUC2, and MUC4 mucins in human neoplasms and their relationship with biological behavior. Proteomics, 2008. 8: 3329-41.
11. Yamada, N., S. Kitamoto, S. Yokoyama, T. Hamada, M. Goto, H. Tsutsumida, et al. Epigenetic regulation of mucin genes in human cancers. Clin Epigenetics, 2011. 2: 85-96.

12. Yamada, N., Y. Nishida, H. Tsutsumida, M. Goto, M. Higashi, M. Nomoto, et al. Promoter CpG methylation in cancer cells contributes to the regulation of MUC4. Br J Cancer, 2009. 100: 344-51.
13. Yamada, N., Y. Nishida, H. Tsutsumida, T. Hamada, M. Goto, M. Higashi, et al. MUC1 expression is regulated by DNA methylation and histone H3 lysine 9 modification in cancer cells. Cancer Res, 2008. 68: 2708-16.
14. Saitou, M., M. Goto, M. Horinouchi, S. Tamada, K. Nagata, T. Hamada, et al. MUC4 expression is a novel prognostic factor in patients with invasive ductal carcinoma of the pancreas. J Clin Pathol, 2005. 58: 845-52.
15. Osako, M., S. Yonezawa, B. Siddiki, J. Huang, J.J. Ho, Y.S. Kim, et al. Immunohistochemical study of mucin carbohydrates and core proteins in human pancreatic tumors. Cancer, 1993. 71: 2191-99.
16. Yonezawa, S., M. Higashi, N. Yamada, S. Yokoyama, S. Kitamoto, S. Kitajima, et al. Mucins in human neoplasms: clinical pathology, gene expression and diagnostic application. Pathol Int, 2011. 61: 697-716.
17. Hinoda, Y., Y. Ikematsu, M. Horinochi, S. Sato, K. Yamamoto, T. Nakano, et al. Increased expression of MUC1 in advanced pancreatic cancer. J Gastroenterol, 2003. 38: 1162-66.
18. Klöppel, G., G. Adler, R. Hruban, S. Kern, D. Longnecker, T. Partanen. Ductal adenocarcinoma of the pancreas. pp. 281-91. In: F. Bosman, F. Carneiro, R. Hruban, N. Theise (Eds.). World Health Organization Classification of Tumours of the Digestive System. Lyon, France. 2010.
19. Luttges, J., B. Feyerabend, T. Buchelt, M. Pacena, G. Kloppel. The mucin profile of noninvasive and invasive mucinous cystic neoplasms of the pancreas. Am J Surg Pathol, 2002. 26: 466-71.
20. Basturk, O., S.M. Hong, L.D. Wood, N.V. Adsay, J. Albores-Saavedra, A.V. Biankin, et al. Consensus: a revised classification system and recommendations from the Baltimore Consensus Meeting for neoplastic precursor lesions in the pancreas. Am J Surg Pathol, 2015. 39: 1730-41.
21. Nagata, K., M. Horinouchi, M. Saitou, M. Higashi, M. Nomoto, M. Goto, et al. Mucin expression profile in pancreatic cancer and the precursor lesions. J Hepatobiliary Pancreat Surg, 2007. 14: 243-54.
22. Swartz, M.J., S.K. Batra, G.C. Varshney, M.A. Hollingsworth, C.J. Yeo, J.L. Cameron, et al. MUC4 expression increases progressively in pancreatic intraepithelial neoplasia. Am J Clin Pathol, 2002. 117: 791-6.
23. Zrihan-Licht, S., M. Weiss, I. Keydar, D.H. Wreschner. DNA methylation status of the MUC1 gene coding for a breast-cancer-associated protein. Int J Cancer, 1995. 62: 245-51.
24. Rajabi, H., C. Jin, R. Ahmad, C. McClary, M.D. Joshi, D. Kufe. Mucin 1 oncoprotein expression is suppressed by the miR-125b oncomir. Genes Cancer, 2010. 1: 62-68.
25. Sachdeva, M., Y.Y. Mo. MicroRNA-145 suppresses cell invasion and metastasis by directly targeting mucin 1. Cancer Res, 2010. 70: 378-87.
26. Hanski, C., E. Riede, A. Gratchev, H.D. Foss, C. Bohm, E. Klussmann, et al. MUC2 gene suppression in human colorectal carcinomas and their metastases: in vitro evidence of the modulatory role of DNA methylation. Lab Invest, 1997. 77: 685-95.
27. Mesquita, P., A.J. Peixoto, R. Seruca, C. Hanski, R. Almeida, F. Silva, et al. Role of site-specific promoter hypomethylation in aberrant MUC2 mucin expression in mucinous gastric carcinomas. Cancer Lett, 2003. 189: 129-36.

28. Siedow, A., M. Szyf, A. Gratchev, U. Kobalz, M.L. Hanski, C. Bumke-Vogt, et al. De novo expression of the MUC2 gene in pancreas carcinoma cells is triggered by promoter demethylation. Tumour Biol, 2002. 23: 54-60.
29. Hamada, T., M. Goto, H. Tsutsumida, M. Nomoto, M. Higashi, T. Sugai, et al. Mapping of the methylation pattern of the MUC2 promoter in pancreatic cancer cell lines, using bisulfite genomic sequencing. Cancer Lett, 2005. 227: 175-84.
30. Yamada, N., Y. Nishida, S. Yokoyama, H. Tsutsumida, I. Houjou, S. Kitamoto, et al. Expression of MUC5AC, an early marker of pancreatobiliary cancer, is regulated by DNA methylation in the distal promoter region in cancer cells. J Hepatobiliary Pancreat Sci, 2010. 17: 844-54.
31. Vincent, A., M. Perrais, J.L. Desseyn, J.P. Aubert, P. Pigny, I. Van Seuningen. Epigenetic regulation (DNA methylation, histone modifications) of the 11p15 mucin genes (MUC2, MUC5AC, MUC5B, MUC6) in epithelial cancer cells. Oncogene, 2007. 26: 6566-76.
32. Okudaira, K., S. Kakar, L. Cun, E. Choi, R. Wu Decamillis, S. Miura, et al. MUC2 gene promoter methylation in mucinous and non-mucinous colorectal cancer tissues. Int J Oncol, 2010. 36: 765-75.
33. Kitamoto, S., N. Yamada, S. Yokoyama, I. Houjou, M. Higashi, S. Yonezawa. Promoter hypomethylation contributes to the expression of MUC3A in cancer cells. Biochem Biophys Res Commun, 2010. 397: 333-9.
34. Vincent, A., M.P. Ducourouble, I. Van Seuningen. Epigenetic regulation of the human mucin gene MUC4 in epithelial cancer cell lines involves both DNA methylation and histone modifications mediated by DNA methyltransferases and histone deacetylases. FASEB J, 2008. 22: 3035-45.
35. Zhu, Y., J.J. Zhang, R. Zhu, Y. Zhu, W.B. Liang, W.T. Gao, et al. The increase in the expression and hypomethylation of MUC4 gene with the progression of pancreatic ductal adenocarcinoma. Med Oncol, 2011. 28(Suppl 1): S175-84.
36. Matsubayashi, H., M. Canto, N. Sato, A. Klein, T. Abe, K. Yamashita, et al. DNA methylation alterations in the pancreatic juice of patients with suspected pancreatic disease. Cancer Res, 2006. 66: 1208-17.
37. Yokoyama, S., S. Kitamoto, N. Yamada, I. Houjou, T. Sugai, S. Nakamura, et al. The application of methylation specific electrophoresis (MSE) to DNA methylation analysis of the 5' CpG island of mucin in cancer cells. BMC Cancer, 2012. 12: 67.
38. Kitazono, I., M. Higashi, S. Kitamoto, S. Yokoyama, M. Horinouchi, M. Osako, et al. Expression of MUC4 mucin is observed mainly in the intestinal type of intraductal papillary mucinous neoplasm of the pancreas. Pancreas, 2013. 42: 1120-28.
39. Yokoyama, S., S. Kitamoto, M. Higashi, Y. Goto, T. Hara, D. Ikebe, et al. Diagnosis of pancreatic neoplasms using a novel method of DNA methylation analysis of mucin expression in pancreatic juice. PLoS One, 2014. 9: e93760.
40. Kitamoto, S., S. Yokoyama, M. Higashi, N. Yamada, S. Takao, S. Yonezawa. MUC1 enhances hypoxia-driven angiogenesis through the regulation of multiple proangiogenic factors. Oncogene, 2013. 32: 4614-21.
41. Ko, M., Y. Huang, A.M. Jankowska, U.J. Pape, M. Tahiliani, H.S. Bandukwala, et al. Impaired hydroxylation of 5-methylcytosine in myeloid cancers with mutant TET2. Nature, 2010. 468: 839-43.
42. Tahiliani, M., K.P. Koh, Y. Shen, W.A. Pastor, H. Bandukwala, Y. Brudno, et al. Conversion of 5-methylcytosine to 5-hydroxymethylcytosine in mammalian DNA by MLL partner TET1. Science, 2009. 324: 930-5.

43. Yokoyama, S., M. Higashi, S. Kitamoto, M. Oeldorf, U. Knippschild, M. Kornmann, et al. Aberrant methylation of MUC1 and MUC4 promoters are potential prognostic biomarkers for pancreatic ductal adenocarcinomas. Oncotarget, 2016. 7: 42553-65.
44. Yokoyama, S., M. Higashi, H. Tsutsumida, J. Wakimoto, T. Hamada, E. Wiest, et al. TET1-mediated DNA hypomethylation regulates the expression of MUC4 in lung cancer. Genes Cancer, 2017. 8: 517-27.

CHAPTER 8

Receptor Tyrosine Kinase Signaling Pathways as a Goldmine for Targeted Therapy in Head and Neck Cancers

Muzafar A. Macha[1,2*], Satyanarayana Rachagani[1], Sanjib Chaudhary[1], Zafar Sayed[2], Dwight T. Jones[2] and Surinder K. Batra[1,3,4*]

[1] Department of Biochemistry and Molecular Biology
[2] Department of Otolaryngology/Head & Neck Surgery
[3] Eppley Institute for Research in Cancer and Allied Diseases
[4] Buffett Cancer Center, University of Nebraska Medical Center (UNMC), Omaha, NE, 68198-5870, USA.

Introduction

Head and neck squamous cell carcinoma (HNSCC) includes tumors of the oral cavity, nasopharynx, oropharynx, hypopharynx, larynx and paranasal sinuses. It is one of the six most common cancers accounting for over 600,000 new cases and 350,000 deaths annually worldwide [1]. In the United States alone, 53,260 (38,380 males and 14,880 females) new cases and over 10,750 deaths are estimated in 2020 [2], together accounting for about 3.6% of new cancer cases and 2% of cancer related deaths. Squamous cell carcinoma (OSCC) of tongue and mouth are the major subtypes of HNSCC which accounts for two-third of the cases represented in developing countries and approximately 31,980 cases in the United States annually [2]. The other anatomical distribution of the OSCC includes approximately 17,950 cases of the pharynx and about 3,330 cases of other oral cavity cancers [2]. Smoking and excessive alcohol consumption are the major risk factors in OSCC development in the western countries, while use of smokeless tobacco products such as gutkha, pan masala and betel quid remain the major risk factors in the Asian countries [3-6]. However, the human papilloma virus (HPV) associated oropharyngeal (OPC) cancers, particularly in younger patients with little or no smoking or alcohol history [3] is on surge. At the current rate of infection, the annual incidence of HPV-positive OPCs is expected to surpass cervical cancer in the United States by 2020 [7]. Of the 40

---

*Corresponding author: muzafar.macha@unmc.edu; sbatra@unmc.edu

HPV types known to infect the mucosal surfaces, only 16, 18, 31, 33, 35, 39, 45, 51, 52 and 58 are oncogenic or high-risk types, while 6, 11, 40, 42, 43, 44 and 54 are non-oncogenic/low-risk types [8]. In addition, chronic inflammation associated with periodontal disease and autoimmune conditions such as erosive lichen planus are other potential risk factors.

Of all the tumors in different anatomic sites of OSCC, tumors occurring in oral tongue is the most aggressive and exhibit high rates of lymph node metastases [9]. In addition to the impact of anatomic site on prognosis, size of primary tumor, poor differentiation status, infiltrated tumors (compared to erosive pattern), presence of perineural invasion and lymphocytic host response are associated with worse prognosis. Moreover, the depth of invasion of the primary tumor (oral tongue and floor of mouth) have also been reported to predict a higher risk of nodal metastases, with the presence of nodal metastases being the most significant prognostic factor for oral cavity cancer.

Despite significant advances in surgery and chemotherapy, the poor prognosis and low survival rates of HNSCC is primarily due to loco-regional (lymph nodes) and distant (lungs, liver and bones) metastasis [10, 11]. Currently, conventional prognostic factors for HNSCC patients are histological tumor grade, tumor stage and involvement of regional lymph nodes at the time of diagnosis; however, these parameters are infrequently used to guide treatment decisions due to lack of sensitivity and accuracy. Therefore, there is an urgent need to identify novel diagnostic and therapeutic targets in HNSCC that could accurately predict patients at risk for treatment failure and disease recurrence.

ErbB Signaling Pathway Regulation in Cancer

Differential expression of tumor specific molecules is not only useful as markers for diagnosis and prognosis, but also may be used as a potential target for therapeutic interventions. Our recent advances in understanding the pathobiology of HNSCC have revealed several genetic, epigenetic and metabolic alterations including over-expression of oncogenes (ErbB, Myc, cyclin D1), mutational activation (Ras, PI3K), inactivation of tumor suppressor genes (TP53 and p16), DNA ploidy and loss of heterozygosity at numerous chromosomal locations [12-15] that result in the appearance of abnormal histological phenotypes.

The receptor tyrosine kinase ErbB family includes ErbB-1/EGFR, ErbB-2/HER2/neu, ErbB-3/HER3 and ErbB-4/HER4 with an extracellular ligand-binding domain, a hydrophobic transmembrane domain and a tyrosine kinase containing cytoplasmic domain. Binding of ligands to extracellular domain results in the homo or hetero dimerization of ErbB family receptors or with other tyrosine kinase receptor (IGF-1R or c-Met) (reviewed in [16]). This dimerization causes trans-phosphorylation of tyrosine residues on the cytoplasmic domain leading to activation of several downstream pathways, including Ras/Raf/MAPK/ERK, PI3K/Akt, STAT and PLC-γ signaling pathways [17, 18]. While no ligand is identified yet for ErbB2, 11 different ligands are known to bind to ErbB1 (EGF, heparin binding EGF (HB-EGF), transforming growth factor-α (TGF-α), amphi-regulin, epigen, beta cellulin (BTC) and epiregulin (EPR), ErbB3 (neuregulin/hereregulin (NRG1, NRG2) and ErbB4 (BTC, HB-EGF, EPR, NRG-1, NRG-2, NRG-3 and NRG-4) (reviewed in [19]). In addition to these ligands, insulin-like growth factor receptor (IGF1-R), G-protein-coupled receptors (GPCRs) and cell adhesion molecules like E-cadherin and integrins can potentially activate ErbB signaling (reviewed in [20]). These studies have shown that GPCRs, either by activating matrix metalloproteinases (MMPs) and cleavage of membrane-tethered ligands for ErbB binding or by indirect activation of Src to phosphorylate the intracellular receptor tyrosine residues, activate the ErbB signaling. In addition, steroid hormones by activating the transcriptional upregulation of EGFR ligands like TGFα are known to transactivate ErbB signaling [21]. A recent study has shown the involvement of urokinase plasminogen activator receptor (uPAR) in EGFR activation by modulating α5β1 integrin, Src and MMPs and results in enhanced cancer cell invasion, in contrast to proliferation by EGF stimulation [22].

Deregulated ErbB Signaling in HNSCC

Deregulation of ErbB signaling pathway in HNSCC progression and metastasis is well established. Xia et al. have reported over-expression of EGFR1, HER-2, HER-3 and HER-4, and their association with shortened HNSCC patient survival [23]. Furthermore, co-expression of EGFR1, HER-2 and HER-3 (but not HER-4) is associated with lymph node metastasis and aggressive HNSCC tumors [23]. While EGFR over-expression is considered as an early event in HNSCC development, its expression progressively increases from hyperplasia to dysplasia to invasive carcinoma [24]. Many studies have reported EGFR1 over expression in ~90% of HNSCCs [25, 26] and its involvement in aggressive tumor behavior [27], radio-resistance [28] and poor prognosis of patient [29]. While therapeutic benefits of targeting membrane-bound EGFR in many cancers including HNSCC have been reported, currently the pathobiological importance of nuclear EGFR expression as observed in cancer cells and primary tumor specimens of various origins is being investigated (reviewed in [19]).

Apart from EGFR and HER2 over-expression in primary HNSCC tumors, these molecules are also upregulated in a subset of metastatic non-small cell lung [30], breast [31] and gastric cancers [32] compared to matched primary tumors. While HER-2 expression increases progressively with HNSCC pathogenesis [33, 34], its pathobiological importance in HNSCC has received less attention compared to breast or ovarian tumors. Although HER-2/-3 heterodimers potentially induce PI3K/AKT pathway [35] through direct interaction of HER-3 with p85 subunit of PI3K, distinct distribution of HER-3 over-expression has been observed in HNSCC tumors. O-Charoenrat et al. reported elevated expression of HER-4 in newly developed cell lines compared to well established HNSCC cell lines [36]; however, no significant association of HER-4 over-expression was observed with invasion, angiogenesis and metastasis in HNSCC patients [37] or survival [23].

Genetic variations including mutations or alternate splicing of EGFR play an important role in expression, protein stability, and function which defines therapy outcome in HNSCC patients [38]. The incidence of EGFR mutations in HNSCC is low ranging from 0-4% and 7% in non-Hispanic whites and Asians respectively [39], while EGFRvIII variants is expressed in 42% of HNSCC patients [40]. This splice variant (145-kDa) has a deletion of amino acids (6-273) in the extracellular domain that renders its constitutively active function even in the absence of ligands and also bypasses the feedback regulatory mechanisms.

Role of ErbB Signaling in Resistance to Therapy in HNSCC

Surgery along with chemo-radiation therapy (CRT) and/or targeted therapy (TT) is the treatment of choice for locally advanced (LA) HNSCC. Despite recent improvements in therapy and pathological understanding of the disease, the five-year patient survival and prognosis has not improved much over decades [10, 41, 42] rendering new therapeutic avenues in dire need [43]. The reasons of poor therapy response includes treatment associated toxicities [44] and inherent resistance of HNSCC to CRT and TT. While the underlying mechanism of resistance is multifactorial, the recent molecular studies have shown the modulation of expression of drug target, pro- survival and apoptotic pathways by tumor cells [45]. Various studies have shown EGFR activation following radiation therapy (RT) in many tumors, including HNSCC (reviewed in [46]). While RT causes cytotoxicity by damaging DNA that results in cell cycle arrest, senescence and tumor regression [47], it can also activate DNA repair machinery and pro-survival signaling pathways like EGFR/ATM/ATR/BRCA1/Chk1 and PI3K/Akt or RAS/Raf/ERK$_{1/2}$ signaling imparting radio-resistant (RR) phenotype to this cancer cells (reviewed in [46]). Although the underlying mechanism for ErbB activation by RT is not completely known, certain studies suggest the role of RT-induced reactive oxygen/nitrogen species (ROS/RNS) in inhibiting receptor protein tyrosine phosphatases (PTPs) [48]. However, studies have shown that RT causes translocation of EGFR inside the nucleus and interacts with DNA-PKCs, Ku70 and Ku80 proteins to repair DNA double strand breaks (DSB), thus resulting in increased cell survival (reviewed in [49] (Fig. 1). Studies have recently shown that EGFR blockade by monoclonal antibody (mAb) (cetuximab/IMC C225) and tyrosine kinase inhibitors TKIs (gefetinib and afatinib) increases the radiosensitivity of the tumors including HNSCC by inhibiting RT-induced nuclear translocation of EGFR, its interaction with DNA-PKc (reviewed in [49], and also by inhibiting EGFR/ATM/ATR/BRCA1/Chk1 mediated DNA DSB repair [50, 51] (Fig. 1). In contrast, some

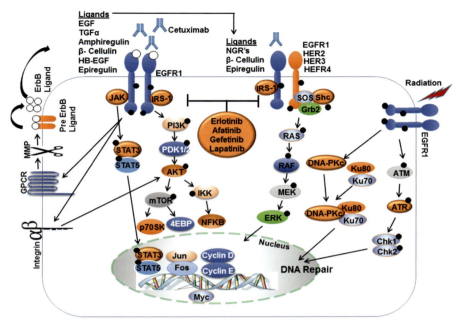

Fig. 1: ErbB signaling pathway and its inhibition and activation. ErbB signaling is activated by the binding of specific ligands to the members of ErbB family that results in homo/hetero-dimerization, activation and recruitment of docking proteins resulting in activation of downstream signaling pathways including RAS/Raf/MAPK, PI-3K/Akt and Jak/STAT. Use of anti-EGFR MAb's (cetuximab) or small molecular inhibitors like erlotinib, gefetinib, afatinib, lapitinib inhibits ErbB signaling and therefore abrogates downstream signaling pathways. However, radiation treatment induces EGFR activation in absence of its ligand and results in activation of DNA double strand break (DSB) repair signaling pathway like ATM/ATR/Chk1/Chk2/BRCA and DNA-KPc/Ku70/Ku80 leading to radio-resistance (RR) of HNSCC. Further, upregulation of GPCR activates matrix metallo-proteinases that cleave the membrane-tethered ligands for ErbB signaling activation. In addition, urokinase plasminogen activator receptor (uPAR) mediated upregulation of α5β1 integrin activates Akt resulting in resistance to ErbB inhibition.

studies revealed impaired cell cycle progression, proliferation and increased apoptosis, and downregulation of cell survival pathways via Ras-MAPK and PI3K-AKT signaling as mechanisms of RS induced by EGFR blockers (reviewed in [49]).*

Role of ErbB Signaling in HNSCC Cancer Stem Cells

Like other solid tumors, HNSCC tumors are heterogeneous and characterized by the presence of tumor cells, cancer stem cells (CSCs), desmoplasia, i.e. stromal component consisting of cells (fibroblasts, endothelial cells, immune cells) and ECM (collagen and fibronectin), collectively called tumor microenvironment (TME) (Fig. 2). CSCs, also termed as tumor initiating

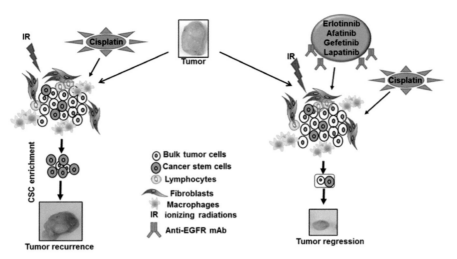

Fig. 2: Cancer stem cell (CSC) enrichment in HNSCC. Use of conventional chemo-radiation therapy (CRT) kills bulk of the tumor cells leaving behind viable CSCs that result in tumor recurrence. Combination of ErbB inhibitors including monoclonal antibody (cetuximab) or small monoclonal antibody like erlotinib, gefetinib, afatinib, lapitinib with CRT kills both the CSCs and bulk tumor cells.

cells or side population, are responsible for tumor initiation, disease aggressiveness and also serve as a reservoir for refractory tumors [52]. The CSCs are a small sub-population of undifferentiated and self-renewing cells among the bulk of tumor cells. CSCs have the ability to differentiate to heterogeneous tumor cells that are highly tumorigenic and resistant to CRT which are responsible for tumor recurrence in various cancers [53, 54]. Conventional CRT kills the bulk population of rapidly proliferating tumor cells, but subsequently enriches or selects CSCs which are inherently CR resistant [55]. Although the underlying cause of CRT resistance and disease recurrence is poorly understood, recent studies have shown TME to protect CSCs from CRT interventions which results in tumor recurrence and poor patient prognosis (Fig. 2). In addition, though less explored, the CRT resistance is also expected to be due to intrinsic capabilities like DNA repair, survival pathways and extrinsic cues from TME. It is also speculated that the ability of CSCs to remain quiescent also make them resistant to CRT (reviewed in [56]). Therefore, understanding the underlying biology of CSC resistance to CRT is important to establish CSC targeted therapy for effective management of HNSCCs.

CSCs have been identified in various types of tumors including HNSCC pathogenesis [57, 58]. Although there is no universal CSC marker for each cancer type, CSCs are identified either by enzyme activity of aldehyde dehydrogenase 1 (ALDH1), biomolecules including, CD24, Bmi-1 (B-cell specific Moloney murine leukemia virus insertion site-1), CD133 (prominin-1), CD44, Notch/Hedgehog family members, nestin etc., or by their ability to efflux vital dyes and to form tumor spheres *in vitro* (reviewed in [59]). ALDH1, CD44 and CD133 being the predominant biomarkers for identifying HNSCC CSCs (reviewed in [59]), recent studies have shown greater tumor forming capability and RR of ALDH1+ cells from HNSCC [60]. Similarly, double positive CD44+ and ALDH+ expressing tumors tend to be more advanced [61], and single positive CD44+ positive tumors predict disease recurrence after RT in laryngeal cancers [62]. In addition, HNSCC CD133+ cells are more tumorigenic and resistant to paclitaxel treatment [63].

Many signaling pathways such as Hedgehog (Hh), Notch and Wnt regulate the self-renewal properties of CSCs and RR in multiple cancers (reviewed in [56]). Besides tumors of epithelial cell origin, recent studies have identified the importance of EFGR signaling in regulating and maintaining nasopharyngeal cancer (NPC) CSCs [64] and treatment of NPC cells with EGF leads to increased CSC population. PI3K/AKT-β-catenin signaling pathway causes activation and over-expression of EGFR in HNSCC cells to promote acquisition of cancer stemness properties including sphere-formation and increased expression of cancer stem cell markers like CD44, Oct-4, Sdf-1 and Nanog [65]. However, gefetinib (EGFR inhibitor) decrease the stemness properties, downregulate CSC markers and sensitize the HNSCC cells to chemotherapeutic agents [64, 65]. In addition, our studies also demonstrated that the use of pan-EGFR inhibitor, afatinib, on HNSCC cell lines can significantly decrease the CSCs, inhibition of their self-renewal property and radio-sensitization of HNSCC cells [51]. Thus EGFR does not only plays an important role in cancer cell survival and metastasis, but also regulates CSCs. Therefore, treatment modalities combining EGFR inhibitors with chemotherapeutic drugs will not only target the bulk of tumor cells, but also CSCs, thereby decreasing tumor growth and inhibiting HNSCC recurrence.

ErbB Signaling Inhibitors in HNSCC

EGFR plays a key role in the pathogenesis of HNSCC and is one of the most promising therapeutic target for HNSCCs (reviewed in [66]). Many EGFR targeted therapies including tyrosine kinase inhibitors (TKIs), mAbs and antisense oligonucleotides etc., have been investigated for their therapeutic potential in HNSCC. Therapeutic mAbs like cetuximab (C225, Erbitux[R]; Bristol-Myers Squibb, Princeton, NJ), panitumumab (Vectibix®; Amgen Inc., Thousand Oaks, CA), zalutumumab (formerly HuMax-EGFr; Genmab, Princeton, NJ) and nimotuzumab (Oncoscience AG, Germany and YM Biosciences Inc, Ontario, Canada) target the extracellular EGFR domain III and prevent receptor dimerization mediated phosphorylation and activation of downstream signaling pathways (Fig. 1). These therapeutic antibodies normally compete for binding of the autocrine ligands like amphiregulin and TGF-alpha, and result in either abrogation of EGFR mediated signal transduction or degradation and internalization of the EGFR leading to antibody-dependent cell-mediated cytotoxicity (ADCC). In addition to mAbs, small molecular inhibitors (TKIs) are also used that compete with

ATP to bind at the intracellular ATP binding cleft of ErbB and abrogate the downstream signaling pathways (Fig. 1). Some of these TKIs like, gefitinib (Iressa®; AstraZeneca, Wilmington, DE) and erlotinib (Tarceva®; Genentech, Inc., South San Francisco, CA) bind reversibly to only EGFR, while other inhibitors like lapatinib (reversible) (Tykerb®; GlaxoSmithKline, Research Triangle Park, NC), dacomitinib (irreversible) (PF00299804; Pfizer Inc., New York, NY) and afatinib (Gilotrif™; Boehringer Ingelheim, Ingelheim, Germany) bind irreversibly to EGFR and other ErbB family members. Unfortunately, the use of these mAbs and TKIs against EGFR in HNSCC is often challenged by either intrinsic and acquired resistance or lower therapeutic response when used as a single agent (reviewed in [66]). Nevertheless, combination of conventional cytotoxic drugs with EGFR inhibitors has shown promising results in the clinical settings [67]. However, there is an urgent need to understand the underlying cause of anti-EGFR therapeutic resistance so as to identify efficacy biomarkers, develop new treatment strategies to overcome resistance, improve efficacy and prevent tumor's recurrence [68-70]. But, recent molecular studies have identified the involvement of GPCRs in EGFR transactivation as one of the important mechanism of resistance to anti-EGFR therapies [71]. Other studies have reported increased expression of HER2 and HER3 in cetuximab resistance in HNSCC cells *in vitro* [72] and downregulation of HER3 leads to sensitization to cetuximab [72]. Furthermore, increased EGFR copy number also mediated cetuximab and gefitinib resistance (reviewed in [66]). Epithelial-mesenchymal transition (EMT) promotes CRT resistance [54] and several studies have reported EGFR induced EMT as one of the mechanism of CRT [73] and cetuximab resistance in HNSCCs [74]. In addition, activation of signal transducer and activator of transcription 3 (STAT3) and upregulation of vascular endothelial growth factor (VEGF) are other mechanisms for anti-EGFR therapy resistance (reviewed in [75]). While mutations in the EGFR kinase domain are common in many cancers which are associated with decreased response to EGFR inhibitors, such mutations are rarely observed in HNSCC [76].

EGFR as a Therapeutic Target in HNSCC: Clinical Studies

EGFR and its family members are the therapeutic targets in various cancers including HNSCC. Many EGFR targeted therapeutics are in various phases of clinical development and all the phase III clinical trials are summarized in Table 1. The FDA approved cetuximab mAb has been shown to inhibit EGFR downstream signaling as determined by decreased phosphorylated ERK$_{1/2}$ [77]. Interestingly, use of mAbs targeting many epitopes of EGFR caused a synergistic downregulation of downstream signaling pathways [78, 79]. While the use of some of these mAbs as a monotherapy in HNSCC has not improved the overall survival (OS) of the patients [80], the combination with chemotherapy (CT) and RT improved response of patient with refractory/metastatic (R/M) HNSCC disease (R/M) [80, 81]. In a phase I and II clinical trial, combination of cetuximab with platinum and 5FU increased the median progression free survival to 5.6 months compared to 3.3 months in platinum and fluorouracil therapy only [82, 83]. Based on successful completion of the above trials, several phase III clinical trials were initiated to use cetuximab in combination with either RT or CT. The results from these trials showed significant improvement in the OS when combined with RT in LA HNSCC or with cisplatin/5FU in R/M disease [67, 81, 84-86]. Currently, many trials to test the efficacy of cetuximab with concurrent CRT [66] and as a chemo- and radiation sensitizer are underway [84]. While the use of nimotuzumab (humanized mAb) is still not approved in the United States for HNSCC, phase III clinical trials of nimotuzumab in combination with CT, RT or CRT are underway (NCT01345084, NCT00957086) in other countries. Similarly, clinical studies on Panitumumab (ABX-EGF, humanized mAb) and Zalitumumab (HuMax-EGFR, fully humanized mAb) in combination with CT, RT or CRT for HNSCC are ongoing and discussed elsewhere [66] and summarized in Table 1. In addition, mAb MEHD7945A (Genentech, Inc., South San Francisco, CA) that targets both EGFR and HER3 are being investigated in phase I/II (NCT01577173), but with limited success [80].

While all these above studies show much of the clinical benefits in combination, unfortunately development of therapeutic resistance due to upregulation of other ErbB family members renders them less effective [72]. Therefore, strategies to target other family members including EGFR2, EGFR3 and EGFR4 or development of pan-EGFR inhibitors are being tested with adjuvant or concurrent CRT in HNSCC. Though many clinical trials were done with these inhibitors as a monotherapy, most of these studies only showed a modest response in HNSCC [80]. However, gefitinib with methotrexate for R/M HNSCC (phase III trial) failed to improve OS resulting in the early termination of the trial [87], whereas gefitinib with RT improved the 3-year OS to 71% and distant metastatic control in 88% in phase II trial [88]. Phase II clinical trial with erlotinib demonstrated a modest overall therapy response with a median OS of six months [89]. Combining erlotinib with cisplatin and RT to increase the efficacy in LA HNSCC did not increase the toxicity of cisplatin and RT in the patients but failed to significantly increase CRR or PFS [90]. Both afatinib and dacomitinib are pan-EGFR irreversible TKIs targeting EGFR, HER2, and HER4 [80]. Many pre-clinical [91-93] and clinical [94-96] studies have shown that afatinib significantly inhibit growth of many cancers overexpressing either wild type EGFR and/or HER2 or EGFR L858R/T790M double mutations. While the phase II study of afatinib in HNSCC showed comparable efficacy to cetuximab (NCT00514943) [97], interim results of the phase III trial (LUX-Head & Neck1; NCT01345682) in R/M HNSCC patients showed improved PFS with afatinib compared to methotrexate treated patients [98, 99]. A recent phase II trial of dacomitinib as a monotherapy in R/M HNSCC patients showed modest response [100] and many phase I/II trials (NCT01449201, NCT01116843) are currently ongoing.

Lapatinib, that targets EGFR and HER2 showed no response as a monotherapy in R/M HNSCC [101], whereas 17% response rate was observed in LA HNSCC patients when given prior to CRT [102]. However, when combined with CRT with subsequent maintenance of monotherapy in LA HNSCC patients, it showed 53% response rate compared to only 36% with placebo [61]. More interestingly, the progression-free survival (PFS) was longer in HPV negative patients treated with lapatinib compared to placebo arm [102]. While many phase II clinical trials are ongoing, including (a) lapatinib into induction CT and concurrent CRT for high-risk patients (NCT01612351), (b) with RT in LA HNSCC patients that do not tolerate CT (NCT00490061) and (c) with capecitabine in R/M HNSCC patients (NCT01044433), a recent phase III trial of lapatinib with adjuvant CRT for high-risk resected SCCHN (NCT00424255) has been completed with no efficacy benefits and associated toxicity [103].

Besides using mAbs and TKIs, many other strategies aimed at inhibiting the ErbB signaling pathway have been investigated in HNSCC and other cancers. Antisense oligonucleotides are ~20 nucleotides, single stranded DNA molecules that bind to the mRNA and prevent translation of protein. Using this strategy, EGFR was downregulated and seen to be associated with DNA fragmentation, induction of apoptosis *in vitro* and significantly reduced growth of xenograft tumors *in vivo* [104]. In addition, targeting EGFR ligand using the oligonucleotides also resulted in tumor regression in HNSCC xenograft mouse model [105]. Like EGFR inhibitors in combination with cytotoxic drugs showed improved response, combination of EGFR antisense oligonucleotides with docetaxel significantly reduced xenograft tumors compared to docetaxel alone [106]. While a moderate response with no cytotoxicity using the antisense oligonucleotides was observed in phase I trial [107], its combination with RT and cetuximab in advanced locoregional HNSCC was investigated (NCT00903461). Many of the clinical trials using this technology in HNSCC and their outcomes are summarized in Table 1. Further, siRNA (double stranded 20nt RNA molecules) mediated down regulation of EGFR in combination with cisplatin also significantly reduced the tumor volume in mouse xenografts models [108].

Conclusion

ErbB receptor tyrosine kinases (EGFR, ErbB2, ErbB-3 and ErbB-4) play an important role in HNSCC pathogenesis and regulate key processes like cell proliferation, invasion, metastasis, angiogenesis, survival, CSCs maintenance and induce CRT resistance. Therefore, targeted therapies against ErbBs are clinically relevant to advance treatment that simultaneously inhibit different aspects of HNSCC phenotypes. A subset of LA HNSCC patients have benefited from combining EGFR inhibitors such as mAbs (cetuximab, panitumumab or nimotuzumab) and TKIs (lapatinib, erlotinib, gefitinib and afatinib) with RT (reviewed in [77]). However, recent studies have shown the existence of alternate downstream escape mechanisms which includes mutational

Table 1: Phase III clinical anti-ErbB trials in HNSCC

S.No	Target	Identifier no.	Inhibitor used	Primary/ Secondary end point	Type of patients	Treatment	Outcome	References
1	EGFR	NCT00003809	Cetuximab	PFS/OS/RR	R/M	Cetuximab + cisplatin vs cisplatin	No improvement in PFS and OS; Improved RR	[67]
2	EGFR	NCT00265941	Cetuximab	PFS/OS	Stage III or IV	Cetuximab + cisplatin + RT vs cisplatin + RT	No improvement	[109, 110]
3	EGFR	NCT01969877	Cetuximab	OS	LA stage III or IV	Cetuximab + RT vs cisplatin + RT	Recruiting	
4	EGFR	NCT01012258	Cetuximab	BOR/PFS	LA	Cetuximab + RT vs RT	Completed	
5	EGFR	NCT01302834	Cetuximab	OS/PFS	HPV +ve oropharyngeal patients	Cetuximab + RT vs Cisplatin + RT	Active, not recruiting	
6	EGFR	NCT01810913	Cetuximab	DFS/loco-regional control	LA stage III or IV	Cetuximab + RT vs Cisplatin + Docetaxel + RT	Active, not recruiting	
7	EGFR	NCT00956007	Cetuximab	OS	LA stage III or IV	Cetuximab + RT vs RT	Active, not recruiting	
8	EGFR		Cetuximab	FFS/OS	R/M	Cetuximab + methotrexate	Recruiting	
9	EGFR	NCT01884623	Cetuximab	FFS/OS	R/M Elderly unfit patients	Cetuximab vs methotrexate	Recruiting	

(Contd.)

Table 1: (Contd.)

S.No	Target	Identifier no.	Inhibitor used	Primary/ Secondary end point	Type of patients	Treatment	Outcome	References
10	EGFR	NCT01177956	Cetuximab	ORR	R/M patients	Cetuximab and Concomitant cisplatin + 5FU	Marginal response	[111]
11	EGFR	NCT02383966	Cetuximab	PFS/OS	R/M patients	Cetuximab +Cisplatin + 5-FU vs Cisplatin + 5-FU vs	Recruiting	
12	EGFR	NCT01302834	Cetuximab	OS/PFS	HPV+ve oropharyngeal cancers	Cetuximab + RT vs CRT	Active, not recruiting	
13	EGFR	NCT01855451	Cetuximab	SS	LA HPV+ve oropharyngeal	Cetuximab + RT vs Cisplatin + RT	Recruiting	
14	EGFR	NCT03258554	Cetuximab	Efficacy & toxicity	Stage III-IVB who cannot take cisplatin	Durvalumab or Cetuximab + RT	Recruiting	
15	EGFR	NCT00004227	Cetuximab	LR disease control	Pharyngeal stage III/IV	Cetuximab + RT vs RT	Terminated	[85, 112]
16	EGFR	NCT01086826	Cetuximab	OS/PFS	LA	Neoadjuvant Docetaxel+Cisplatin and 5-fluorouracil (TPF) followed by RT + Concomitant Chemo or Cetuximab Versus RT + Concomitant Chemo or Cetuximab	Completed with improved patient outcome	[113]

(Contd.)

17	EGFR	NCT01233843	Cetuximab	CRR/OS	LA	CT + RT + Cetuximab vs CRT	Active, not recruiting	
18	EGFR	NCT02633176	Cetuximab	PFS/EFS	Metastatic Nasopharyngeal Carcinoma	Cetuximab + Cisplatin + Docetaxel + RT vs Cisplatin + Docetaxel + RT	Recruiting	
19	EGFR	NCT00820248	Panitumumab	PFS	LA stage III or IV	Pan + acc. RT vs CT + acc RT	Active, not recruiting	
20	EGFR	NCT01142414	Panitumumab	DFS	Resected with high risk of recurrence	Panitumumab + CRT vs CRT (cisplatin+5FU)	Withdrawn	
21	EGFR	NCT01345084	Nimotuzumab	OS/PFS	LA	Nimotuzumab + CRT vs CRT	Withdrawn	
22	EGFR	NCT00957086	Nimotuzumab	DFS	Stage III or IV	Nimotuzumab + CRT vs CRT	Recruiting	
23	EGFR	NCT00496652	Zalutumumab	LRC	-	Zalutumumab + CRT vs CRT	Ongoing	
24	EGFR	NCT00088907	Gefitinib	PFS	R/M patients	Gefitinib with or without Docetaxel	No improvement in patient prognosis	[87]
25	EGFR	NCT00903461	Antisense DNA	DP	LA	Cetuximab + RT + antisense DNA vs	Terminated (drug supply issues)	
26	HER1, 2	NCT00422255	Lapatinib	DFS/OS	Stage II, III or IVa	Post-surgery Lapatinib or placebo + Adjuvant concurrent CRT	Toxicity with no benefits	[103]

(Contd.)

Table 1: (Contd.)

S.No	Target	Identifier no.	Inhibitor used	Primary/ Secondary end point	Type of patients	Treatment	Outcome	References
27	HER1, 2, 4	NCT01345682 NCT01856478*	Afatinib	PFS	R/M patients after PT based CT	Afatinib vs methotrexate	Improved PFS with manageable safety profile	[98, 99]
28	HER1, 2, 4	NCT01427478	Afatinib	DFS	Non metastatic	Afatinib vs placebo after post-operative CRT	Recruiting	
29	HER1, 2, 4	NCT01856478	Afatinib	PFS/OS	R/M patients	Afatinib vs methotrexate after PT based CT	Active, not recruiting	
30	HER1, 2, 4	NCT01345669	Afatinib	DFS	unresectable stage III-IVb	Adjuvant afatinib vs placebo after PT based CRT	Terminated	[12]
31	HER1, 2, 4	NCT02131155	Afatinib	DFS/OS	unresectable stage III-IVb	Adjuvant afatinib vs placebo after PT based CRT	Terminated	

*Active but not recruiting, LR – Loco regional, RR – Response rate, FFS – Failure free survival, ORR – Overall response rate, LRC – Loco regional control, BOR – Best overall response, CRR – Complete response rate, EFS – Event free survival, DP – Disease progression, SS – Symptom severity activation of Ras, PI3K, ERK or upregulation of Met, STAT, IGF-1R signaling pathways that result in refractory tumors (Fig. 1). While combination of EGFR inhibitors along with RT and conventional chemotherapeutic drugs like cisplatin, docetaxel and 5FU have shown promise, but associated toxicities have limited their use (reviewed in [77]). Although combination of EGFR inhibitors with either STAT3, Met, or IGF-1R inhibitors has shown promise in preclinical HNSCC models (reviewed in [78]), however efficacy of these combination treatments is still under clinical investigation. Therefore, future studies aimed to understand the pathways associated with resistance mechanism shall determine the success of combination therapies.

References

1. Leemans, C.R., P.J. Snijders, R.H. Brakenhoff. The molecular landscape of head and neck cancer. Nat Rev Cancer, 2018. 18(5): 269.
2. Siegel, R.L., K.D. Miller, A. Jemal. Cancer statistics. CA Cancer J Clin, 2020. 70(1): 7-30.
3. Fakhry, C., M.L. Gillison. Clinical implications of human papillomavirus in head and neck cancers. J Clin Oncol, 2006. 24(17): 2606-11.
4. Hashibe, M., P. Brennan, S.-C. Chuang, S. Boccia, X. Castellsague, C. Chen. Interaction between tobacco and alcohol use and the risk of head and neck cancer: pooled analysis in the International Head and Neck Cancer Epidemiology Consortium. Cancer Epidemiol Biomarkers Prev, 2009. 18(2): 541-50.
5. Maier, H., A. Dietz, U. Gewelke, W. Heller, H. Weidauer. Tobacco and alcohol and the risk of head and neck cancer. Clin Investig, 1992. 70(3-4): 320-7.
6. Sturgis, E.M., Q. Wei. Genetic susceptibility—molecular epidemiology of head and neck cancer. CurrOpinOncol, 2002. 14(3): 310-7.
7. Chaturvedi, A.K., E.A. Engels, R.M. Pfeiffer, B.Y. Hernandez, W. Xiao, E. Kim, et al. Human papillomavirus and rising oropharyngeal cancer incidence in the United States. J Clin Oncol, 2011. 29(32): 4294-301.
8. Kreimer, A.R., G.M. Clifford, P. Boyle, S. Franceschi. Human papillomavirus types in head and neck squamous cell carcinomas worldwide: a systematic review. Cancer Epidemiol Biomarkers Prev, 2005. 14(2): 467-75.
9. Regezi, J.A., R.C. Jordan. Oral cancer in the molecular age. J Calif Dent Assoc, 2001. 29(8): 578-84.
10. Bernier, J. Adjuvant treatment of head and neck cancers: advances and challenges. Bull Cancer, 2007. 94(9): 823-7.
11. Warnakulasuriya, K.A., R. Ralhan. Clinical, pathological, cellular and molecular lesions caused by oral smokeless tobacco—a review. J Oral Pathol Med, 2007. 36(2): 63-77.
12. Ginos, M.A., G.P. Page, B.S. Michalowicz, K.J. Patel, S.E. Volker, S.E. Pambuccian, et al. Identification of a gene expression signature associated with recurrent disease in squamous cell carcinoma of the head and neck. Cancer Res, 2004. 64(1): 55-63.
13. Agrawal, N., M.J. Frederick, C.R. Pickering, C. Bettegowda, K. Chang, R.J. Li, et al. Exome sequencing of head and neck squamous cell carcinoma reveals

inactivating mutations in NOTCH1. Science (New York, NY), 2011. 333(6046): 1154-7.
14. Martin, D., M.C. Abba, A.A. Molinolo, L. Vitale-Cross, Z. Wang, M. Zaida, et al. The head and neck cancer cell oncogenome: a platform for the development of precision molecular therapies. Oncotarget, 2014. 5(19): 8906-23.
15. Stransky, N., A.M. Egloff, A.D. Tward, A.D. Kostic, K. Cibulskis, A. Sivachenko, et al. The mutational landscape of head and neck squamous cell carcinoma. Science (New York, NY), 333(6046): 2011. 1157-60.
16. Ohnishi, Y., Y. Minamino, K. Kakudo, M. Nozaki. Resistance of oral squamous cell carcinoma cells to cetuximab is associated with EGFR insensitivity and enhanced stem cell-like potency. Oncol Rep. 2014. 32(2): 780-6.
17. Bussink, J., J.H. Kaanders, A.J. van der Kogel. Microenvironmental transformations by VEGF- and EGF-receptor inhibition and potential implications for responsiveness to radiotherapy. Radiother Oncol, 2007. 82(1): 10-7.
18. Rodemann, H.P., K. Dittmann, M. Toulany. Radiation-induced EGFR-signaling and control of DNA-damage repair. J Radiat Biol, 2007. 83(11-12): 781-91.
19. Ribeiro, F.A.P., J. Noguti, C.T.F. Oshima, D.A. Ribeiro. Effective targeting of the epidermal growth factor receptor (EGFR) for treating oral cancer: a promising approach. Anticancer Res, 2014. 34(4): 1547-52.
20. Rogers, S.J., K.J. Harrington, P. Rhys-Evans, O. Pornchai, S.A. Eccles. Biological significance of c-erbB family oncogenes in head and neck cancer. Cancer Metastasis Rev, 2005. 24(1): 47-69.
21. Bates, S.E., N.E. Davidson, E.M. Valverius, C.E. Freter, R.B. Dickson, J.P. Tam, et al. Expression of transforming growth factor alpha and its messenger ribonucleic acid in human breast cancer: its regulation by estrogen and its possible functional significance. Mol Endocrinol, 1988. 2(6): 543-55.
22. Guerrero, J., J.F. Santibanez, A. González, J. Martınez. EGF receptor transactivation by urokinase receptor stimulus through a mechanism involving Src and matrix metalloproteinases. Exp Cell Res, 2004. 292(1): 201-8.
23. Xia, W., Y.-K. Lau, H.-Z. Zhang, F.-Y. Xiao, D.A. Johnston, A.-R. Liu, et al. Combination of EGFR, HER-2/neu, and HER-3 is a stronger predictor for the outcome of oral squamous cell carcinoma than any individual family members. Clin Cancer Res, 1999. 5(12): 4164-74.
24. Grandis, J.R., D.J. Tweardy, M.F. Melhem Asynchronous modulation of transforming growth factor alpha and epidermal growth factor receptor protein expression in progression of premalignant lesions to head and neck squamous cell carcinoma. Clin Cancer Res. 1998. 4(1): 13-20.
25. Grandis, J.R., D.J. Tweardy. Elevated levels of transforming growth factor alpha and epidermal growth factor receptor messenger RNA are early markers of carcinogenesis in head and neck cancer. Cancer Res, 1993. 53(15): 3579-84.
26. Ishitoya, J., M. Toriyama, N. Oguchi, K. Kitamura, M. Ohshima, K. Asano, et al. Gene amplification and overexpression of EGF receptor in squamous cell carcinomas of the head and neck. Br J Cancer, 1989. 59(4): 559-62.
27. Olayioye, M.A., R.M. Neve, H.A. Lane, N.E. Hynes. The ErbB signaling network: receptor heterodimerization in development and cancer. EMBO J, 2000. 19(13): 3159-67.
28. Liang, K., K.K. Ang, L. Milas, N. Hunter, Z. Fan. The epidermal growth factor receptor mediates radioresistance. Int J Radiat Oncol Biol Phys, 2003. 57(1): 246-54..

29. Ang, K.K., B.A. Berkey, X. Tu, H.-Z. Zhang, R. Katz, E.H. Hammond, et al. Impact of epidermal growth factor receptor expression on survival and pattern of relapse in patients with advanced head and neck carcinoma. Cancer Res, 2002. 62(24): 7350-6.
30. Italiano, A., F.B. Vandenbos, J. Otto, J. Mouroux, D. Fontaine, P.-Y. Marcy, et al. Comparison of the epidermal growth factor receptor gene and protein in primary non-small-cell-lung cancer and metastatic sites: implications for treatment with EGFR-inhibitors. Ann Oncol, 2006. 17(6): 981-5.
31. Zidan, J., I. Dashkovsky, C. Stayerman, W. Basher, C. Cozacov, A. Hadary. Comparison of HER-2 overexpression in primary breast cancer and metastatic sites and its effect on biological targeting therapy of metastatic disease. Br J Cancer, 2005. 93(5): 552-6.
32. Kim, J.H., M.A. Kim, H.S. Lee, W.H. Kim. Comparative analysis of protein expressions in primary and metastatic gastric carcinomas. Hum Pathol, 2009. 40(3): 314-22.
33. Wilkman, T.S., J.H. Hietanen, M.J. Maimström, Y.T. Konttinen. Immunohistochemical analysis of the oncoprotein c-erbB-2 expression in oral benign and malignant lesions. Int J Oral Maxillofac Surg, 1998. 27(3): 209-12.
34. Hou, L., D. Shi, S.-M. Tu, H.-Z. Zhang, M.-C. Hung, D. Ling. Oral cancer progression and c-erbB-2/neu proto-oncogene expression. Cancer Lett, 1992. 65(3): 215-20.
35. Hellyer, N.J., M.-S. Kim, J.G. Koland. Heregulin-dependent activation of phosphoinositide 3-kinase and Akt via the ErbB2/ErbB3 co-receptor. J Biol Chem, 2001. 276(45): 42153-61.
36. O-Charoenrat, P., P. Rhys-Evans, S. Eccles. Characterization of ten newly-derived human head and neck squamous carcinoma cell lines with special reference to c-erbB proto-oncogene expression. Anticancer Res, 2001. 21(3B): 1953-63.
37. O-Charoenrat, P., P.H. Rhys-Evans, D.J. Archer, S.A. Eccles. C-erbB receptors in squamous cell carcinomas of the head and neck: clinical significance and correlation with matrix metalloproteinases and vascular endothelial growth factors. Oral Oncol, 2002. 38(1): 73-80.
38. Gebhardt, F., H. Bürger, B. Brandt. Modulation of EGFR gene transcription by a polymorphic repetitive sequence—a link between genetics and epigenetics. Int J Biol Markers, 2000. 15(1): 105-10.
39. Schwentner, I., M. Witsch-Baumgartner, G.M. Sprinzl, J. Krugmann, A. Tzankov, S. Jank, et al. Identification of the rare EGFR mutation p.G796S as somatic and germline mutation in white patients with squamous cell carcinoma of the head and neck. Head & Neck, 2008. 30(8): 1040-4.
40. Wikstrand, C.J., C. Reist, G. Archer, M. Zalutsky, D. Bigner D. The class III variant of the epidermal growth factor receptor (EGFRvIII): characterization and utilization as an immunotherapeutic target. J Neurovirol, 1998. 4(2): 148-58.
41. Begg, A.C. Predicting recurrence after radiotherapy in head and neck cancer. Semin Radiat Oncol, 2012. 22(2): 108-18.
42. Haddad, R.I., D.M. Shin. Recent advances in head and neck cancer. N Engl J Med, 2008. 359(11): 1143-54.
43. Martin, L., M. Zoubir, C.T. Le. Recurrence of upper aerodigestive tract tumors. Bull Cancer, 2014. 101(5): 511-20.
44. Rivelli, T.G., M. Mak, R.E. Martins, e Silva VTdC, G. de Castro Jr. Cisplatin based chemoradiation late toxicities in head and neck squamous cell carcinoma patients. Discov Med, 2015. 20(108): 57-66.

45. Thariat, J., L. Milas, K.K. Ang. Integrating radiotherapy with epidermal growth factor receptor antagonists and other molecular therapeutics for the treatment of head and neck cancer. Int J Radiat Oncol Biol Phys, 2007. 69(4): 974-84.
46. Hein, A.L., M.M. Ouellette, Y. Yan. Radiation-induced signaling pathways that promote cancer cell survival (review). Int J Oncol, 2014. 45(5): 1813-9.
47. Park, S.Y., Y.M. Kim, H. Pyo. Gefitinib radiosensitizes non-small cell lung cancer cells through inhibition of ataxia telangiectasia mutated. Mol Cancer, 2010. 9(1): 222.
48. Leach, J.K., G. Van Tuyle, P.-S. Lin, R. Schmidt-Ullrich, R.B. Mikkelsen. Ionizing radiation-induced, mitochondria-dependent generation of reactive oxygen/nitrogen. Cancer Res, 2001. 61(10): 3894-901.*
49. Baumann, M., M. Krause, E. Dikomey, K. Dittmann, W. Dörr, U. Kasten-Pisula, et al. EGFR-targeted anti-cancer drugs in radiotherapy: preclinical evaluation of mechanisms. Radiother Oncol, 2007. 83(3): 238-48.
50. Huang, S.M., P.M. Harari. Modulation of radiation response after epidermal growth factor receptor blockade in squamous cell carcinomas: inhibition of damage repair, cell cycle kinetics, and tumor angiogenesis. Clin Cancer Res, 2000. 6(6): 2166-74.
51. Macha, M.A., S. Rachagani, A.K. Qazi, R. Jahan, S. Gupta, A. Patel, et al. Afatinib radiosensitizes head and neck squamous cell carcinoma cells by targeting cancer stem cells. Oncotarget, 2017. 8(13): 20961-73.
52. Wang, X.K., J.-h. He, J.-h. Xu, S. Ye, F. Wang, H. Zhang, et al. Afatinib enhances the efficacy of conventional chemotherapeutic agents by eradicating cancer stem-like cells. Cancer Res, 2014. 74(16): 4431-45.
53. Ojha, R., S. Bhattacharyya, S.K. Singh. Autophagy in cancer stem cells: a potential link between chemoresistance, recurrence, and metastasis. Biores Open Access, 2015. 4(1): 97-108.
54. Chen, Y.C., C.-J. Chang, H.-S. Hsu, Y.-W. Chen, L.-K. Tai, L.-M. Tseng, et al. Inhibition of tumorigenicity and enhancement of radiochemosensitivity in head and neck squamous cell cancer-derived ALDH1-positive cells by knockdown of Bmi-1. Oral Oncology, 2010. 46(3): 158-65.
55. Nor, C., Z. Zhang, K.A. Warner, L. Bernardi, F. Visioli, J.I. Helman, et al. Cisplatin induces Bmi-1 and enhances the stem cell fraction in head and neck cancer. Neoplasia (New York, NY), 2014. 16(2): 137-46.
56. Skvortsova, I., P. Debbage, V. Kumar, S. Skvortsov. Radiation resistance: cancer stem cells (CSCs) and their enigmatic pro-survival signaling. Semin Cancer Biol, 2015. 35: 39-44.
57. Ailles, L.E., I.L. Weissman. Cancer stem cells in solid tumors. Curr Opin Biotechnol, 2007. 18(5): 460-6.
58. Diehn, M., R.W. Cho, M.F. Clarke. Therapeutic implications of the cancer stem cell hypothesis. Semin Radiat Oncol, 2009. 19(2): 78-86.
59. Szafarowski, T., M.J. Szczepanski. Cancer stem cells in head and neck squamous cell carcinoma. Otolaryngol Pol, 2014. 68(3): 105-11.
60. Chen, Y.C., Y.-W. Chen, H.-S. Hsu, L.-M. Tseng, P.-I. Huang, K.-H. Lu, et al. Aldehyde dehydrogenase 1 is a putative marker for cancer stem cells in head and neck squamous cancer. Biochem Biophys Res Commun, 2009. 385(3), 307-13.
61. Okamoto, A., K. Chikamatsu, K. Sakakura, K. Hatsushika, G. Takahashi, K. Masuyama. Expansion and characterization of cancer stem-like cells in squamous cell carcinoma of the head and neck. Oral Oncol, 2009. 45(7): 633-9.

62. de Jong, M.C., J. Pramana, J.E. van der Wal, M. Lacko, C.J. Peutz-Kootstra, J.M. de Jong, et al. CD44 expression predicts local recurrence after radiotherapy in larynx cancer. Clin Cancer Res, 2010. 16(21): 5329-38.
63. Zhang, Q., S. Shi, Y. Yen, J. Brown, J.Q. Ta, A.D. Le .A subpopulation of CD133(+) cancer stem-like cells characterized in human oral squamous cell carcinoma confer resistance to chemotherapy. Cancer Lett, 2010. 289(2): 151-60.
64. Ma, L., G. Zhang, X.B. Miao, X.B. Deng, Y. Wu, Y. Liu, et al. Cancer stem-like cell properties are regulated by EGFR/AKT/beta-catenin signaling and preferentially inhibited by gefitinib in nasopharyngeal carcinoma. FEBS Journal, 2013. 280(9): 2027-41.
65. Abhold, E.L., A. Kiang, E. Rahimy, S.Z. Kuo, J. Wang-Rodriguez, J.P. Lopez, et al. EGFR kinase promotes acquisition of stem cell-like properties: a potential therapeutic target in head and neck squamous cell carcinoma stem cells. PloS One, 2012. 7(2): e32459.
66. Cassell, A., J.R. Grandis. Investigational EGFR-targeted therapy in head and neck squamous cell carcinoma. Expert Opin Investig Drugs, 2010. 19(6): 709-22.
67. Burtness, B., M.A. Goldwasser, W. Flood, B. Mattar, A.A. Forastiere. Phase III randomized trial of cisplatin plus placebo compared with cisplatin plus cetuximab in metastatic/recurrent head and neck cancer: an Eastern Cooperative Oncology Group study. J Clin Oncol, 2005. 23(34): 8646-54.
68. Schaaij-Visser, T.B., R.H. Brakenhoff, C.R. Leemans, A.J. Heck, M. Slijper M.Protein biomarker discovery for head and neck cancer. J Proteomics, 2010. 73(10): 1790-803.
69. Chang, S.S., J. Califano. Current status of biomarkers in head and neck cancer. J Surg Oncol, 2008. 97(8): 640-3.
70. Ferreira, M.B., J.A. De Souza, E.E. Cohen. Role of molecular markers in the management of head and neck cancers. Curr Opin Oncol, 2011. 23(3): 259-64.
71. Zhang, Q., S.M. Thomas, V.W.Y. Lui, S. Xi, J.M. Siegfried, H. Fan, et al. Phosphorylation of TNF-alpha converting enzyme by gastrin-releasing peptide induces amphiregulin release and EGF receptor activation. Proc Natl Acad Sci USA, 2006. 103(18): 6901-6.
72. Wheeler, D.L., S. Huang, T.J. Kruser, M.M. Nechrebecki, E.A. Armstrong, S. Benavente, et al. Mechanisms of acquired resistance to cetuximab: role of HER (ErbB) family members. Oncogene, 2008. 27(28): 3944-56.
73. Singh, A., J. Settleman, EMT, cancer stem cells and drug resistance: an emerging axis of evil in the war on cancer. Oncogene, 2010. 29(34): 4741-51.
74. Holz, C., F. Niehr, M. Boyko, T. Hristozova, L. Distel, V. Budach, et al. Epithelial-mesenchymal-transition induced by EGFR activation interferes with cell migration and response to irradiation and cetuximab in head and neck cancer cells. Radiother Oncol, 2011. 101(1): 158-64.
75. Rabinowits, G., R.I. Haddad. Overcoming resistance to EGFR inhibitor in head and neck cancer: a review of the literature. Oral Oncol, 2012. 48(11): 1085-9.
76. Loeffler-Ragg, J., M. Witsch-Baumgartner, A. Tzankov, W. Hilbe, I. Schwentner, G.M. Sprinzl, et al. Low incidence of mutations in EGFR kinase domain in Caucasian patients with head and neck squamous cell carcinoma. Eur J Cancer, 2006. 42(1): 109-11.
77. Yoshida, T., I. Okamoto, T. Iwasa, M. Fukuoka, K. Nakagawa. The anti-EGFR monoclonal antibody blocks cisplatin-induced activation of EGFR signaling mediated by HB-EGF. FEBS Lett, 2008. 582(30): 4125-30.

78. Friedman, L.M., A. Rinon, B. Schechter, L. Lyass, S. Lavi, S.S. Bacus, et al. Synergistic down-regulation of receptor tyrosine kinases by combinations of mAbs: implications for cancer immunotherapy. Proc Natl AcadSci USA, 2005. 102(6): 1915-20.
79. Pedersen, M.W., H.J. Jacobsen, K. Koefoed, A. Hey, C. Pyke, J.S. Haurum, et al. Sym004: a novel synergistic anti-epidermal growth factor receptor antibody mixture with superior anticancer efficacy. Cancer Res, 2010. 70(2): 588-97.
80. Price, K.A., E.E. Cohen. Mechanisms of and therapeutic approaches for overcoming resistance to epidermal growth factor receptor (EGFR)-targeted therapy in squamous cell carcinoma of the head and neck (SCCHN). Oral Oncol, 2015. 51(5): 399-408.
81. Vermorken, J.B., R. Mesia, F. Rivera, E. Remenar, A. Kawecki, S. Rottey, et al. Platinum-based chemotherapy plus cetuximab in head and neck cancer. N Engl J Med, 2008. 359(11): 1116-27.
82. Robert, F., M.P. Ezekiel, S.A. Spencer, R.F. Meredith, J.A. Bonner, M. Khazaeli, et al. Phase I study of anti-epidermal growth factor receptor antibody cetuximab in combination with radiation therapy in patients with advanced head and neck cancer. J Clin Oncol, 2001. 19(13): 3234-43.
83. Shin, D.M., N.J. Donato, R. Perez-Soler, H.J.C. Shin, J.Y. Wu, P. Zhang, et al. Epidermal growth factor receptor-targeted therapy with C225 and cisplatin in patients with head and neck cancer. Clin Cancer Res, 2001. 7(5): 1204-13.
84. Bonner, J.A., P.M. Harari, J. Giralt, N. Azarnia, D.M. Shin, R.B. Cohen, et al. Radiotherapy plus cetuximab for squamous-cell carcinoma of the head and neck. N Engl J Med, 2006. 354(6): 567-78.
85. Bonner, J.A., P.M. Harari, J. Giralt, R.B. Cohen, C.U. Jones, R.K. Sur, et al. Radiotherapy plus cetuximab for locoregionally advanced head and neck cancer: 5-year survival data from a phase 3 randomised trial, and relation between cetuximab-induced rash and survival. Lancet Oncol, 2010. 11(1): 21-8.
86. Vermorken, J.B., J. Trigo, R. Hitt, P. Koralewski, E. Diaz-Rubio, F. Rolland, et al. Open-label, uncontrolled, multicenter phase II study to evaluate the efficacy and toxicity of cetuximab as a single agent in patients with recurrent and/or metastatic squamous cell carcinoma of the head and neck who failed to respond to platinum-based therapy. J Clin Oncol, 2007. 25(16): 2171-7.
87. Argiris, A., M. Ghebremichael, J. Gilbert, J.-W. Lee, K. Sachidanandam, J.M. Kolesar, et al. Phase III randomized, placebo-controlled trial of docetaxel with or without gefitinib in recurrent or metastatic head and neck cancer: an eastern cooperative oncology group trial. J Clin Oncol, 2013. 31(11): 1405-14.
88. Rodriguez, C.P., D.J. Adelstein, L.A. Rybicki, J.P. Saxton, R.R. Lorenz, B.G. Wood, et al. Single-arm phase II study of multiagent concurrent chemoradiotherapy and gefitinib in locoregionally advanced squamous cell carcinoma of the head and neck. Head & Neck, 2012. 34(11): 1517-23.
89. Soulieres, D., N.N. Senzer, E.E. Vokes, M. Hidalgo, S.S. Agarwala, L.L. Siu. Multicenter phase II study of erlotinib, an oral epidermal growth factor receptor tyrosine kinase inhibitor, in patients with recurrent or metastatic squamous cell cancer of the head and neck. J Clin Oncol, 2004. 22(1): 77-85.
90. Martins, R.G., U. Parvathaneni, J.E. Bauman, A.K. Sharma, L.E. Raez, M.A. Papagikos, et al. Cisplatin and radiotherapy with or without erlotinib in locally advanced squamous cell carcinoma of the head and neck: a randomized phase II trial. J Clin Oncol, 2013. 31(11): 1415-21.

91. Harbeck, N., F. Solca, T.C. Gauler. Preclinical and clinical development of afatinib: a focus on breast cancer and squamous cell carcinoma of the head and neck. Future Oncol, 2014. 10(1): 21-40.
92. Mihaly, Z., B. Gyorffy. HER2-positive breast cancer: available targeted agents and biomarkers for therapy response. Magy Onkol, 2013. 57(3): 147-56.
93. Walter, A.O., R.T.T. Sjin, H.J. Haringsma, K. Ohashi, J. Sun, K. Lee, et al. Discovery of a mutant-selective covalent inhibitor of EGFR that overcomes T790M-mediated resistance in NSCLC. Cancer Discov, 2013. 3(12): 1404-15.
94. Ferrarotto, R., K.A. Gold. Afatinib in the treatment of head and neck squamous cell carcinoma. Expert Opin Investig Drugs, 2014. 23(1): 135-43.
95. Ioannou, N., A.M. Seddon, A. Dalgleish, D. Mackintosh, H. Modjtahedi. Treatment with a combination of the ErbB (HER) family blocker afatinib and the IGF-IR inhibitor, NVP-AEW541 induces synergistic growth inhibition of human pancreatic cancer cells. BMC Cancer, 2013. 13(1): 41.
96. Roskoski, R., Jr. The ErbB/HER family of protein-tyrosine kinases and cancer. Pharmacol Res, 2014. 79: 34-74.
97. Seiwert, T.Y., J. Fayette, D. Cupissol, J. Del Campo, P. Clement, R. Hitt, et al. A randomized, phase II study of afatinib versus cetuximab in metastatic or recurrent squamous cell carcinoma of the head and neck. Ann Oncol, 2014. 25(9): 1813-20.
98. Machiels, J.-P.H., L.F. Licitra, R.I. Haddad, M. Tahara, E.E. Cohen. Rationale and design of LUX-Head & Neck 1: a randomised, Phase III trial of afatinib versus methotrexate in patients with recurrent and/or metastatic head and neck squamous cell carcinoma who progressed after platinum-based therapy. BMC Cancer, 2014. 14(1): 473.
99. Machiels, J.-P.H., R.I. Haddad, J. Fayette, L.F. Licitra, M. Tahara, J.B. Vermorken, et al. Afatinib versus methotrexate as second-line treatment in patients with recurrent or metastatic squamous-cell carcinoma of the head and neck progressing on or after platinum-based therapy (LUX-Head & Neck 1): an open-label, randomised phase 3 trial. Lancet Oncol, 2015. 16(5): 583-94.
100. Abdul Razak, A., D. Soulieres, S. Laurie, S. Hotte, S. Singh, E. Winquist, et al. A phase II trial of dacomitinib, an oral pan-human EGF receptor (HER) inhibitor, as first-line treatment in recurrent and/or metastatic squamous-cell carcinoma of the head and neck. Ann Oncol, 2013. 24(3): 761-9.
101. de Souza, J.A., D.W. Davis, Y. Zhang, A. Khattri, T.Y. Seiwert, S. Aktolga, et al. A phase II study of lapatinib in recurrent/metastatic squamous cell carcinoma of the head and neck. Clin Cancer Res, 2012. 18(8): 2336-43.
102. Del Campo, J.M., R. Hitt, P. Sebastian, C. Carracedo, D. Lokanatha, J. Bourhis, et al. Effects of lapatinib monotherapy: results of a randomised phase II study in therapy-naive patients with locally advanced squamous cell carcinoma of the head and neck. Br J Cancer, 2011. 105(5): 618-27.
103. Harrington, K., S. Temam, H. Mehanna, A. D'Cruz, M. Jain, I. D'Onofrio, et al. Postoperative adjuvant lapatinib and concurrent chemoradiotherapy followed by maintenance lapatinib monotherapy in high-risk patients with resected squamous cell carcinoma of the head and neck: a phase III, randomized, double-blind, placebo-controlled study. J Clin Oncol, 2015. 33(35): 4202-9.
104. He, Y., Q. Zeng, S.D. Drenning, M.F. Melhem, D.J. Tweardy, L. Huang, et al. Inhibition of human squamous cell carcinoma growth in vivo by epidermal growth factor receptor antisense RNA transcribed from the U6 promoter. J Natl Cancer Inst, 1998. 90(14): 1080-7.

105. Endo, S., Q. Zeng, N. Burke, Y. He, M. Melhem, S. Watkins, et al. TGF-alpha antisense gene therapy inhibits head and neck squamous cell carcinoma growth in vivo. Gene Ther, 2000. 7(22): 1906-14.
106. Thomas, S.M., M.J. Ogagan, M.L. Freilino, S. Strychor, D.R. Walsh, W.E. Gooding, et al. Antitumor mechanisms of systemically administered epidermal growth factor receptor antisense oligonucleotides in combination with docetaxel in squamous cell carcinoma of the head and neck. Mol Pharmacol, 2008. 73(3): 627-38.
107. Lai, S.Y., P. Koppikar, S.M. Thomas, E.E. Childs, A.M. Egloff, R.R. Seethala, et al. Intratumoral epidermal growth factor receptor antisense DNA therapy in head and neck cancer: first human application and potential antitumor mechanisms. J Clin Oncol, 2009. 27(8): 1235-42.
108. Nozawa, H., T. Tadakuma, T. Ono, M. Sato, S. Hiroi, K. Masumoto, et al. Small interfering RNA targeting epidermal growth factor receptor enhances chemosensitivity to cisplatin, 5-fluorouracil and docetaxel in head and neck squamous cell carcinoma. Cancer Sci, 2006. 97(10): 1115-24.
109. Ang, K.K., Q. Zhang, D.I. Rosenthal, P.F. Nguyen-Tan, E.J. Sherman, R.S. Weber, et al. Randomized phase III trial of concurrent accelerated radiation plus cisplatin with or without cetuximab for stage III to IV head and neck carcinoma: RTOG 0522. J Clin Oncol, 2014. 32(27): 2940-50.
110. Truong, M.T., Q. Zhang, D.I. Rosenthal, M. List, R. Axelrod, E. Sherman, et al. Quality of life and performance status from a substudy conducted within a prospective phase 3 randomized trial of concurrent accelerated radiation plus cisplatin with or without cetuximab for locally advanced head and neck carcinoma: NRG oncology radiation therapy oncology group 0522. Int J Radiat Oncol Biol Phys, 2017. 97(4): 687-99.
111. Guo, Y., M. Shi, A. Yang, J. Feng, X. Zhu, Y.J. Choi, et al. Platinum-based chemotherapy plus cetuximab first-line for Asian patients with recurrent and/or metastatic squamous cell carcinoma of the head and neck: results of an open-label, single-arm, multicenter trial. Head & Neck, 2015. 37(8): 1081-7.
112. Bonner, J., J. Giralt, P. Harari, S. Spencer, J. Schulten, A. Hossain, et al. Cetuximab and radiotherapy in laryngeal preservation for cancers of the larynx and hypopharynx: a secondary analysis of a randomized clinical trial. JAMA Otolaryngol Head Neck Surg, 2016. 142(9): 842-9.
113. Ghi, M.G., A. Paccagnella, D. Ferrari, P. Foa, D. Alterio, C. Codecà, et al. Induction TPF followed by concomitant treatment versus concomitant treatment alone in locally advanced head and neck cancer. A phase II-III trial. Ann Oncol, 2017. 28(9): 2206-12.

CHAPTER 9

Molecular Drivers in Lung Adenocarcinoma: Therapeutic Implications

Imayavaramban Lakshmanan[1] and Apar Kishor Ganti[2*]

[1] Department of Biochemistry and Molecular Biology, University of Nebraska Medical Center, Omaha, NE, USA
[2] VA Nebraska Western Iowa Health Care System, University of Nebraska Medical Center, Omaha, NE 68198-6840

Introduction

Lung cancer is the leading cause of cancer related deaths not just in the US, but worldwide [1]. Adenocarcinoma (38.5%) is the most common subtype of non-small cell lung cancer [2]. Our knowledge of the pathogenesis of lung cancer, especially lung adenocarcinoma has undergone a rapid change over the last decade. This has led to the emergence of targeted agents that have revolutionized the management of lung adenocarcinoma. Targeted therapies against pathways, such as the epidermal growth factor receptor (EGFR), anaplastic lymphoma kinase (ALK) and C-ros oncogene 1, receptor tyrosine kinase (ROS-1) rearrangements have increased progression free survival (PFS) in patients with the respective mutations, with much less toxicity as compared to cytotoxic chemotherapy [3, 4]. Given this, the National Comprehensive Cancer Network recommends routine testing for these mutations for all patients with primary lung adenocarcinomas and any lung cancer with an adenocarcinoma component [5].

The most common abnormalities seen in lung adenocarcinoma include K-RAS mutations, but unfortunately, attempts to identify therapeutic agents targeting this pathway have not been successful thus far. Of the targetable abnormalities, activating mutations in the EGFR gene are the most common, followed by ALK translocations (Fig. 1). A study from the Lung Cancer Mutation Consortium evaluated 1007 metastatic lung adenocarcinoma patients and identified a number of less common abnormalities [6]. ERBB2 (formerly HER2) mutation was found in 3%, BRAF

---

*Corresponding author: aganti@unmc.edu

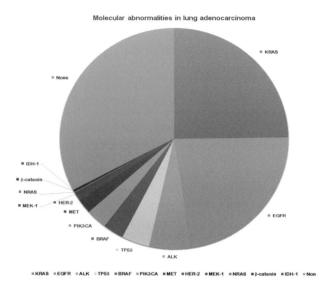

Fig. 1: Frequency of various molecular drivers in lung adenocarcinoma.

in 2%, MET amplification in <1% and NRAS in <1% cases. Interestingly, patients who received molecularly targeted therapy against the less common genetic abnormalities had a slightly longer median overall survival [2.2 years (95% CI, 1.3 to 2.7 years)] than those without actionable mutations [1.9 years (95% CI, 1.6 to 2.1 years); P<0.001] [7]. Thus, targeting potential targets, for which drugs are currently available, could improve outcomes.

KRAS

KRAS is a member of the Ras family that promotes cell growth and division. This pathway is activated by the binding of guanosine triphosphate (GTP) and phosphorylates downstream signaling proteins. The process continues until the GTP is converted to guanosine diphosphate (GDP) through an intrinsic GTPase activity that is present in the Ras family enzymes. While the Ras family includes both HRAS and NRAS, in addition to KRAS, KRAS alone has been implicated in the lung carcinogenesis.

In lung cancer, mutations in KRAS (exons 12 [80%], 13, and 61) arise, that decrease its intrinsic GTPase activity. This causes a marked upregulation of kinase activity and downstream growth and mitotic signaling. These mutations have been demonstrated in approximately 15-20% of all NSCLC and appear to be associated with a smoking history and adenocarcinoma histology (30-50%) [8].

Clinical Evidence

KRAS mutations have been shown to be associated with worse outcomes and shown to be associated with reduced benefits from adjuvant cytotoxic chemotherapy in resected non-small cell lung cancer [9-11]. Unlike colorectal cancer, in NSCLC, the presence of KRAS mutations does not necessarily confer resistance to cetuximab, an anti-EGFR monoclonal antibody [12].

Unfortunately, there are currently no therapies that can target the KRAS pathway and management of patients with KRAS-mutated NSCLC is a major unmet need today. A major reason for this appears to be the high affinity binding to the GTP substrate that has hindered the development of direct KRAS inhibitors.

Epidermal Growth Factor Receptor (EGFR)

The EGF pathway plays a critical role in the pathogenesis and progression of NSCLC. In the normal epithelium, EGFR is expressed predominantly in the basal layer, but as progressive dyplasia develops, the bronchial epithelium develops additional cell layers with nuclear atypia, and widespread EGFR expression [13]. EGFR structure is composed of an extracellular ligand-binding domain comprising 621 amino acids, a transmembrane anchoring region that is 23 amino acids long, and an intracellular tyrosine kinase that is made up of 542 amino acids. Ligand binding results in the dimerization of the receptor subunits, which in turns leads to the autophosphorylation of intracellular tyrosine residues. This process creates docking sites for intracellular effector proteins, thereby generating multiple signal transduction cascades. These include the Ras-Raf-MEK (mitogen-activated and extracellular-signal regulated kinase kinase), PI3K (phosphatidylinositol-3-kinase)-Akt, and STAT (signal transducer and activator of transcription) pathways. Phosphorylation of the tyrosine residues in the EGFR results in the activation of multiple down-stream pathways that promote cell proliferation, inhibition of apoptosis and angiogenesis within the tumor [14] (Fig. 2). In addition, the EGFR pathway can lead to modification of gene transcription and modulation of DNA repair mechanisms independent of the aforementioned pathways. This is a result of internalization of the EGFR complex and subsequent translocation into the nucleus [15, 16].

The presence of somatic mutations in the tyrosine kinase domain of the EGFR gene appears to predict for sensitivity to EGFR tyrosine kinase inhibitors [17, 18]. EGFR mutations are not only predicts TKI response but also act as prognostic indicator [19]. The prevalence of these mutations seems to be highest in Asians and patients from the Indian subcontinent (45-50%), while Hispanic and African-Americans had the lowest prevalence rates (15-20%) [20]. Another factor that affects the presence of these activating mutations is the smoking history.

A number of methods of testing for EGFR mutations are available [21]. Allele-specific mutation uses polymerase chain reaction targeting specific mutations. The main advantage of this technique is efficiency, but the sensitivity is low. The amplification refractory mutation system (ARMS) requires specially engineered primers for the detection of a single mutation. The EGFR ScorpionsIM Kit (DxS, Manchester, United Kingdom)

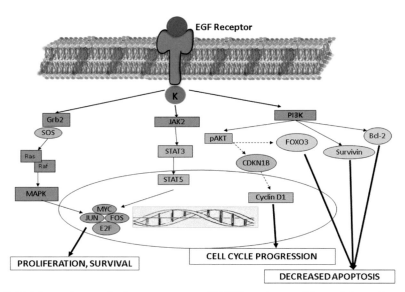

Fig. 2: Epidermal growth factor receptor (EGFR) pathway. The epidermal growth factor activates its receptor tyrosine kinase which in turn results in the activation of multiple pathways including the Ras/Raf/mitogen-activated kinase-like protein (MAPK), phosphatidylinositol 3-kinase (PI3K) and the JAK/STAT pathways. These pathways eventually result in increased cell proliferation, and evasion of apoptosis leading to increased survival and increased cell cycle progression.

system is a commercialized kit that can detect 29 common EGFR mutations. The application is relatively easy and the sensitivity is higher than that of standard sequencing [22]. Nowadays, EGFR testing is routinely performed as part of a next generation sequencing platform, in conjunction with tests for other actionable mutations.

Clinical Evidence

The IPASS (Iressa Pan-Asia Study) randomized study analyzed the EGFR mutation status between treatment-naive non smokers or former light smokers with advanced lung adenocarcinoma to either gefitinib or chemotherapy (carboplatin-paclitaxel) [19]. In patients who harbored an activating EGFR mutation, gefitinib was associated with higher response rates (71.2% vs. 1.1%) and a significantly prolonged PFS (HR, 0.48; 95% CI, 0.36-0.64; P<0.001). In contrast, chemotherapy was superior in patients without the mutation (HR for gefitinib, 2.85; 95% CI, 2.05-3.98; P<0.001). Similar findings have been observed with different EGFR TKIs in patients whose tumors harbor activating EGFR mutations (Table 1). A weighted pooled analysis of these studies of EGFR-mutated NSCLC patients showed that both EGFR TKIs were associated with a longer median PFS, but there does not seem to be a survival benefit [23]. A meta-analysis of data on 2620 patient from 13 phase III trials comparing an EGFR-TKI with platinum-based chemotherapy demonstrated a significantly longer progression-free survival with the TKI (HR 0.43, 95% CI 0.38-0.49), but there was no difference in overall survival (HR 1.01, 95% CI 0.87-1.18) [3].

Table 1: Trials of EGFR TKIs in EGFR-mutation-positive patients

Study	Treatment	n	Median PFS (months)	Median OS (months)
IPASS [19, 92]	Gefitinib#	132	9.5	22
	Carboplatin-paclitaxel	129	6.3	22
NEJ002 [24]	Gefitinib	114	10.8	30.5
	Carboplatin-paclitaxel	110	5.4*	23.6
WJTOG3405 [25]	Gefitinib	86	9.2	30.9
	Cispatin-docetaxel	86	6.3*	Not reached
OPTIMAL [93]	Erlotinib	82	13.1	22.8
	Carboplatin-gemcitabine	72	4.6*	27.2
EURTAC [26]	Erlotinib	86	23.9	19.3
	Chemotherapy	87	6*	19.5
LUX-Lung 3 [94]	Afatinib	230	11.1	Not reached
	Cisplatin-pemetrexed	115	6.9	
LUX-Lung 6 [95]	Afatinib	242	11	22.1
	Cisplatin-gemcitabine	122	5.6	22.2
[96]	Gefitinib	128	10.4	20.1
	Erlotinib	128	13	22.9
FLAURA [30]	Osimertinib	279	18.9	Not reached
	Standard of Care TKI	277	10.2	

OS – Overall survival; PFS – Progression-free survival
* Statistically significant
# Data for mutation positive patients only

Resistance to EGFR TKI

Despite the impressive response rates, the majority of patients stop responding to the initial EGFR TKI within approximately 10-12 months [19, 24-26]. The most common mechanisms of development of resistance to EGFR TKIs include development of resistance mutations in the EGFR gene; T790M is the most common resistance mutation accounting for almost half of the patients who develop resistance [27]. Other mechanisms include MET amplification (5%), EGFR overexpression (5%), small cell transformation (8%) and PIK3CA mutations (5%).

Osimertinib is an EGFR TKI that is active against T790M, the most common resistance mutation seen in patients who have progressed following first line EGFR-TKI therapy. In a phase I/II study of 253 patients who had progressed on an EGFR-TKI, osimertinib demonstrated a response rate of 51% [28]. Among 127 patients with a confirmed EGFR T790M, the response rate was 61%. The median progression-free survival was 9.6 months in EGFR T790M-positive patients. The benefit of osimertinib in EGFRT790M positive lung cancer was confirmed in a phase III trial against chemotherapy following first-line EGFR TKI therapy [29]. In this trial of 419 patients, osimertinib had a PFS of 10.1 months vs. 4.4 months with platinum-pemetrexed (HR 0.30; 95% CI 0.23 to 0.41; P<0.001). The corresponding response rates were 71% and 31%, respectively. In a phase III trial in treatment naïve patients with activating EGFR mutations, osimertinib had a superior PFS compared to standard of care TKIs (18.9 vs. 10.2 months; HR 0.46 (0.37, 0.57); p<0.0001) [30].

Anaplastic Lymphoma Kinase (ALK)

The anaplastic lymphoma kinase (ALK) gene encodes a tyrosine kinase receptor. The carcinogenic effect of ALK is often seen when it forms a fusion gene with any of several other genes such as, Nucleophosmin (NPM) or Echinoderm microtubule-associated protein-like 4 (EML-4). In addition, increased gene copy number or mutations within the gene can also lead to carcinogenesis. As the name suggests, the pathogenic role of ALK was originally identified in a subset of anaplastic large cell lymphomas with a t(2; 5) (p23; q35) translocation (NPM-ALK).

Since then, ALK translocation has been implicated in the pathogenesis of a small subgroup of lung adenocarcinoma as well. In lung cancer, inversion of chromosome 2p results in fusion of the protein encoded by EML-4 gene with the intracellular signaling portion of the ALK receptor tyrosine kinase. EML4-ALK fusions result in constitutive tyrosine kinase activity resulting in the activation of downstream pathways (including AKT, ERK, and STAT3) [31].

The incidence of ALK translocation is rarely seen in 3-5 percent of cases. The patients with ALK translocated lung cancer seem to have some characteristic features including younger age, no or light smoking history, solid or acinar type adenocarcinoma with signet ring differentiation and prominent intracellular mucin. In addition, these tumors stain positively for thyroid transcription factor-1 (TTF-1) [32, 33].

ALK rearrangements can be identified by immunohistochemistry (IHC), fluorescence *in situ* hybridization (FISH), or polymerase chain reaction (PCR). Of these, FISH appears to be the most clinically applicable, but IHC can be used as a screening technique since it is readily available. In addition, the combination of IHC and FISH may increase the detection of these cases [33]. The commonly used *EML4-ALK* FISH technique employs differently labeled break-apart (split signal) probes on the 5′ and 3′ ends of the ALK gene.

Normal ALK generates a fused (yellow) signal, while ALK rearrangements appear as separate signals. A positive result is confirmed when a split signal (red and green) is seen in >15% of cells examined [34].

Clinical Evidence

Since the discovery of the role of ALK in lung cancer in 2007 [35], there have been a number of agents that have shown to inhibit ALK tyrosine kinase. Similar to the results seen with EGFR inhibitors, ALK inhibitors have been shown to be superior to chemotherapy in terms of PFS, but not OS, for these patients. Currently, crizotinib, ceritinib, alectinib and brigatinib have been approved for clinical use in this setting. The randomized clinical trials of some of these agents are described in Table 2. The recent results of the ALEX study establish alectinib as the preferred agent for the management of treatment naïve ALK positive lung cancer [36].

Resistance to ALK TKI

Similar to the experience with EGFR TKIs, most patients develop resistance to ALK TKIs, especially crizotinib, after a while. This phenomenon appears due to an acquired secondary mutation within the ALK tyrosine kinase domain [37]. The most common mutation seen after crizotinib is the L1196M mutation, while other common mutations include G1269A, C1156Y and I1171T/N/S. Ceritinib use is associated with the development of G1202R, F1174C/L and V1180L mutations and deletion of G1202, while S1206Y, and E1210K are commonly seen after alectinib use. Lorlatinib, an ALK inhibitor currently under investigation, appears to have activity against many of these mutations, including the G1202R, which confers resistance to ceritinib, alectinib and brigatinib [38]. Hence, it is important to re-biopsy at disease progression and check for mutation status, in order to choose the most sensitive treatment option. There has been a single report of a patient who progressed on sequential crizotinib and lorlatinib, but a biopsy following lorlatinib showed that the tumor now had a mutation that paradoxically increased the sensitivity to crizotinib. The patient derived clinical benefit with a rechallenge of crizotinib [39].

Another mechanism of ALK resistance is the amplification of the ALK fusion gene, either alone or in combination with one of the aforementioned secondary resistance mutations. Other mechanisms of ALK TKI resistance include abnormalities in pathways such as EGFR, KIT, and insulin-like growth factor-1 receptor.

BRAF Mutations

BRAF is a serine/threonine kinase downstream from KRAS in the RAS–RAF–MEK–ERK–Mitogen-activated protein kinase (MAPK) pathway that is often altered in cancer [40]. RAS is a critical downstream effector of the

Table 2: Randomized trials of ALK inhibitors in ALK positive NSCLC

Study	Patient population	Treatment	n	Median PFS (months)	Median OS (months)
[4]	Prior platinum-based chemotherapy	Crizotinib	173	7.7	20.3
		Docetaxel/pemetrexed	174	3	22.8
PROFILE 1014 [97]	Treatment naïve	Crizotinib	172	10.9	No difference
		Platinum-pemetrexed	171	7	
ASCEND-4 [98]	Untreated stage IIIB/IV	Ceritinib	189	16.6	Not reached
		Platinum-pemetrexed	187	8.1	26.2
J-ALEX [99]	Chemotherapy naïve or one prior regimen	Alectinib	103	NR	Not reached
		Crizotinib	104	10.2	
ALEX [36]	Previously untreated	Alectinib	152	25.7	Not reached
		Crizotinib	151	10.4	

epidermal growth factor receptor (EGFR). Stimulation of this pathway leads to a transient formation of the active, GTP-bound RAS. Mutant RAS proteins are insensitive to proteins that hydrolyze GTP to GDP and hence constitutionally activate downstream pathways. Activated RAS-GTP binds to RAF kinases leading to subsequent activation of MEK1 and MEK2 protein kinases, which in turn activate ERK1 and ERK2 MAPKs. Activated ERKs affect regulation of normal cell proliferation, survival and differentiation. ERK activation also promotes an autocrine growth loop that upregulates cellular expression of EGFR ligands resulting in tumor growth [40].

Mutated BRAF proteins have increased kinase activity resulting in increased cell proliferation and survival BRAF mutations have been described in up to 5% of patients with lung adenocarcinoma [41]. Of the different mutations, V600E (50%) and G469A (39%) are the most common genotypes [42]. BRAF mutations, especially non-V600E genotypes, are more common in patients with a smoking history [41, 42]. BRAF V600E positive tumors seem to have a more aggressive phenotype and a worse prognosis as compared to non-V600E tumors [41].

Clinical Evidence

Dabrafenib, a reversible, ATP-competitive inhibitor that selectively inhibits BRAF V600E kinase, increases progression-free survival in BRAF V600E-mutated metastatic melanoma compared to conventional therapy [43]. The BRF113928 study (NCT01336634) was an international, multicenter, open-label, trial in patients with BRAF V600E mutation-positive metastatic NSCLC. Of the 93 patients who were treated with combination of dabrafenib and trametinib, 36 had received no prior systemic therapy for metastatic NSCLC and 57 had progressed following at least one platinum-based regimen. In the previously treated group, the response rate (ORR) for the combination was 63% (95% CI: 49%, 76%) with a median duration of response of 12.6 months (95% CI: 5.8, not estimable [NE]). In the treatment-naïve group, the ORR was 61% (95% CI: 44%, 77%) and median duration of response was not reached (https://www.fda.gov/drugs/informationondrugs/approveddrugs/ucm564331.htm). Based on these findings, the US FDA granted approval for the combination of dabrafenib and trametinib for the treatment of BRAF V600E positive NSCLC.

Vemurafenib is a selective oral inhibitor of the BRAF V600 kinase, approved for the treatment of BRAF mutated metastatic melanoma. In a prospective study, vemurafenib had a response rate of 42% (95% CI, 20 to 67) in the cohort of patients with NSCLC [44]. The median PFS in these patients was 7.3 months (95% CI, 3.5 to 10.8).

Different mechanisms of resistance to BRAF V600 targeted therapies have been described in melanoma, leading to disease progression after few months of use [45]. In a comparison of pre- and post-tumor mutational profiles in a case of lung adenocarcinoma that progressed after eight months on dabrafenib, Rudin et al. found three new mutations including KRAS and

Tp53 in the post-dabrafenib specimen. This raises the possibility of acquired KRAS mutation upstream from BRAF causing resistance [46]. Gautschi et al. described a case of metastatic lung adenocarcinoma with BRAF G469L mutation, which was refractory to vemurafenib. It is possible that secondary mutations or impaired binding of vemurafenib to non-V600E mutated BRAF proteins could lead to resistance [47].

MET Pathway

MET [also known as, hepatocyte growth factor receptor (HGFR)] is a transmembrane receptor protein tyrosine kinase. Hepatocyte growth factor (HGF), secreted by fibroblasts and smooth muscle cells, binds to the MET receptor and induces MET dimerization and autophosphorylation, leading to the activation of tyrosine kinase [48]. MET mediates activation of downstream signaling pathways, including phosphoinositide 3-kinase (PI3K)/AKT, Ras-Rac/Rho, mitogen-activated protein kinase, and phospholipase C, that stimulate pathways involved in cell growth, apoptosis, motility, and invasiveness [49]. MET pathway is dysregulated in multiple human cancers, including lung adenocarcinoma [48]. Several mechanisms of MET pathway overexpression have been identified, namely, receptor overexpression, constitutive kinase activation, gene amplification, paracrine/autocrine activation via HGF, MET mutation or epigenetic mechanisms (tumor secreted growth factors, hypoxia, and other oncogenes) [48, 50].

MET pathway overexpression is an independent negative prognostic factor in NSCLC [51, 52]. MET amplification was found in up to 20% of lung cancer specimens that had developed resistance to EGFR-TKI [27, 53]. Aberrant MET signaling results in activation of downstream signaling pathways, especially the PI3K-AKT pathway. This allows cancer cells to maintain growth in the presence of the EGFR TKI, thus conferring resistance to EGFR-TKIs. Several pre-clinical studies in animal models have suggested possible synergistic benefits of dual inhibition of MET and EGFR pathways [54, 55].

Clinical Evidence

Several MET pathway inhibitors like HGF inhibitors (ficlatuzumab), anti-MET monoclonal antibodies (onartuzumab), and MET tyrosine kinase inhibitors (crizotinib, tivantinib) are currently being studied in combination with EGFR-TKIs [50, 56] (Table 3). A study by Mok et al. involving 188 Asian patients with stage IIIB/IV lung adenocarcinoma, who received either ficlatuzumab and gefitinib or gefitinib alone, did not find any statistically significant improvement with the combination [56]. However, preliminary results in a subset of patients with both EGFR sensitizing mutations and low c-Met biomarker levels showed that the combination had a trend towards ORR and PFS improvement, and for prolonged OS in those with high stromal HGF (P = 0.03) [56]. In another phase II study, onartuzumab, a monovalent

Table 3: Clinical trials evaluating targeted therapies against MET pathways in NSCLC

Trials	Study population	Treatment	n	Response rate (%)	Median PFS (months)	Median OS (months)
[56]	Treatment Naïve Asian patients with Stage IIIB/IV lung adenocarcinoma	Ficlatuzumab + gefitinib	94	43	5.6	NA
		Gefitinib	94	40	4.7	NA
[57]	Stage IIIB/IV NSCLC treated with ≥1 systemic regimens – MET positive	Erlotinib	31	3.2	1.5	3.8
		Onartuzumab + erlotinib	35	8.6	2.9	12.6
	NSCLC with ≥1 systemic regimens – MET negative group	Placebo + erlotinib	31	6.5	2.7	15.3
		Onartuzumab + erlotinib	31	3.2	1.4	8.1
[100]	Previously treated MET-positive NSCLC	Placebo + erlotinib	499	9.6	2.6	9.1
		Onartuzumab + erlotinib		8.4	2.7	6.8
[58]	Previously treated patients, but EGFR TKI-naïve NSCLC	Erlotinib + tivantinib	84	10	3.8	8.5
		Erlotinib + placebo	83	7	2.3	6.9
[62]	Advanced c-Met-amplified NSCLC	Crizotinib	13	33	8.25	NA

NA – Not available

antibody against c-MET-, was associated with improved PFS and OS in the MET-positive population while the MET-negative patients had worse outcomes [57]. Sequist et al. compared erlotinib alone with the combination of erlotinib and tivantinib (ARQ 197), a non-patients in the study had MET gene copy number ≥ 4, equally distributed between the treatment arms. There were no statistically significant differences in PFS or OS, though the combination of erlotinib and tivantinib had a trend towards benefit in patients with increased MET gene copy number [58] (Table 3).

Interestingly, while the efficacy of crizotinib in ALK-positive NSCLC is well established [4], it was initially developed as a potent MET inhibitor [59]. Ou et al. described a case of ALK negative but MET amplified stage IV adenocarcinoma of lung where crizotinib was used after the progression of disease on chemotherapy with an excellent partial response and a progression-free interval of at least six months [60]. In another case, crizotinib was used after radiation in c-MET amplified squamous cell lung cancer with significant partial response for at least eight weeks [61]. In an ongoing phase 1 study (NCT00585195), crizotinib was reported to have some antitumor activity in patients with c-Met-amplified NSCLC, with the response rate as high as 50% in the subgroup with high MET/CEP7 ratio [62].

ROS1 Rearrangement

ROS1 is a human receptor tyrosine kinase, encoded by the ROS1 gene closely related to ALK gene [63]. ROS1 is believed to play a role in epithelial cell differentiation during the development of a variety of organs. However, a ligand for this receptor has not been identified and its exact role in human cells is not well established [64]. Oncogenic activation of ROS as a result of different chromosomal rearrangements has been identified in a variety of human tumors including NSCLC, where it has been reported in 1.2-1.7% of the cases [65, 66]. ROS1-rearranged NSCLC resembles ALK-rearranged NSCLC in terms of clinicopathologic characteristics: young never-smokers with adenocarcinoma [66].

At least 12 fusion genes have been identified in NSCLC, including CD74-ROS1 and SLC34A2-ROS1; the ROS1 kinase gene is retained in all of these fusion events, and the expressed fusion genes are believed to play a role in carcinogenesis [66, 67]. The mechanisms of oncogenic transformation resulting from these fusion genes are believed to be the upregulation of the phosphatase SHP-2, the PI3K/AKT/mTOR pathway, the JAK/STAT pathway, and the MAPK/ERK pathway [65]. In a study of 162 never-smokers with adenocarcinoma, patients with ALK or ROS fusion gene (n = 19) were found to have significantly worse disease-free survival than fusion-negative patients (hazard ratio – 2.11; 95% confidence interval, 1.19-4.30; p = 0.022), thus implying possible prognostic role [68].

A phase I study of the efficacy of crizotinib in ROS1-rearranged NSCLC showed marked antitumor activity, with an objective response rate of 72% (95% CI, 58 to 84), and median PFS of 19.2 months (95% CI, 14.4 to not reached).

Though several types of ROS1 re-arrangements have been described, this study did not find any correlation between the type of rearrangement and response to therapy (Table 4) [69]. A recent retrospective study in Europe in 31 patients with previously treated stage IV lung adenocarcinoma who were positive for ROS1 rearrangement found an overall RR of 80% and median PFS of 9.1 months in patients receiving crizotinib (Table 4) [70]. Until further results become available, the evidence to establish the efficacy of crizotinib in NSCLC patients with ROS1 rearrangement is limited to these two single-arm studies (Table 4).

Table 4. Clinical trials targeting ROS re-arrangement in NSCLC

Trials	Treatment regimens	Patients (n)	Response rate (%)	Median PFS (months)
[69]	Crizotinib	50	72	19.2
[70]	Crizotinib	32	80	9.1

RET Rearrangement

RET is another receptor tyrosine kinase, encoded by the RET gene. It is believed to be required for development of the kidneys and enteric system, as well as for the differentiation and survival of neurons [71, 72]. The ligand for RET receptors are glial-derived neurotrophic factor (GDNF) family of neurotrophins, which include neurturin (NTN), artemin (ART) and persephin (PSP) [72]. After ligand binding, the intracellular kinase domain is activated, followed by autophosphorylation of intracellular tyrosine residues. These phospho-tyrosine residues then serve as platform where downstream signaling proteins carrying SRC homology 2 (SH2) or phosphotyrosine-binding (PTB) domains bind and transmit signals into the cell, leading to the activation of RAS/ERK1/2 and PI3K/AKT pathways [73]. Rearrangements in RET gene result in formation of fusion proteins (with the tyrosine kinase domain of Ret receptor) which are capable of undergoing constitutive dimerization leading to subsequent ligand independent kinase activation, potentially resulting in neoplastic transformation [72]. Although data from *in vitro* and xenograft models support the potential of RET inhibitors in the treatment of RET fusion-positive NSCLC, information about mechanisms by which these oncogenic fusion proteins exert their transforming action is lacking [73].

RET rearrangements have been reported in approximately 1-2% of all NSCLC patients [74-76], and in up to 16% of NSCLC cases negative for other mutations [77]. RET rearrangements are largely mutually exclusive with genetic alterations in other oncogenic drivers, such as EGFR, KRAS, ALK, and ROS1 [76]. One particular driver mutation, KIF5B-RET, has been identified in majority of lung carcinomas with RET rearrangement [74].

A variety of targeted agents against RET pathways has been studied in other malignancies. These include vandetanib, cabozantinib, sunitinib, sorafenib, fostamatinib and ponatinib with good response, especially in RET mutated tumors [73]. However, studies with RET inhibitors in NSCLC are lacking. Vandetanib has been shown to have some *in vitro* activity against cancer cells with RET rearrangements [74]. Gautschi et al. described a case of lung adenocarcinoma, positive for RET-KIF5B, where vandetanib was used at disease progression following chemotherapy, surgery and radiation, with the development of disease remission after four weeks of vandetanib treatment [78]. A prospective phase II trial studying the role of cabozantinib, a multi-tyrosine kinase inhibitor and a potent inhibitor of RET, in NSCLC patients with RET rearrangements also showed partial responses in two out of three cases and prolonged stable disease after eight months in the third case, with no progression of the disease reported in any of the three [77]. There are ongoing phase II trials to evaluate the efficacy of lenvatinib, vandetanib, and sunitinib in NSCLC with RET rearrangements [79].

HER-2

HER-2 is a member of the Human epidermal growth factor receptor (HER) family of transmembrane receptor tyrosine kinases. Other members of this family include HER1 (epidermal growth factor receptor [EGFR]), HER3 (erbB3), and HER4 (erbB4). HER2 is constitutively active and has no ligand. Overexpression, gene amplification or mutation of HER-2 affects the downstream PI3K-Akt and MAPK pathways, leading to the disruption of the regulation of cell-cycle progression and apoptosis [80, 81]. Approximately 2-4% of NSCLC have HER2 mutations, which seem to be mutually exclusive from EGFR/KRAS/ALK-mutations. HER-2 gene amplification and HER-2 over expression are more prevalent (20% and 35% of NSCLC cases) [82]. HER-2 overexpression [by immunohistochemistry (IHC)] was found to be a marker for poor prognosis in NSCLC (HR 1.48; 95% CI: 1.22-1.80) in a meta-analysis of 6135 patients [83].

Besides a possible role in carcinogenesis, HER2 over expression has been postulated as a possible mechanism of resistance to treatment with EGFR tyrosine kinase inhibitors [84]. *In vitro* studies have found that these patients are more responsive to agents with combined HER-2 and EGFR inhibitors than EGFR inhibitors alone, suggesting some role of HER-2 inhibition in their overall efficacy [85]. The prognostic and predictive significance of HER-2 mutations, gene amplifications and overexpression requires further validation, with fairly good evidence supporting the response to HER-2 targeted agents in at least HER-2 gene mutated cases.

Clinical Evidence

Initial studies of trastuzumab (a monoclonal antibody against HER-2) combined with cytotoxic chemotherapy had mixed results in patients with HER-2 overexpressing tumors [86-88]. A phase II trial of gemcitabine-cisplatin +/- trastuzumab in HER2-positive (IHC) found inferior results with trastuzumab [86]. The response rate with trastuzumab was 36% as compared to 41% without it; similarly, the median PFS was 6.1 vs. 7 months. Another phase II study of carboplatin, paclitaxel, and trastuzumab in 53 HER-2/neu positive (IHC) advanced NSCLC patients found a response rate of 24.5% (95% CI, 13.8 to 38.3), median PFS of 3.3 months and median survival of 10.1 months [88]. However, there were suggestions of potential benefit in tumors with 3+ HER-2/neu expression. However, a study evaluating single agent trastuzumab in 24 patients with 2+ or 3+ HER-2 expressing tumors (IHC) did not find any benefit [89].

Capuzzo et al. described a case with HER-2 amplification determined by fluorescent *in situ* hybridization (FISH) that responded to paclitaxel and trastuzumab with partial response up to 4 months after therapy. DNA sequencing revealed an EGFR exon 21 mutation (A859T) and a HER2 exon 20 mutation (G776L) [90]. De Greve et al. described three cases of exon 20 HER2-mutant relapsed lung adenocarcinomas, where afatinib was associated with an objective response [91]. A study of 16 advanced NSCLC patients with HER-2 mutations showed a RR of 50%, median PFS of 5.1 months and median OS of 22.9 months with addition of HER-2 targeted therapies (trastuzumab and afatinib) to conventional therapies [82]. Hence, HER2 mutation is a potential therapeutic target, contrary to the evidence in breast cancer [82, 90]. In light of the current evidence, the usefulness of HER2 targeted agents in patients with relapsed advanced NSCLC with HER2 mutations needs to be studied further before routine use.

Conclusions

With increasing number of mutations being identified in NSCLC, the use of targeted agents against these pathways offers a promising therapeutic option. While it is possible that these agents will help NSCLC patients, the differential predictive ability of EGFR expression in lung, colorectal and squamous cell carcinoma of the head and neck serves as a cautionary tale. Patients found to have these molecular targets should be offered targeted therapies, but the optimal choice and timing of treatment is unclear.

References

1. Siegel, R.L., K.D. Miller, A. Jemal. Cancer statistics, 2015. CA Cancer J Clin, 2015. 65: 5-29.
2. Dela Cruz, C.S., L.T. Tanoue, R.A. Matthay. Lung cancer: epidemiology, etiology, and prevention. Clin Chest Med, 2011. 32: 605-44.
3. Lee, C.K., C. Brown, R.J. Gralla, V. Hirsh, S. Thongprasert, C.M. Tsai, et al. Impact of EGFR inhibitor in non-small cell lung cancer on progression-free and overall survival: a meta-analysis. J Natl Cancer Inst, 2013. 105: 595-605.
4. Shaw, A.T., D.W. Kim, K. Nakagawa, T. Seto, L. Crino, M.J. Ahn, et al. Crizotinib versus chemotherapy in advanced ALK-positive lung cancer. N Engl J Med, 2013. 368: 2385-94.

5. Ettinger, D.S., D.E. Wood, D.L. Aisner, W. Akerley, J. Bauman, L.R. Chirieac, et al. Non-small cell lung cancer, Version 5.2017, NCCN Clinical Practice Guidelines in Oncology. J Natl Compr Canc Netw, 2017. 15: 504-35.
6. Kris, M.G., B.E. Johnson, L.D. Berry, D.J. Kwiatkowski, A.J. Iafrate, I.I. Wistuba, et al. Using multiplexed assays of oncogenic drivers in lung cancers to select targeted drugs. Jama, 2014. 311: 1998-2006.
7. Lopez-Chavez, A., A. Thomas, A. Rajan, M. Raffeld, B. Morrow, R. Kelly, et al. Molecular profiling and targeted therapy for advanced thoracic malignancies: a biomarker-derived, multiarm, multihistology phase II basket trial. J Clin Oncol, 2015. 33: 1000-7.
8. Rodenhuis, S., R.J. Slebos. Clinical significance of ras oncogene activation in human lung cancer. Cancer Res, 1992. 52: 2665s-9s.
9. Coate, L.E., T. John, M.S. Tsao, F.A. Shepherd. Molecular predictive and prognostic markers in non-small-cell lung cancer. Lancet Oncol, 2009. 10: 1001-10.
10. Miller, V.A., G.J. Riely, M.F. Zakowski, A.R. Li, J.D. Patel, R.T. Heelan, et al. Molecular characteristics of bronchioloalveolar carcinoma and adenocarcinoma, bronchioloalveolar carcinoma subtype, predict response to erlotinib. J Clin Oncol, 2008. 26: 1472-8.
11. Shepherd, F.A., J. Rodrigues Pereira, T. Ciuleanu, E.H. Tan, V. Hirsh, S. Thongprasert, et al. Erlotinib in previously treated non-small-cell lung cancer. N Engl J Med, 2005. 353: 123-32.
12. Gatzemeier, U., L. Paz-Ares, J. Rodrigues Pereira, J. Von Pawel, R. Ramlau, J.K. Roh, et al. Molecular and clinical biomarkers of cetuximab efficacy: data from the phase III FLEX study in non-small cell lung cancer (NSCLC). J Thorac Oncol, 2009. 4: S324 (abstract B2.3).
13. Franklin, W.A., R. Veve, F.R. Hirsch, B.A. Helfrich, P.A. Bunn Jr. Epidermal growth factor receptor family in lung cancer and premalignancy. Semin Oncol, 2002. 29: 3-14.
14. Ganti, A.K., A. Potti. Epidermal growth factor inhibition in solid tumours. Expert Opin Biol Ther, 2005. 5: 1165-74.
15. Lin, S.Y., K. Makino, W. Xia, A. Matin, Y. Wen, K.Y. Kwong, et al. Nuclear localization of EGF receptor and its potential new role as a transcription factor. Nat Cell Biol, 2001. 3: 802-8.
16. Dittmann, K., C. Mayer, B. Fehrenbacher, M. Schaller, U. Raju, L. Milas, et al. Radiation-induced epidermal growth factor receptor nuclear import is linked to activation of DNA-dependent protein kinase. J Biol Chem, 2005. 280: 31182-9.
17. Lynch, T.J., D.W. Bell, R. Sordella, S. Gurubhagavatula, R.A. Okimoto, B.W. Brannigan, et al. Activating mutations in the epidermal growth factor receptor underlying responsiveness of non-small-cell lung cancer to gefitinib. N Engl J Med, 2004. 350: 2129-39.
18. Bell, D.W., T.J. Lynch, S.M. Haserlat, P.L. Harris, R.A. Okimoto, B.W. Brannigan, et al. Epidermal growth factor receptor mutations and gene amplification in non-small-cell lung cancer: molecular analysis of the IDEAL/INTACT gefitinib trials. J Clin Oncol, 2005. 23: 8081-92.
19. Mok, T.S., Y.L. Wu, S. Thongprasert, C.H. Yang, D.T. Chu, N. Saijo, et al. Gefitinib or carboplatin-paclitaxel in pulmonary adenocarcinoma. N Engl J Med, 2009. 361: 947-57.
20. Lindeman, N.I., P.T. Cagle, M.B. Beasley, D.A. Chitale, S. Dacic, G. Giaccone, et al. Molecular testing guideline for selection of lung cancer patients for EGFR and ALK tyrosine kinase inhibitors: guideline from the College of American

Pathologists, International Association for the Study of Lung Cancer, and Association for Molecular Pathology. J Thorac Oncol, 2013. 8: 823-59.
21. Mok, T., Y.L. Wu, L. Zhang. A small step towards personalized medicine for non-small cell lung cancer. Discov Med, 2009. 8: 227-31.
22. Ellison, G., E. Donald, G. McWalter, L. Knight, L. Fletcher, J. Sherwood, et al. A comparison of ARMS and DNA sequencing for mutation analysis in clinical biopsy samples. J Exp Clin Cancer Res, 2010. 29: 132.
23. Paz-Ares, L., D. Soulieres, I. Melezinek, J. Moecks L. Keil, T. Mok, et al. Clinical outcomes in non-small-cell lung cancer patients with EGFR mutations: pooled analysis. J Cell Mol Med, 14: 51-69.
24. Maemondo, M., A. Inoue, K. Kobayashi, S. Sugawara, S. Oizumi, H. Isobe, et al. Gefitinib or chemotherapy for non-small-cell lung cancer with mutated EGFR. N Engl J Med, 2010. 362: 2380-8.
25. Mitsudomi, T., S. Morita, Y. Yatabe, S. Negoro, I. Okamoto, J. Tsurutani, et al. Gefitinib versus cisplatin plus docetaxel in patients with non-small-cell lung cancer harbouring mutations of the epidermal growth factor receptor (WJTOG3405): an open label, randomised phase 3 trial. Lancet Oncol, 2010. 11: 121-8.
26. Rosell, R., E. Carcereny, R. Gervais, A. Vergnenegre, B. Massuti, E. Felip, et al. Erlotinib versus standard chemotherapy as first-line treatment for European patients with advanced EGFR mutation-positive non-small-cell lung cancer (EURTAC): a multicentre, open-label, randomised phase 3 trial. Lancet Oncol, 2012. 13: 239-46.
27. Sequist, L.V., B.A. Waltman, D. Dias-Santagata, S. Digumarthy, A.B. Turke, P. Fidias, et al. Genotypic and histological evolution of lung cancers acquiring resistance to EGFR inhibitors. Sci Transl Med, 2011. 3: 75ra26.
28. Janne, P.A., J.C. Yang, D.W. Kim, D. Planchard, Y. Ohe, S.S. Ramalingam, et al. AZD9291 in EGFR inhibitor-resistant non-small-cell lung cancer. N Engl J Med, 2015. 372: 1689-99.
29. Mok, T.S., Y.L. Wu, M.J. Ahn, M.C. Garassino, H.R. Kim, S.S. Ramalingam, et al. Osimertinib or Platinum-Pemetrexed in EGFR T790M-Positive Lung Cancer. N Engl J Med, 2017. 376: 629-40.
30. Ramalingam, S., T. Reungwetwattana, B. Chewaskulyong, A. Dechaphunkul, K. Lee, F. Imamura, et al. Osimertinib vs standard of care (SoC) EGFR-TKI as first-line therapy in patients (pts) with EGFRm advanced NSCLC: FLAURA. ESMO 2017 Congress, 2017: LBA2_PR.
31. Gerber, D.E., J.D. Minna. ALK inhibition for non-small-cell lung cancer: from discovery to therapy in record time. Cancer Cell, 2010. 18: 548-51.
32. Shaw, A.T., B.Y. Yeap, M. Mino-Kenudson, S.R. Digumarthy, D.B. Costa, R.S. Heist, et al. Clinical features and outcome of patients with non-small-cell lung cancer who harbor EML4-ALK. J Clin Oncol, 2009. 27: 4247-53.
33. Rodig, S.J., M. Mino-Kenudson, S. Dacic, B.Y. Yeap, A. Shaw, J.A. Barletta, et al. Unique clinicopathologic features characterize ALK-rearranged lung adenocarcinoma in the western population. Clin Cancer Res, 2009. 15: 5216-23.
34. Camidge, D.R., S.A. Kono, A. Flacco, A.C. Tan, R.C. Doebele, Q. Zhou, et al. Optimizing the detection of lung cancer patients harboring anaplastic lymphoma kinase (ALK) gene rearrangements potentially suitable for ALK inhibitor treatment. Clin Cancer Res, 2010.
35. Soda, M., Y.L. Choi, M. Enomoto, S. Takada, Y. Yamashita, S. Ishikawa, et al. Identification of the transforming EML4-ALK fusion gene in non-small-cell lung cancer. Nature, 2007. 448: 561-6.

36. Peters, S., D.R. Camidge, A.T. Shaw, S. Gadgeel, J.S. Ahn, D.W. Kim, et al. Alectinib versus crizotinib in untreated ALK-positive non-small-cell lung cancer. N Engl J Med, 2017. 377: 829-38.
37. Gainor, J.F., L. Dardaei, S. Yoda, L. Friboulet, I. Leshchiner, R. Katayama, et al. Molecular mechanisms of resistance to first- and second-generation ALK inhibitors in ALK-rearranged lung cancer. Cancer Dis, 2016. 6: 1118-33.
38. Colwell J. Lorlatinib is active in drug-resistant NSCLC. Cancer Dis, 2016. 6: OF1.
39. Shaw, A.T., L. Friboulet, I. Leshchiner, J.F. Gainor, S. Bergqvist, A. Brooun, et al. Resensitization to crizotinib by the lorlatinib ALK resistance mutation L1198F. N Engl J Med, 2016. 374: 54-61.
40. Roberts, P., C. Der. Targeting the Raf-MEK-ERK mitogen-activated protein kinase cascade for the treatment of cancer. Oncogene, 2007. 26: 3291-310.
41. Marchetti, A., L. Felicioni, S. Malatesta, M. Grazia Sciarrotta, L. Guetti, A. Chella, et al. Clinical features and outcome of patients with non-small-cell lung cancer harboring BRAF mutations. J Clin Oncol, 2011. 29: 3574-9.
42. Paik, P.K., M.E. Arcila, M. Fara, C.S. Sima, V.A. Miller, M.G. Kris, et al. Clinical characteristics of patients with lung adenocarcinomas harboring BRAF mutations. J Clin Oncol, 2011. 29: 2046-51.
43. Hauschild, A., J.J. Grob, L.V. Demidov, T. Jouary, R. Gutzmer, M. Millward, et al. Dabrafenib in BRAF-mutated metastatic melanoma: a multicentre, open-label, phase 3 randomised controlled trial. Lancet, 2012. 380: 358-65.
44. Hyman, D.M., I. Puzanov, V. Subbiah, J.E. Faris, I. Chau, J.Y. Blay, et al. Vemurafenib in multiple nonmelanoma cancers with BRAF V600 mutations. N Engl J Med, 2015. 373: 726-36.
45. Sullivan, R.J., K.T. Flaherty. Resistance to BRAF-targeted therapy in melanoma. European J Cancer (Oxford, England: 1990), 2013. 49: 1297-304.
46. Rudin, C.M., K. Hong, M. Streit. Molecular characterization of acquired resistance to the BRAF inhibitor dabrafenib in a patient with BRAF-mutant non-small-cell lung cancer. J Thorac Oncol, 2013. 8: e41-2.
47. Gautschi, O., S. Peters, V. Zoete, F. Aebersold-Keller, K. Strobel, B. Schwizer, et al. Lung adenocarcinoma with BRAF G469L mutation refractory to vemurafenib. Lung Cancer (Amsterdam, Netherlands), 2013. 82: 365-7.
48. Cipriani, N.A., O.O. Abidoye, E. Vokes, R. Salgia. MET as a target for treatment of chest tumors. Lung Cancer (Amsterdam, Netherlands), 2009. 63: 169-79.
49. Robinson, K.W., A.B. Sandler. The role of MET receptor tyrosine kinase in non-small-cell lung cancer and clinical development of targeted anti-MET agents. The Oncologist, 2013. 18: 115-22.
50. Gelsomino, F., F. Facchinetti, E. Haspinger, M. Garassino, L. Trusolino, F. De Braud, et al. Targeting the MET gene for the treatment of non-small-cell lung cancer. Crit Rev Oncol Hemat, 2014. 89: 284-99.
51. Okuda, K., H. Sasaki, H. Yukiue, M. Yano, Y. Fujii. MET gene copy number predicts the prognosis for completely resected non-small-cell lung cancer. Cancer Sci, 2008. 99: 2280-5.
52. Cappuzzo, F., A. Marchetti, M. Skokan, E. Rossi, S. Gajapathy, L. Felicioni, et al. Increased MET gene copy number negatively affects survival of surgically resected non-small-cell lung cancer patients. J Clin Oncol, 2009. 27: 1667-74.
53. Engelman, J.A., K. Zejnullahu, T. Mitsudomi, Y. Song, C. Hyland, J.O. Park, et al. MET amplification leads to gefitinib resistance in lung cancer by activating ERBB3 signaling. Science (New York, NY), 2007. 316: 1039-43.

54. Stabile, L.P., M.E. Rothstein, P. Keohavong, D. Lenzner, S.R. Land, A.L. Gaither-Davis, et al. Targeting of both the c-MET and EGFR pathways results in additive inhibition of lung tumorigenesis in transgenic mice. Cancers, 2010. 2: 2153-70.
55. Zhang, Y.W., B. Staal, C. Essenburg, Y. Su, L. Kang, R. West, et al. MET kinase inhibitor SGX523 synergizes with epidermal growth factor receptor inhibitor erlotinib in a hepatocyte growth factor-dependent fashion to suppress carcinoma growth. Cancer Res, 2010. 70: 6880-90.
56. Mok, T., K. Park, S. Geater, S. Agarwal, M. Han, P. Komarnitsky, et al. A randomized phase 2 study with exploratory biomarker analysis of ficlatuzumab, a humanized hepatocyte growth factor (HGF) inhibitory monoclonal antibody, in combination with gefitinib versus gefitinib alone in Asian patients with lung adenocarcinoma. Ann Oncol, 2012. 23: ix391.
57. Spigel, D.R., T.J. Ervin, R.A. Ramlau, D.B. Daniel, J.H. Goldschmidt Jr., G.R. Blumenschein Jr., et al. Randomized phase II trial of Onartuzumab in combination with erlotinib in patients with advanced non-small-cell lung cancer. J Clin Oncol, 2013. 31: 4105-14.
58. Sequist, L.V., J. von Pawel, E.G. Garmey, W.L. Akerley, W. Brugger, D. Ferrari, et al. Randomized phase II study of erlotinib plus tivantinib versus erlotinib plus placebo in previously treated non-small-cell lung cancer. J Clin Oncol, 2011. 29: 3307-15.
59. Zou, H.Y., Q. Li, J.H. Lee, M.E. Arango, S.R. McDonnell, S. Yamazaki, et al. An orally available small-molecule inhibitor of c-Met, PF-2341066, exhibits cytoreductive antitumor efficacy through antiproliferative and antiangiogenic mechanisms. Cancer Res, 2007. 67: 4408-17.
60. Ou, S.H., E.L. Kwak, C. Siwak-Tapp, J. Dy, K. Bergethon, J.W. Clark, et al. Activity of crizotinib (PF02341066), a dual mesenchymal-epithelial transition (MET) and anaplastic lymphoma kinase (ALK) inhibitor, in a non-small-cell lung cancer patient with de novo MET amplification. J Thorac Oncol, 2011. 6: 942-6.
61. Schwab, R., I. Petak, M. Kollar, F. Pinter, E. Varkondi, A. Kohanka, et al. Major partial response to crizotinib, a dual MET/ALK inhibitor, in a squamous cell lung (SCC) carcinoma patient with de novo c-MET amplification in the absence of ALK rearrangement. Lung Cancer (Amsterdam, Netherlands), 2014. 83: 109-11.
62. Camidge, D., S. Ou, G. Shapiro, G. Otterson, L. Villaruz, M. Villalona-Calero, et al. Efficacy and safety of crizotinib in patients with advanced c-MET-amplified non-small-cell lung cancer (NSCLC). J Clin Oncol, 2014. 32.
63. Robinson, D.R., Y.M. Wu, S.F. Lin. The protein tyrosine kinase family of the human genome. Oncogene, 2000. 19: 5548-57.
64. Acquaviva, J., R. Wong, A. Charest. The multifaceted roles of the receptor tyrosine kinase ROS in development and cancer. Biochimica et Biophysica Acta, 2009. 1795: 37-52.
65. Davies, K.D., A.T. Le, M.F. Theodoro, M.C. Skokan, D.L. Aisner, E.M. Berge, et al. Identifying and targeting ROS1 gene fusions in non-small-cell lung cancer. Clin Cancer Res, 2012. 18: 4570-9.
66. Bergethon, K., A.T. Shaw, S.H. Ou, R. Katayama, C.M. Lovly, N.T. McDonald, et al. ROS1 rearrangements define a unique molecular class of lung cancers. J Clin Oncol, 2012. 30: 863-70.
67. Ou, S.H., C.H. Bartlett, M. Mino-Kenudson, J. Cui, A.J. Iafrate. Crizotinib for

the treatment of ALK-rearranged non-small-cell lung cancer: a success story to usher in the second decade of molecular targeted therapy in oncology. The Oncologist, 2012. 17: 1351-75.
68. Kim, M.H., H.S. Shim, D.R. Kang, J.Y. Jung, C.Y. Lee, D.J. Kim, et al. Clinical and prognostic implications of ALK and ROS1 rearrangements in never-smokers with surgically resected lung adenocarcinoma. Lung Cancer (Amsterdam, Netherlands), 2014. 83: 389-95.
69. Shaw, A.T., S.H. Ou, Y.J. Bang, D.R. Camidge, B.J. Solomon, R. Salgia, et al. Crizotinib in ROS1-rearranged non-small-cell lung cancer. N Engl J Med, 2014. 371: 1963-71.
70. Mazieres, J., G. Zalcman, L. Crino, P. Biondani, F. Barlesi, T. Filleron, et al. Crizotinib therapy for advanced lung adenocarcinoma and a ROS1 rearrangement: results from the EUROS1 Cohort. J Clin Oncol, 2015. 33(9): 972-4.
71. Eng, C. RET proto-oncogene in the development of human cancer. J Clin Oncol, 1999. 17: 380-93.
72. Blume-Jensen, P., T. Hunter. Oncogenic kinase signalling. Nature, 2001. 411: 355-65.
73. Plaza-Menacho, I., L. Mologni, N.Q. McDonald. Mechanisms of RET signaling in cancer: current and future implications for targeted therapy. Cellular Signalling, 2014. 26: 1743-52.
74. Takeuchi, K., M. Soda, Y. Togashi, R. Suzuki, S. Sakata, S Hatano, et al. RET, ROS1 and ALK fusions in lung cancer. Nature Med, 2012. 18: 378-81.
75. Lipson, D., M. Capelletti, R. Yelensky, G. Otto, A. Parker, M. Jarosz, et al. Identification of new ALK and RET gene fusions from colorectal and lung cancer biopsies. Nature Med, 2012. 18: 382-4.
76. Kohno, T., H. Ichikawa, Y. Totoki, K. Yasuda, M. Hiramoto, T. Nammo, et al. KIF5B-RET fusions in lung adenocarcinoma. Nature Med, 2012. 18: 375-7.
77. Drilon, A., L. Wang, A. Hasanovic, Y. Suehara, D. Lipson, P. Stephens, et al. Response to cabozantinib in patients with RET fusion-positive lung adenocarcinomas. Cancer Discov, 2013. 3: 630-5.
78. Gautschi, O., T. Zander, F.A. Keller, K. Strobel, A. Hirschmann, S. Aebi, et al. A patient with lung adenocarcinoma and RET fusion treated with vandetanib. J Thorac Oncol, 2013. 8: e43-4.
79. Song, M. Progress in discovery of KIF5B-RET kinase inhibitors for the treatment of non-small-cell lung cancer. J Med Chem, 2015. 58: 3672-81.
80. Spector, N.L., K.L. Blackwell. Understanding the mechanisms behind trastuzumab therapy for human epidermal growth factor receptor 2-positive breast cancer. J Clin Oncol, 2009. 27: 5838-47.
81. Mar, N., J.J. Vredenburgh, J.S. Wasser. Targeting HER2 in the treatment of non-small-cell lung cancer. Lung Cancer (Amsterdam, Netherlands), 2015. 87: 220-5.
82. Mazieres, J., S. Peters, B. Lepage, A.B. Cortot, F. Barlesi, M. Beau-Faller, et al. Lung cancer that harbors an HER2 mutation: epidemiologic characteristics and therapeutic perspectives. J Clin Oncol, 2013. 31: 1997-2003.
83. Liu, L., X. Shao, W. Gao, J. Bai, R. Wang, P. Huang, et al. The role of human epidermal growth factor receptor 2 as a prognostic factor in lung cancer: a meta-analysis of published data. J Thorac Oncol, 2010. 5: 1922-32.
84. Takezawa, K., V. Pirazzoli, M.E. Arcila, C.A. Nebhan, X. Song, E. de Stanchina, et al. HER2 amplification: a potential mechanism of acquired resistance to EGFR inhibition in EGFR-mutant lung cancers that lack the second-site EGFRT790M mutation. Cancer Discov, 2012. 2: 922-33.

85. Wang, S.E., A. Narasanna, M. Perez-Torres, B. Xiang, F.Y. Wu, S. Yang, et al. HER2 kinase domain mutation results in constitutive phosphorylation and activation of HER2 and EGFR and resistance to EGFR tyrosine kinase inhibitors. Cancer Cell, 2006. 10: 25-38.
86. Gatzemeier, U., G. Groth, C. Butts, N. Van Zandwijk, F. Shepherd, A. Ardizzoni, et al. Randomized phase II trial of gemcitabine-cisplatin with or without trastuzumab in HER2-positive non-small-cell lung cancer. Ann Oncol, 2004. 15: 19-27.
87. Lara, P.N., Jr., L. Laptalo, J. Longmate, D.H. Lau, R. Gandour-Edwards, P.H. Gumerlock, et al. Trastuzumab plus docetaxel in HER2/neu-positive non-small-cell lung cancer: a California Cancer Consortium screening and phase II trial. Clin Lung Cancer, 2004. 5: 231-6.
88. Langer, C.J., P. Stephenson, A. Thor, M. Vangel, D.H. Johnson. Trastuzumab in the treatment of advanced non-small-cell lung cancer: is there a role? Focus on Eastern Cooperative Oncology Group study 2598. J Clin Oncol, 2004. 22: 1180-7.
89. Clamon, G., J. Herndon, J. Kern, R. Govindan, J. Garst, D. Watson, et al. Lack of trastuzumab activity in non-small-cell lung carcinoma with overexpression of erb-B2: 39810: a phase II trial of cancer and leukemia group B. Cancer, 2005. 103: 1670-5.
90. Cappuzzo, F., L. Bemis, M. Varella-Garcia. HER2 mutation and response to trastuzumab therapy in non-small-cell lung cancer. N Engl J Med, 2006. 354: 2619-21.
91. De Greve, J., E. Teugels, C. Geers, L. Decoster, D. Galdermans, J. De Mey, et al. Clinical activity of afatinib (BIBW 2992) in patients with lung adenocarcinoma with mutations in the kinase domain of HER2/neu. Lung Cancer (Amsterdam, Netherlands), 2012. 76: 123-7.
92. Fukuoka, M., Y.L. Wu, S. Thongprasert, P. Sunpaweravong, S.S. Leong, V. Sriuranpong, et al. Biomarker analyses and final overall survival results from a phase III, randomized, open-label, first-line study of gefitinib versus carboplatin/paclitaxel in clinically selected patients with advanced non-small-cell lung cancer in Asia (IPASS). J Clin Oncol, 2011. 29: 2866-74.
93. Zhou, C., Y.L. Wu, G. Chen, J. Feng, X.Q. Liu, C. Wang, et al. Final overall survival results from a randomised, phase III study of erlotinib versus chemotherapy as first-line treatment of EGFR mutation-positive advanced non-small-cell lung cancer (OPTIMAL, CTONG-0802). Ann Oncol, 2015. 26: 1877-83.
94. Sequist, L.V., J.C. Yang, N. Yamamoto, K. O'Byrne, V. Hirsh, T. Mok, et al. Phase III study of afatinib or cisplatin plus pemetrexed in patients with metastatic lung adenocarcinoma with EGFR mutations. J Clin Oncol, 2013. 31: 3327-34.
95. Wu, Y.L., C. Zhou, C.P. Hu, J. Feng, S. Lu, Y. Huang, et al. Afatinib versus cisplatin plus gemcitabine for first-line treatment of Asian patients with advanced non-small-cell lung cancer harbouring EGFR mutations (LUX-Lung 6): an open-label, randomised phase 3 trial. Lancet Oncol, 2014. 15: 213-22.
96. Yang, J.J., Q. Zhou, H.H. Yan, X.C. Zhang, H.J. Chen, H.Y. Tu, et al. A phase III randomised controlled trial of erlotinib vs gefitinib in advanced non-small-cell lung cancer with EGFR mutations. Br J Cancer, 2017. 116: 568-74.
97. Solomon, B.J., T. Mok, D.W. Kim, Y.L. Wu, K. Nakagawa, T. Mekhail, et al. First-line crizotinib versus chemotherapy in ALK-positive lung cancer. N Engl J Med, 2014. 371: 2167-77.
98. Soria, J.C., D.S.W. Tan, R. Chiari, Y.L. Wu, L. Paz-Ares, J. Wolf, et al. First-line ceritinib versus platinum-based chemotherapy in advanced ALK-rearranged

non-small-cell lung cancer (ASCEND-4): a randomised, open-label, phase 3 study. Lancet, 2017. 389: 917-29.
99. Hida, T., H. Nokihara, M. Kondo, Y.H. Kim, K. Azuma, T. Seto, et al. Alectinib versus crizotinib in patients with ALK-positive non-small-cell lung cancer (J-ALEX): an open-label, randomised phase 3 trial. Lancet, 2017. 390: 29-39.
100. Spigel, D., M. Edelman, K. O'Byrne. Onartuzumab plus erlotinib versus erlotinib in previously treated stage IIIb or IV NSCLC. Presentation at American Society of Clinical Oncology Annual Meeting, Chicago, Illinois. May 30–June 3, 2014. J Clin Oncol, 2014. 32: 128045-144.

CHAPTER

10

Molecular Mediator of Prostate Cancer Progression and Its Implication in Therapy

Samikshan Dutta, Navatha Shree Sharma, Ridwan Islam and Kaustubh Datta*
Department of Biochemistry and Molecular Biology, University of Nebraska Medical Center, Omaha, NE

Introduction

Back in 1853, histological analysis by J. Adams confirmed the possible reported case of prostate cancer [1]. Since then, there is a tremendous progress in identifying and treatment of prostate cancer. However, in the United States alone, with 174,650 new cases and nearly 31,620 deaths estimated annually, it is the most common cancer in men and stands second in the leading cause of cancer deaths in men next to lung cancer [2]. Prostate cancer is multifocal and multifactorial disease. There is a considerable degree of heterogeneity in prostate tumors as evaluated at genetic, histological or cell signaling level. While many prostate cancer patients are detected with localized disease, which can be cured by surgery, the disease for some patients can be evolved into metastatic castration resistant prostate cancer (mCRPC). Despite the advent for new therapeutic options such as abiraterone acetate, enzalutamide, radium-223, sipuleucel-T and taxane-based chemotherapies, mCRPC patients are generally resistant to these therapies and die due to the complications from metastatic disease. Despite the significant advancement and knowledge for initiation and progression of prostate cancer generated by the next generation sequencing technology, there is no definite molecular markers that can distinguish indolent verses aggressive phenotype [2-8]. Some studies had been performed to differentiate the genetic expression profiles between benign vs prostatic intraepithelial neoplasia (PIN) or between localized and castration resistance prostate cancer (CRPC); however, specific genetic profiling among the various prostate cancer grades is still

---

*Corresponding author: Kaustubh.datta@unmc.edu.

lacking [9]. In this chapter, we will update the genetic mediators involved in the progression of prostate cancer.

Prostate and Its Cell Types

Prostate is a walnut-sized secretory gland, involved in the male reproductive system by secreting various components of seminal fluids and maintain alkalinity. Architecturally, adult prostate consists of three distinct zones, named peripheral, central and transition. Among these zones, prostate cancer mainly arises from the peripheral zones [10-13]. Histopathologically, prostate consists of three different epithelial layers of cell, namely luminal, basal and neuroendocrine [14-18]. The highly polarized columnar type luminal cells are mainly responsible for the production of secretory proteins. These cells are characterized by the expression cytokeratins 8 and 18 and produce high levels of androgen receptors (AR) [19-22]. On the other hand, basal cells are characterized by the expression of proteins like p63, cytokeratins 5 and 14; however, AR expression is low or often undetectable [22-24]. The basal cells surround the luminal columnar epithelial cells. Neuroendocrine cells are the rarest cells among the prostate epithelial cell types and are AR-negative but express the marker proteins like chromogranin A and synaptophysin [25-27].

Determination of Prostate Cancer Stages

Identification and staging of prostate cancer is the most critical step for choosing the treatment options to be given to the patients. There are two types of staging associated with cancer determination:

1. Clinical staging: before surgery
2. Pathological staging: after surgery

Clinical staging is based mainly on histopathological evaluation of biopsy specimen (Gleason score), TNM classification, PSA level and imaging study. On the other hand, during or after surgery, this clinical staging often changes due to the up or down gradation on pathological staging. Histologic grades such as Gleason scoring remains an important prognostic factor in PCa [28, 29]. A high Gleason score predicts rapid progression of the disease and thereby indicates the requirement of a severe treatment regimen. However, the Gleason score does not provide information on the type of treatment strategy to be used on a specific patient leading to over or non-optimum treatment. Patients with low Gleason score usually have indolent, low risk cancer [30-32]. Unfortunately, as previously mentioned, cancer in some patients do progress and becomes a lethal disease. Because of the fear of having aggressive disease, patients with low Gleason grades are often treated with radical therapy leading to morbidity and poor quality of life [33, 34]. The risk stratification strategy for localized PCa patients is therefore utmost necessary, with an underlying objective to identify and categorize patient groups who can be benefited from aggressive treatments versus those with indolent tumors who can be good candidates for either active surveillance or with no or limited intervention. However, the challenges remain in the identification of molecular subtypes that drive differential prognoses in localized PCa.

Prostate Cancer Subtypes

Majority of the prostate cancers are pathologically classified as adenocarcinoma with luminal prototype. It is diagnosed by the absence of the basal cell markers like cytokeratin 5/14 and p63 and elevated luminal proteins such as AR and a-methylacyl-CoA racemase [35-38]. Other subtypes of prostate cancers like ductal adenocarcinoma, neuroendocrine prostate cancer, mucinous carcinoma, and signet ring carcinoma account for very low percentage [39-42].

In most occasions, prostate cancer contains very low level of somatic and germline mutations. Copy number variation or random gene fusion events are more frequent in primary prostate cancer. Epigenetic changes within the genetic environment has also played a strong role in development of prostate cancer [43, 44]. The genetic landscape of prostate cancer was intensely explored in the last few years with Next Generation Sequencing (NGS), whole genome expression analyses and analyses of epigenetic alterations. These findings, along with the results from genetically engineered mouse models (GEMM) for PCa initiation and progression, confirmed a number of features that define prostate cancer. These are:

- A relatively low rate of mutations in PCa compared to other tumors [45, 46].
- Prevalence of non-random copy number variations (CNV) in most PCa tumors involving well-known prostate oncogenes or tumor suppressors [47, 48].
- Recurrent chromosomal rearrangements involving ETS transcription factors, most frequently ERG, in ~50% of PCa. The genomic arrangements observed in PCa are complex and are thought to evolve in a punctuated manner with translocations and deletions occurring interdependently, via a process known as "chromoplexy" [49-56].
- Involvement of pathways that govern prostate embryonic development during the initiation and particularly during progression to CRPC [47, 57, 58].
- The importance of epigenetic changes such as chromatin remodeling, DNA methylation and histone acetylation [59, 60].
- The whole scale alterations in transcriptional programs, particularly those governed by androgen receptor (AR), and their prominent role in driving DNA rearrangements and co-opting developmental pathways [51, 55, 61, 62].

TMPRSS2-ERG and Other Rearrangements Involving ETS Family

One of the most common recurrent genomic rearrangement in prostate cancer is the fusion of androgen regulated gene *TMPRSS2* to ETS family of transcription factors, the most common one being ERG [61, 63, 64]. Gene fusion is detected in ~50% of prostate cancer patients and occurs due to improper repair of double stranded breaks. Because of TMPRSS2-ERG fusion, the expression of this family of transcription factors is regulated by androgen axis in prostate cancer cells. It has been suggested that AR itself increases the probability of this specific fusion event to occur in cancer cells [51, 65, 66]. Interestingly, ~90 % patients with early onset of prostate cancer have been detected with the fusion and the prostate cancer in these patients is also dependent on AR axis suggesting an AR-dependent fusion mechanism [67-69]. Fusion of other ETS factors such as ETV1, ETV4, and ETV5 with binding factors such as TMPRSS2, HERV-K, KLK2, SLC45A3, CANT1, HERV-K17, DDX5 and FOXP1 have also been reported [51, 61, 63-69]. These ETS transcription factors promote the expression of genes such as MYC, EZH2 and SOX9 involved in proliferation, dedifferentiation, migration, invasion and oncogenesis [70-72]. TMPRSS2-ERG fusion has been detected in both low and high-grade intraepithelial neoplasia (PIN) and is therefore thought to be an early event in prostate cancer [51, 63, 66]. PIN is also detected in mouse prostate when overexpressing prostate-specific ETV1 in transgenic mice. Interestingly, no tumor is developed in this mouse model, suggesting the requirement of other genetic alterations for oncogenic transformation [73-75]. Some studies also reported increased expression of ETS factors in hormone-naïve and castration-resistant metastatic prostate cancer indicating potential roles of fusion genes in advanced cancer [54, 75, 76]. Studies have indicated co-operation between TMPRSS2-ERG fusion and other oncogenic events such as PTEN loss, AKT activation and AR overexpression, which promotes the emergence of advanced prostate cancer [77, 78]. The TMPRSS2-ERG fusion can activate RAS-MAPK and is associated with downregulation of WNT and TGFβ signaling pathways [79-83]. Moreover, removal of these fusion proteins reduces tumor growth in nude mice highlighting the potential of ETS transcription factors as target for prostate cancer [84-86]. Inhibiting upstream signaling kinases and downstream targets of ETS transcription factors are also considered as effective therapeutic target as an alternative of directly targeting the ETS factors [73, 87, 88]. Because of its cancer-specific expression and potential impact on tumor initiation and growth, many studies investigated the prognostic importance of ETS gene rearrangement. The results are conflicting as they reported association of ETS fusions with both aggressive and indolent disease [64, 89-91]. There could be several reasons for this anomaly, including heterogeneity of the prostate tumors, study cohorts and the impact of sampling.

Somatic Copy Number Alterations (SCNAs)

Gains or losses in genetic materials occur in SCNAs. 90% of prostate cancer shows SCNAs and is responsible for the activation of oncogenes and inactivation of tumor suppressor genes. The event is particularly prominent in metastatic prostate cancer [92-94]. Deletions of chromosomes 6q, 8p, 10q and 13q are frequently observed in prostate cancer, which result in loss of tumor suppressor genes such as NKX3.1, BRCA2, PTEN and RB1 [95-98]. Amplification of chromosomes X, 7, 8q and 9q are also observed in metastatic castration resistant prostate cancer leading to increased expression of genes involved in androgen receptor pathway and MYC oncogene [97, 99].

NKX3.1

Down-regulation of the NKX3.1 homeobox gene is a frequent event in prostate cancer and is likely to involve multiple mechanisms [100]. Loss-of-heterozygosity (LOH) of NKX3.1 at chromosome 8p21.2 occurs in nearly 85% of high-grade PIN lesions and adenocarcinomas [101-105]. However, although LOH of 8p21 progressively increases in frequency with cancer grade, the remaining allele of NKX3.1 remains as wild type [102, 105-107]. In addition, there is substantial evidence that NKX3.1 undergoes epigenetic down-regulation especially during later stages of prostate cancer progression, probably through its promoter methylation [108]. During development, NKX3.1 is expressed in all epithelial cells of the nascent prostate buds from the urogenital sinus, and represents the earliest known marker for the prostate epithelium [109]. In the absence of NKX3.1, there is a significant decrease in prostatic ductal branching, as well as in production of secretory proteins [109-111]. Notably, young adult mice with NKX3.1 heterozygous and homozygous mutants frequently display prostate epithelial hyperplasia and dysplasia, and often develop intraductal neoplasia (PIN) by 1 year of age [109-113]. These findings are consistent with the tumor suppressor activity of NKX3.1 in cell culture and xenograft assays [113, 114]. Analyses of NKX3.1 function in human tumor cells and genetically engineered mice have provided insights into its potential roles in cancer. In particular, NKX3.1 inactivation in mice results in a defective response to oxidative damage, while its expression in human prostate cancer cell lines protects against DNA damage and is regulated by inflammation [100, 115-117]. Although earlier studies had suggested that NKX3.1 expression is completely lost in advanced cancers, recent analyses using a highly sensitive antibody indicate that low levels of NKX3.1 expression can be demonstrated in nearly all prostate cancers and metastases examined [118, 119]. Thus, there appears to be a selection for reduction, but not loss, of NKX3.1 expression throughout prostate cancer progression. These findings are highly suggestive, since NKX3.1 has been shown to be a critical regulator of prostate epithelial differentiation and stem cell function in mouse models. Recent work has shown that Nkx3.1 expression in the androgen-deprived prostate marks a rare population of prostate epithelial stem cells that is a cell of origin for prostate cancer in mouse models [120]. A causal role for Nkx3.1 in these processes has been suggested by analyses of genes that are dysregulated following perturbation of Nkx3.1 expression in mouse models or human cell lines [115, 121-123].

Myc

Amplification in Myc oncogene has been detected during the initiation (PIN lesions) phase of prostate cancer and also during advanced stage [97, 99, 124, 125]. Myc can be amplified in prostate cancer by long-range regulatory regions in cancer cells or due to the inactivating mutation of the transcriptional repressor, FOXP3 [99]. Myc overexpression has been shown to immortalize prostate cells and together with Pim1, it can promote carcinoma with neuroendocrine differentiation [126, 127].

PTEN

Phosphatase and Tensin homolog (PTEN) mainly acts as phosphatidylinositol-3,4,5-triphosphate (PIP3) phosphatase and is a known tumor suppressor protein. It is frequently mutated or deleted in prostate cancer [128]. Chromosome loci 10q23, where PTEN resides, often undergoes allelic loss in prostate cancer. Reduction or loss of protein expression of PTEN are also observed in prostate tumors [129-132]. Studies have indicated that PTEN undergoes copy number loss as an early event in prostate carcinogenesis, and is correlated with progression to aggressive, castration-resistant disease. Germline loss of Pten in heterozygous mutants or conditional deletion in the prostate epithelium results in PIN and/or adenocarcinoma [129, 132-135]. Analyses of Pten deletion in genetically engineered mouse models have uncovered its cooperativity with inactivation of other key genes that are deregulated in prostate tumorigenesis, and have also provided insights into new therapeutic options for the treatment of prostate cancer. Inactivation of Pten has been shown to cooperate with loss of function of the Nkx3.1 homeobox gene, upregulation of the c-Myc proto-oncogene, or the TMPRSS2-ERG fusion [113, 136-138]. PTEN loss also cooperates with cell cycle regulators such as p27, p18ink4c, and p14arf, or components of key signaling pathways such as Rheb, TSC2, and Rictor [139-144]. Complete inactivation of Pten in mouse prostate tumors leads to cellular senescence, highlighting PTEN as a target in prostate cancer [145]. PTEN reduction or loss in prostate cancer predisposes to the emergence of castration-resistant prostate cancer and has been experimentally proved by mouse model of prostate cancer [50, 146-149].

Loss of Chromosome 13q and Rb [150]

Retinoblastoma (Rb) gene is located in chromosome 13q, which is lost in 50% of prostate cancer patients [151-153]. Mutations of the Rb gene and loss of Rb protein expression have been also reported in clinically localized and more advanced prostate carcinomas [154-156] . Experiments with embryonic prostatic rescue and tissue recombination approaches have shown that homozygous loss of Rb results in dysplasia and invasive carcinoma, which can be aggravated by hormonal stimulation [157]. Reintroduction of Rb into Rb-negative prostate carcinoma cell lines inhibited tumorigenicity, further suggesting Rb as a tumor suppressor protein. Rb also regulates apoptosis in prostate cells, particularly in response to androgens [158-161]. In addition, loss of RB1 has been found to be associated with neuroendocrine trans-differentiation of prostate epithelium, where luminal epithelial cells undergo lineage switch to behave as neuroendocrine-like basal cells [25, 26]. This lineage switch allows the prostate tumor to evade androgen receptor targeted therapy and renders an aggressive therapy resistant phenotype [26].

Altered Cell-cycle Regulatory Genes

Altered cell-cycle regulation plays a major role in progression of clinically localized prostate cancer. Among cell cycle regulatory genes, loss of function of the CDK4 inhibitor p27kip1 is frequently observed in prostate tumors, and can be considered as a prognostic marker [162, 163] . p27kip1 inactivation occurs through loss of expression or altered subcellular localization because of aberrant phosphorylation and/or ubiquitinylation [109, 164-168]. Synergism between losses of p27kip1 with loss of Rb in prostate tumorigenesis has been reported [168]. Another cell cycle regulator is p16, whose loss is known in bypassing senescence [169-171]. p16 is rarely mutated in primary prostate carcinomas but is instead mutated in advanced metastatic disease [172-177].

Somatic Mutations in DNA Repair Genes

TCGA database suggested mutations in many DNA repair genes like BRCA1, BRCA2, FANCD2, CKD12 and ATM in primary prostate cancer, although it is to be noted that the frequency of DNA repair mutations in localized prostate cancer is low [108, 178, 179]. Interestingly, mutations in DNA repair enzymes (BRCA2, ATM, BRCA1, FANCA, RAD51B, RAD51C, MLH1, and MSH2) are the major disease-specific mutations in mCRPC [180, 181]. The understanding that a quarter of men with mCRPC have mutations in DNA repair pathway genes such as BRCA genes has led to clinical trials with poly ADP ribose polymerase (PARP) inhibitors. PARPs are required to repair the DNA single-strand breaks through base-excision repair. It has been shown that cancer cells with BRCA gene mutations become significantly more sensitive to PARP inhibitors as a therapeutic approach popularly known as synthetic lethal strategy [180, 181]. It is further noted that mutations of other DNA repair pathway genes can be sensitive to PARP inhibitors. These genes are ATM, BRIP1, BARD1, CDK12, CHEK2, FANCA, NBN, PALB2, and RAD51 [182, 183].

Germline Mutations

Family history of prostate cancer is a known risk factor. Familial prostate cancer is often defined as having one first degree relative with prostate cancer. Sometimes, the term hereditary prostate cancer is also used especially when a family with either three affected generations, three affected first-degree relatives or two affected relatives before 55 years of age develops prostate cancer. The factors responsible for familial or heritable prostate cancer can be either single nucleotide polymorphisms identified through genome-wide association studies which are usually associated with modest increased risk of prostate cancer or some rare mutants that can substantially increase the risk [98, 184-186]. The genome wide association studies indicated chromosomal loci such as 8q24, 17p with familial risk of prostate cancer [98, 184-186]. On the other hand, rare pathogenic germline mutations such as mutations in DNA repair genes such as BRCA2 and BRCA1 not only increase the risk of developing prostate cancer and earlier onset, but also lead to aggressive cancers with higher rate of recurrence following primary therapy. These mutations therefore are responsible for cancer-associated mortality [98, 185, 187]. Recent whole exome sequencing of autosomal dominant cancer risk genes isolated from tumor tissues of mCRPC indicated 11.8% of men were detected to have germline mutations in one of the 16 DNA repair genes in addition to BRAC2 and BRCA1 (ATM , CHEK2, PALB2, RAD51C, RAD51D); as a result, multigene cancer screening panels have been increasingly suggested [178, 188, 189]. These screenings would help identify men with advanced prostate cancer harboring deficiency in somatic homologous recombination DNA repair, as these patients would be benefitted from PARP inhibitor and platinum chemotherapy [179, 190, 191]. However, elaborate studies in this area should be performed to link the genetic testing with the identification of familial prostate cancer predisposition and design specific treatment strategies to treat this high-risk prostate cancer patient. High-risk prostate cancer predisposition genes other than DNA repair genes also exist. One such gene is HOXB13, which carries a rare missense mutation (G84E). This mutation is specifically found in young prostate cancer patient with a family history [192-194]. RNaseL, MSR1 and ELAC2/HPC2 are tumor suppressor genes and have been identified in hereditary prostate cancer [195, 196].

Signaling Pathways—Akt/mTOR and MAPK Signaling

Up-regulation of the Akt/mTOR signaling pathway primarily through activation of Akt1 is common in prostate cancer because of loss of Pten function, activation mutations of Akt1 and activation of the p110b isoform of PI3K [50, 146, 197-201]. The Akt/mTOR pathway activation can lead to several tumor promoting functions such as enhanced survival of cancer cells during pathologic and therapeutic stress and increase metabolic activity required for cancer cells to grow. They are particularly relevant for castration-resistant prostate cancer [202-204]. In addition to Akt/mTOR signaling, Erk (p42/44) MAPK signaling is also frequently activated in prostate cancer, particularly in advanced disease, and is often coordinately deregulated together with Akt signaling [197, 198, 205-208]. Simultaneous activation of these signaling pathways promotes tumor progression and castration resistance in prostate cancer cell lines and mouse models, while combinatorial inhibition of these pathways inhibits castration resistant prostate cancer in genetically engineered mice [203, 204, 208]. In contrast to Akt/mTOR signaling, the upstream events that lead to activation of Erk MAPK signaling are less well defined, but are thought to be linked to aberrant growth factor signaling [206]. Although mutations of RAS or RAF, the immediate up-streams of MAPK, are rarely found in human prostate cancer, the Ras/Raf/MAPK pathway is frequently perturbed in advanced prostate cancers [209]. Notably, expression of activated forms of either Raf or Ras in the mouse prostate epithelium results in MAPK activation and promotes cancer formation [210, 211]. Interestingly, a small percentage of aggressive prostate tumors contains a translocation of B-RAF or C-RAF that results in activation, suggesting that perturbations of Ras or Raf signaling may occur in prostate cancer through mechanisms other than activating mutations [212].

Oncogenic Tyrosine Kinases

The deregulated expression of oncogenic tyrosine kinases has been studied extensively in many cancers, since these can represent targets for therapeutic intervention [213]. In prostate cancer, aberrant tyrosine kinase signaling, particularly through Her2/ Neu or SRC tyrosine kinases, has been implicated in aggressive disease, progression to metastasis, and castration resistance, and, consequently, has been implicated as a key therapeutic target in patients with advanced disease [214, 215]. In particular, stimulation of AR signaling leads to activation of SRC in prostate cancer cells, which can lead to phosphorylation of AR, castration resistance, and cellular proliferation and invasiveness [216-218]. However, most functional analyses of SRC and other oncogenic tyrosine kinases have been limited to studies of prostate cancer cell lines in culture or in xenografts, and further insights will require analyses of *in vivo* models and correlative studies of clinical specimens.

Driver Mutations in ETS Fusion-Negative PCa

Until recently, the driver mutations in ETS fusion-negative PCa were unknown. In the last year or two, a number of genomic aberrations that occur selectively in ETS fusion-negative PCa were identified, mostly through use of NGS and analysis of epigenetic alteration.

Mutations in spop (Speckle-type POZ protein) gene (6-15% of PCa) represents a subclass of fusion-negative PCa patients with poor prognosis [219-221]. SPOP is a POZ domain adaptor protein that modulates the transcriptional repression activities of genes and thus involve an error-prone method to repair broken DNA strands. SPOP forms a complex with CULLIN3 E3 ubiquitin ligase, and ubiquitinates for degradation of SRC-3/AIB1, a cofactor of AR [222]. SPOP mutation therefore can activate AR in castration-resistant prostate cancer [223, 224]. SPOP mutation also activates Hedgehog pathway, polycomb group protein BMI1, which then promotes the progression of prostate cancer [225-227]. SPOP mutations are strongly associated with copy loss of CHD, FOXO3 and PRDM1 and rarely have accompanying mutations in PTEN or PIK3CA or TP53 in localized cancers [228]. SPINK1 overexpression is another molecular event found in 5-10% of PCa, which usually does not show gene rearrangement [229]. Although SPINK expression and ERG-negative status are not always mutually exclusive [230-232], SPINK1 encodes a secreted serine peptidase inhibitor, which may involve EGFR in its tumorigenic effects, and thus promotes an aggressive subtype of PCa [232]. It is strongly associated with copy loss of PTEN [230]. Other genetic events that are detected in fusion-negative prostate cancer are methylation of miR-26a, high expression of EZH2, and deletion of tumor suppressor MAP3K7/TAK1 [233-235].

Androgen Receptor (AR)

Androgen Receptor (AR) is a member of nuclear receptor superfamily. The gene is located at chromosome Xq11-12. AR is an 11 KDa protein with distinct structural domains such as NH2 terminal transactivation domain, a ZN-finger DNA-binding domain (DBD), a hinge region and a ligand binding domain. Ligands such as testosterone and dihydrotestosterone bind to the ligand-binding domain of AR in the cytoplasm, which promotes dimerization and nuclear translocation of AR. In the nucleus, AR functions as a transcription factor and binds androgen response element of the promoter and enhancer of AR-targeted genes through its DBD. AR dimer upon binding to chromosomes recruits co-activator and co-regulatory proteins to initiate transcription of genes [236, 237]. Some of the known AR regulated genes are PSA, KLK2, NKX3.1, FKBP5, FOXP1, UBE2C and fusion gene, TMPRSS2-ERG [238, 239]. AR is also capable of repressing transcription of genes [240, 241]. In normal prostate, AR is detected in luminal epithelial cells, fibro-muscular stroma and endothelial cells with weak expression in basal epithelial cells [19, 242]. Epithelial AR promotes the expression of secretary proteins in the prostate, while stromal AR favors the expression of growth-promoting peptides and are therefore important for the overall growth of the prostate [243-245]. High expression of AR is detected in prostate cancer and is required for the growth and survival of cancer cells [237, 246]. AR in prostate cancer cells stimulates G1 to S progression by inducing cyclin-dependent kinases, cyclins D1 and E [247, 248]. AR also functions with FOXO3a to express anti-apoptotic FLICE-like inhibitory protein (FLIP) and stimulates angiogenesis in prostate tumor [249]. AR-mediated MMP2 and -9 expressions have been linked to its ability to promote the migration and invasion of prostate cancer cells [250, 251].

Phosphorylation of ezrin by AR inactivates its function and is considered to be another mechanism through which AR promotes increased invasion of cancer cells [252]. AR also interacts with filamin A, recruits integrin beta 1 and thus controls focal adhesion kinase, paxilin, and Rac for migration [253]. Because of the several oncogenic functions of AR in prostate cancer, AR axis has been demonstrated to be an effective target especially for locally invasive or metastatic prostate cancer. Androgen deprivation therapy (ADT) through surgical and chemical castration is the standard of care for aggressive prostate cancer. Although ADT is initially effective for the remission of prostate cancer, recurrence inevitably occurs for most patients with the development of castration resistant prostate cancer (CRPC) [254, 255]. CRPC is a lethal disease. Interestingly, activation of AR axis even in the presence of castrated serum level of androgen is one of the major reasons for CRPC and therefore second line of hormonal therapies such as AR antagonists (enzalutamide, apalutamide) and inhibitor of intracrine synthesis of androgen (Abiraterone acetate) can prolong the survival of CRPC patients [256, 257]. AR overexpression (by genomic amplification or transcriptional activation), AR mutations, activation of AR co-regulators, post-translational modification (phosphorylation, acetylation and sumoylation) of AR, non-genomic function of AR and intracrine synthesis of testosterone by resistant prostate cancer cells have been suggested as potential mechanisms of enhanced activation of AR [258, 259].

Unfortunately, resistance to second line of hormonal therapy is also a common event for CRPC patients. Although the tumors resistant to second line of therapies can develop mechanisms that can bypass AR (such as activation of glucocorticoid receptor) or gain independence of AR axis (neuroendocrine like differentiation), AR signaling is once again restored for many of them [260-262]. Mutations at the ligand-binding domain of AR (such as F877L, T878A, H875Y) are observed, which often make the antagonists to behave as agonists or help AR to be activated by adrenal androgens and glucocorticoids [261, 263]. One of the important mechanisms of resistance to second line of hormonal therapy is due to alternative splicing of AR mRNA. These AR variants usually have no carboxy-terminal ligand binding domains and retain only the amino-terminal and DNA binding domain. Because of this specific structural property, the AR variants localize to nucleus without the help from ligand and thus become constitutively active [264].

Summary

Insights gained through cutting edge research is important to classify the molecular subtypes of prostate cancer. It will thus help to identify candidate driver mutations specific to a certain stage of prostate cancer and thus help to develop better treatment strategy. It is crucial to identify the high risk prostate cancer patients early, which will help to start aggressive treatment of those patients with curative intent. At the same time, patients with indolent disease can be spared from morbidity due to unnecessary invasive treatments.

References

1. Adams, J. The case of scirrhous of the prostate gland with corresponding affliction of the lymphatic glands in the lumbar region and in the pelvis. Lancet, 1853. 1: 393.
2. A.C.S. Key Statistics for Prostate Cancer, 2019.
3. Aihara, M., R.M. Lebovitz, T.M. Wheeler, B.M. Kinner, M. Ohori, P.T. Scardino. Prostate specific antigen and gleason grade: an immunohistochemical study of prostate cancer. J Urol, 1994. 151: 1558-64.
4. Bostwick, D.G., A. Pacelli, M. Blute, P. Roche, G.P. Murphy. Prostate specific membrane antigen expression in prostatic intraepithelial neoplasia and adenocarcinoma: a study of 184 cases. Cancer, 1998. 82: 2256-61.
5. McIntosh, T.K., K.E. Saatman, R. Raghupathi, D.I. Graham, D.H. Smith, V.M. Lee, et al. The Dorothy Russell Memorial Lecture. The molecular and cellular sequelae of experimental traumatic brain injury: pathogenetic mechanisms. Neuropathol Appl Neurobiol, 1998. 24: 251-67.
6. Macintosh, C.A., M. Stower, N. Reid, N.J. Maitland. Precise microdissection of human prostate cancers reveals genotypic heterogeneity. Cancer Res, 1998. 58: 23-28.
7. Mehra, R., S.A. Tomlins, R. Shen, O. Nadeem, L. Wang, J.T. Wei, et al. Comprehensive assessment of TMPRSS2 and ETS family gene aberrations in clinically localized prostate cancer. Mod Pathol, 2007. 20: 538-44.
8. Clark, E.L., A. Coulson, C. Dalgliesh, P. Rajan, S.M. Nicol, S. Fleming, et al. The RNA helicase p68 is a novel androgen receptor coactivator involved in splicing and is overexpressed in prostate cancer. Cancer Res, 2008. 68: 7938-46.
9. Martin, N.E., L.A. Mucci, M. Loda, R.A. Depinho. Prognostic determinants in prostate cancer. Cancer J, 2011. 17: 429-37.
10. McNeal, J.E. Origin and development of carcinoma in the prostate. Cancer, 1969. 23: 24-34.
11. McNeal, J.E. The zonal anatomy of the prostate. Prostate, 1981. 2: 35-49.
12. McNeal, J.E., E.A. Redwine, F.S. Freiha, T.A. Stamey. Zonal distribution of prostatic adenocarcinoma. Correlation with histologic pattern and direction of spread. Am J Surg Pathol, 1988. 12: 897-906.
13. Timms, B.G. Prostate development: a historical perspective. Differentiation, 2008. 76: 565-77.
14. Foster, C.S., A. Dodson, V. Karavana, P.H. Smith, Y. Ke. Prostatic stem cells. J Pathol, 2002. 197: 551-65.
15. van Leenders, G.J., J.A. Schalken. Epithelial cell differentiation in the human prostate epithelium: implications for the pathogenesis and therapy of prostate cancer. Crit Rev Oncol Hematol, 2003. 46(Suppl): S3-10.
16. Hudson, D.L. Epithelial stem cells in human prostate growth and disease. Prostate Cancer Prostatic Dis, 2004. 7: 188-94.
17. Shappell, S.B., G.V. Thomas, R.L. Roberts, R. Herbert, M.M. Ittmann, M.A. Rubin, et al. Prostate pathology of genetically engineered mice: definitions and classification. The consensus report from the Bar Harbor meeting of the Mouse Models of Human Cancer Consortium Prostate Pathology Committee. Cancer Res, 2004. 64: 2270-2305.
18. Peehl, D.M. Primary cell cultures as models of prostate cancer development. Endocr Relat Cancer, 2005. 12: 19-47.

19. Xie, Q., Y. Liu, T. Cai, C. Horton, J. Stefanson, Z.A. Wang. Dissecting cell-type-specific roles of androgen receptor in prostate homeostasis and regeneration through lineage tracing. Nat Commun, 2017. 8: 14284.
20. Cunningham, D., Z. You. In vitro and in vivo model systems used in prostate cancer research. J Biol Methods, 2015. 2: 2015. 2: e15
21. Hayward, S.W., R. Del Buono, N. Deshpande, P.A. Hall. A functional model of adult human prostate epithelium. The role of androgens and stroma in architectural organisation and the maintenance of differentiated secretory function. J Cell Sci, 1992. 102(Pt 2): 361-72.
22. Tan, H.L., M.C. Haffner, D.M. Esopi, A.M. Vaghasia, G.A. Giannico, H.M. Ross, et al. Prostate adenocarcinomas aberrantly expressing p63 are molecularly distinct from usual-type prostatic adenocarcinomas. Mod Pathol, 2015. 28: 446-56.
23. Park, J.W., J.K. Lee, J.W. Phillips, P. Huang, D. Cheng, J. Huang, et al. Prostate epithelial cell of origin determines cancer differentiation state in an organoid transformation assay. Proc Natl Acad Sci USA, 2016. 113: 4482-87.
24. Kurita, T., R.T. Medina, A.A. Mills, G.R. Cunha. Role of p63 and basal cells in the prostate. Development, 2004. 131: 4955-64.
25. Ku, S.Y., S. Rosario, Y. Wang, P. Mu, M. Seshadri, Z.W. Goodrich, et al. Rb1 and Trp53 cooperate to suppress prostate cancer lineage plasticity, metastasis, and antiandrogen resistance. Science, 2017. 355: 78-83.
26. Mu, P., Z. Zhang, M. Benelli, W.R. Karthaus, E. Hoover, C.C. Chen, et al. SOX2 promotes lineage plasticity and antiandrogen resistance in TP53- and RB1-deficient prostate cancer. Science, 2017. 355: 84-88.
27. Beltran, H., D. Prandi, J.M. Mosquera, M. Benelli, L. Puca, J. Cyrta, et al. Divergent clonal evolution of castration-resistant neuroendocrine prostate cancer. Nat Med, 2016. 22: 298-305.
28. Gleason, D.F., G.T. Mellinger. Prediction of prognosis for prostatic adenocarcinoma by combined histological grading and clinical staging. J Urol, 1974. 111: 58-64.
29. Gleason, D.F., G.T. Mellinger, G. Veterans. Administration cooperative urological research: prediction of prognosis for prostatic adenocarcinoma by combined histological grading and clinical staging. J Urol, 2017. 197: S134-S139.
30. Irshad, S., M. Bansal, M. Castillo-Martin, T. Zheng, A. Aytes, S. Wenske, et al. A molecular signature predictive of indolent prostate cancer. Sci Transl Med, 2013. 5: 202ra122.
31. Mahal, B.A., M.R. Cooperberg, A.A. Aizer, D.R. Ziehr, A.S. Hyatt, T.K. Choueiri, et al. Who bears the greatest burden of aggressive treatment of indolent prostate cancer? Am J Med, 2015. 128: 609-16.
32. O'Donnell, H., C. Parker. What is low-risk prostate cancer and what is its natural history? World J Urol, 2008. 26: 415-22.
33. Lepor, H. Selecting treatment for high-risk, localized prostate cancer: the case for radical prostatectomy. Rev Urol, 2002. 4: 147-52.
34. Lepor, H. Selecting candidates for radical prostatectomy. Rev Urol, 2000. 2: 182-89.
35. Humphrey, P.A. Diagnosis of adenocarcinoma in prostate needle biopsy tissue. J Clin Pathol, 2007. 60: 35-42.
36. Grisanzio, C., Signoretti, S. p63 in prostate biology and pathology. J Cell Biochem, 2008. 103: 1354-68.
37. Luo, J., S. Zha, W.R. Gage, T.A. Dunn, J.L. Hicks, C.J. Bennett, et al. Alpha-methylacyl-CoA racemase: a new molecular marker for prostate cancer. Cancer Res, 2002. 62: 2220-26.

38. Jiang, W.G., G. Davies, T.A. Martin, C. Parr, G. Watkins, R.E. Mansel, et al. The potential lymphangiogenic effects of hepatocyte growth factor/scatter factor in vitro and in vivo. Int J Mol Med, 2005. 16: 723-28.
39. Grignon, D.J. Unusual subtypes of prostate cancer. Mod Pathol, 2004. 17: 316-27.
40. Kokubo, Y., Y. Nagayama, M. Tsumita, C. Ohkubo, S. Fukushima, P. Vult von Steyern. Clinical marginal and internal gaps of In-Ceram crowns fabricated using the GN-I system. J Oral Rehabil, 2005. 32: 753-8.
41. Kokubo, H., Y. Yamada, Y. Nishio, H. Fukatsu, N. Honda, A. Nakagawa, et al. Immunohistochemical study of chromogranin A in Stage D2 prostate cancer. Urology, 2005. 66: 135-40.
42. Berruti, A., A. Mosca, F. Porpiglia, E. Bollito, M. Tucci, F. Vana, et al. Chromogranin A expression in patients with hormone naive prostate cancer predicts the development of hormone refractory disease. J Urol, 2007. 178: 838-43; quiz 1129.
43. Koochekpour, S. Genetic and epigenetic changes in human prostate cancer. Iran Red Crescent Med J, 2011. 13: 80-98.
44. Wu, Y., M. Sarkissyan, J.V. Vadgama. Epigenetics in breast and prostate cancer. Methods Mol Biol, 2015. 1238: 425-66.
45. Frank, S., P. Nelson, V. Vasioukhin. Recent advances in prostate cancer research: large-scale genomic analyses reveal novel driver mutations and DNA repair defects. F1000Res, 2018. 7(F1000 Faculty Rev):1173.
46. Angeles, A.K., S. Bauer, L. Ratz, S.M. Klauck, H. Sultmann. Genome-based classification and therapy of prostate cancer. Diagnostics (Basel), 2018. 8: 62.
47. Shtivelman, E., T.M. Beer, C.P. Evans. Molecular pathways and targets in prostate cancer. Oncotarget, 2014. 5: 7217-59.
48. Vasmatzis, G., F. Kosari, S.J. Murphy, S. Terra, I.V. Kovtun, F.R. Harris, et al. Large chromosomal rearrangements yield biomarkers to distinguish low-risk from intermediate- and high-risk prostate cancer. Mayo Clin Proc, 2019. 94: 27-36.
49. Rubin, M.A. ETS rearrangements in prostate cancer. Asian J Androl, 2012. 14: 393-9.
50. Shen, M.M., C. Abate-Shen. Pten inactivation and the emergence of androgen-independent prostate cancer. Cancer Res, 2007. 67: 6535-38.
51. Yu, J., J. Yu, R.S. Mani, Q. Cao, C.J. Brenner, X. Cao, et al. An integrated network of androgen receptor, polycomb, and TMPRSS2-ERG gene fusions in prostate cancer progression. Cancer Cell, 2010. 17: 443-54.
52. Flajollet, S., T.V. Tian, A. Flourens, N. Tomavo, A. Villers, E. Bonnelye, et al. Abnormal expression of the ERG transcription factor in prostate cancer cells activates osteopontin. Mol Cancer Res, 2011. 9: 914-24.
53. Baca, S.C., D. Prandi, M.S. Lawrence, J.M. Mosquera, A. Romanel, Y. Drier, et al. Punctuated evolution of prostate cancer genomes. Cell, 2013. 153: 666-77.
54. Wang, K., Y. Wang, C.C. Collins. Chromoplexy: a new paradigm in genome remodeling and evolution. Asian J Androl, 2013. 15: 711-2.
55. Xu, Z., Y. Wang, Z.G. Xiao, C. Zou, X. Zhang, Z. Wang, et al. Nuclear receptor ERRalpha and transcription factor ERG form a reciprocal loop in the regulation of TMPRSS2:ERG fusion gene in prostate cancer. Oncogene, 2018. 37: 6259-74.
56. Hashmi, A.A., E.Y. Khan, M. Irfan, R. Ali, H. Asif, M. Naeem, et al. ERG oncoprotein expression in prostatic acinar adenocarcinoma; clinicopathologic significance. BMC Res Notes, 2019. 12: 35.

57. Prins, G.S., O. Putz. Molecular signaling pathways that regulate prostate gland development. Differentiation, 2008. 76: 641-59.
58. Ceder, Y., A. Bjartell, Z. Culig, M.A. Rubin, S. Tomlins, T. Visakorpi. The molecular evolution of castration-resistant prostate cancer. Eur Urol Focus, 2016. 2: 506-13.
59. Massie, C.E., I.G. Mills, A.G. Lynch. The importance of DNA methylation in prostate cancer development. J Steroid Biochem Mol Biol, 166: 2017. 1-15.
60. Ruggero, K., S. Farran-Matas, A. Martinez-Tebar, A. Aytes. Epigenetic regulation in prostate cancer progression. Curr Mol Biol Rep, 2018. 4: 101-15.
61. White, N.M., F.Y. Feng, C.A. Maher. Recurrent rearrangements in prostate cancer: causes and therapeutic potential. Curr Drug Targets, 2013. 14: 450-59.
62. Kirby, M.K., R.C. Ramaker, B.S. Roberts, B.N. Lasseigne, D.S. Gunther, T.C. Burwell, et al. Genome-wide DNA methylation measurements in prostate tissues uncovers novel prostate cancer diagnostic biomarkers and transcription factor binding patterns. BMC Cancer, 2017. 17: 273.
63. Tomlins, S.A., D.R. Rhodes, S. Perner, S.M. Dhanasekaran, R. Mehra, X.W. Sun, et al. Recurrent fusion of TMPRSS2 and ETS transcription factor genes in prostate cancer. Science, 2005. 310: 644-8.
64. Kumar-Sinha, C., S.A. Tomlins, A.M. Chinnaiyan. Recurrent gene fusions in prostate cancer. Nat Rev Cancer, 2008. 8: 497-511.
65. Augello, M.A., R.B. Den, K.E. Knudsen. AR function in promoting metastatic prostate cancer. Cancer Metastasis Rev, 2014. 33: 399-411.
66. Navaei, A.H., B.A. Walter, V. Moreno, S.D. Pack, P. Pinto, M.J. Merino. Correlation between ERG fusion protein and androgen receptor expression by immunohistochemistry in prostate, possible role in diagnosis and therapy. J Cancer, 2017. 8: 2604-13.
67. Prensner, J.R., M.A. Rubin, J.T. Wei, A.M. Chinnaiyan. Beyond PSA: the next generation of prostate cancer biomarkers. Sci Transl Med, 2012. 4: 127rv123.
68. Tsourlakis, M.C., A. Stender, A. Quaas, M. Kluth, C. Wittmer, A. Haese, et al. Heterogeneity of ERG expression in prostate cancer: a large section mapping study of entire prostatectomy specimens from 125 patients. BMC Cancer, 2016. 16: 641.
69. Spans, L., L. Clinckemalie, C. Helsen, D. Vanderschueren, S. Boonen, E. Lerut, et al. The genomic landscape of prostate cancer. Int J Mol Sci, 2013. 14: 10822-51.
70. Kunderfranco, P., M. Mello-Grand, R. Cangemi, S. Pellini, A. Mensah, V. Albertini, et al. ETS transcription factors control transcription of EZH2 and epigenetic silencing of the tumor suppressor gene Nkx3.1 in prostate cancer. PLoS One, 2010. 5: e10547.
71. Cai, C., H. Wang, H.H. He, S. Chen, L. He, F. Ma, et al. ERG induces androgen receptor-mediated regulation of SOX9 in prostate cancer. J Clin Invest, 2013. 123: 1109-22.
72. Obinata, D., K. Takayama, S. Takahashi, S. Inoue. Crosstalk of the androgen receptor with transcriptional collaborators: potential therapeutic targets for castration-resistant prostate cancer. Cancers (Basel), 2017. 9: 22.
73. Oh, S., S. Shin, R. Janknecht. ETV1, 4 and 5: an oncogenic subfamily of ETS transcription factors. Biochim Biophys Acta, 2012. 1826: 1-12.
74. Higgins, J., M. Brogley, N. Palanisamy, R. Mehra, M.M. Ittmann, J.Z. Li, et al. Interaction of the androgen receptor, ETV1, and PTEN pathways in mouse prostate varies with pathological stage and predicts cancer progression. Horm Cancer, 2015. 6: 67-86.

75. Rezk, M., A. Chandra, D. Addis, H. Moller, M. Youssef, P. Dasgupta, et al. ETS-related gene (ERG) expression as a predictor of oncological outcomes in patients with high-grade prostate cancer treated with primary androgen deprivation therapy: a cohort study. BMJ Open, 2019. 9: e025161.
76. Lian, F., N.V. Sharma, J.D. Moran, C.S. Moreno. The biology of castration-resistant prostate cancer. Curr Probl Cancer, 2015. 39: 17-28.
77. St John, J., K. Powell, M.K. Conley-Lacomb, S.R. Chinni. TMPRSS2-ERG fusion gene expression in prostate tumor cells and its clinical and biological significance in prostate cancer progression. J Cancer Sci Ther, 2012. 4: 94-101.
78. Yang, Y., A.M. Blee, D. Wang, J. An, Y. Pan, Y. Yan, et al. Loss of FOXO1 cooperates with TMPRSS2-ERG overexpression to promote prostate tumorigenesis and cell invasion. Cancer Res, 2017. 77: 6524-37.
79. Hollenhorst, P.C., M.W. Ferris, M.A. Hull, H. Chae, S. Kim, B.J. Graves. Oncogenic ETS proteins mimic activated RAS/MAPK signaling in prostate cells. Genes Dev, 2011. 25: 2147-57.
80. Linn, D.E., K.L. Penney, R.T. Bronson, L.A. Mucci, Z. Li. Deletion of interstitial genes between TMPRSS2 and ERG promotes prostate cancer progression. Cancer Res, 2016. 76: 1869-81.
81. Ratz, L., M. Laible, L.A. Kacprzyk, S.M. Wittig-Blaich, Y. Tolstov, S. Duensing, et al. TMPRSS2:ERG gene fusion variants induce TGF-beta signaling and epithelial to mesenchymal transition in human prostate cancer cells. Oncotarget, 2017. 8: 25115-30.
82. Brase, J.C., M. Johannes, H. Mannsperger, M. Falth, J. Metzger, L.A. Kacprzyk, et al. TMPRSS2-ERG-specific transcriptional modulation is associated with prostate cancer biomarkers and TGF-beta signaling. BMC Cancer, 2011. 11: 507.
83. Li, Y., D. Kong, Z. Wang, A. Ahmad, B. Bao, S. Padhye, et al. Inactivation of AR/TMPRSS-ERG/Wnt signaling networks attenuates the aggressive behavior of prostate cancer cells. Cancer Prev Res (Phila), 2011. 4: 1495-1506.
84. Obinata, D., A. Ito, K. Fujiwara, K. Takayama, D. Ashikari, Y. Murata, et al. Pyrrole-imidazole polyamide targeted to break fusion sites in TMPRSS2 and ERG gene fusion represses prostate tumor growth. Cancer Sci, 2014. 105: 1272-78.
85. Wang, X., Y. Qiao, I.A. Asangani, B. Ateeq, A. Poliakov, M. Cieslik, et al. Development of peptidomimetic inhibitors of the ERG gene fusion product in prostate cancer. Cancer Cell, 2017. 31: 532-48, e537.
86. Mohamed, A.A., C.P. Xavier, G. Sukumar, S.H. Tan, L. Ravindranath, N. Seraj, et al. Identification of a small molecule that selectively inhibits ERG-positive cancer cell growth. Cancer Res, 2018. 78: 3659-71.
87. Tetsu, O., F. McCormick. ETS-targeted therapy: can it substitute for MEK inhibitors? Clin Transl Med, 2017. 6: 16.
88. Turner, D.P., D.K. Watson. ETS transcription factors: oncogenes and tumor suppressor genes as therapeutic targets for prostate cancer. Expert Rev Anticancer Ther, 2008. 8: 33-42.
89. FitzGerald, L.M., I. Agalliu, K. Johnson, M.A. Miller, E.M. Kwon, A. Hurtado-Coll, et al. Association of TMPRSS2-ERG gene fusion with clinical characteristics and outcomes: results from a population-based study of prostate cancer. BMC Cancer, 2008. 8: 230.
90. Mackinnon, A.C., B.C. Yan, L.J. Joseph, H.A. Al-Ahmadie. Molecular biology underlying the clinical heterogeneity of prostate cancer: an update. Arch Pathol Lab Med, 2009. 133: 1033-40.

91. Garcia-Perdomo, H.A., M.J. Chaves, J.C. Osorio, A. Sanchez. Association between TMPRSS2:ERG fusion gene and the prostate cancer: systematic review and meta-analysis. Cent European J Urol, 2018. 71: 410-19.
92. Cheng, I., A.M. Levin, Y.C. Tai, S. Plummer, G.K. Chen, C. Neslund-Dudas, et al. Copy number alterations in prostate tumors and disease aggressiveness. Genes Chromosomes Cancer, 2012. 51: 66-76.
93. Williams, J.L., P.A. Greer, J.A. Squire. Recurrent copy number alterations in prostate cancer: an in silico meta-analysis of publicly available genomic data. Cancer Genet, 2014. 207: 474-88.
94. Robbins, C.M., W.A. Tembe, A. Baker, S. Sinari, T.Y. Moses, S. Beckstrom-Sternberg, et al. Copy number and targeted mutational analysis reveals novel somatic events in metastatic prostate tumors. Genome Res, 2011. 21: 47-55.
95. Stopsack, K.H., C.A. Whittaker, T.A. Gerke, M. Loda, P.W. Kantoff, L.A. Mucci, et al. Aneuploidy drives lethal progression in prostate cancer. Proc Natl Acad Sci USA, 2019. 116: 11390-11395.
96. Qin, L.X. Chromosomal aberrations related to metastasis of human solid tumors. World J Gastroenterol, 2002. 8: 769-76.
97. Schoenborn, J.R., P. Nelson, M. Fang. Genomic profiling defines subtypes of prostate cancer with the potential for therapeutic stratification. Clin Cancer Res, 2013. 19: 4058-66.
98. Wallis, C.J., R.K. Nam. Prostate cancer genetics: a review. EJIFCC, 2015. 26: 79-91.
99. Koh, C.M., C.J. Bieberich, C.V. Dang, W.G. Nelson, S. Yegnasubramanian, A.M. De Marzo. MYC and prostate cancer. Genes Cancer, 2010. 1: 617-28.
100. Abate-Shen, C., M.M. Shen, E. Gelmann. Integrating differentiation and cancer: the Nkx3.1 homeobox gene in prostate organogenesis and carcinogenesis. Differentiation, 2008. 76: 717-27.
101. Emmert-Buck, M.R., C.D. Vocke, R.O. Pozzatti, P.H. Duray, S.B. Jennings, C.D. Florence, et al. Allelic loss on chromosome 8p12-21 in microdissected prostatic intraepithelial neoplasia. Cancer Res, 1995. 55: 2959-62.
102. Vocke, C.D., R.O. Pozzatti, D.G. Bostwick, C.D. Florence, S.B. Jennings, S.E. Strup, et al. Analysis of 99 microdissected prostate carcinomas reveals a high frequency of allelic loss on chromosome 8p12-21. Cancer Res, 1996. 56: 2411-16.
103. Haggman, M.J., J.A. Macoska, K.J. Wojno, J.E. Oesterling. The relationship between prostatic intraepithelial neoplasia and prostate cancer: critical issues. J Urol, 1997. 158: 12-22.
104. Swalwell, J.I., C.D. Vocke, Y. Yang, J.R. Walker, L. Grouse, S.H. Myers, et al. Determination of a minimal deletion interval on chromosome band 8p21 in sporadic prostate cancer. Genes Chromosomes Cancer, 2002. 33: 201-5.
105. Bethel, C.R., D. Faith, X. Li, B. Guan, J.L. Hicks, F. Lan, et al. Decreased NKX3.1 protein expression in focal prostatic atrophy, prostatic intraepithelial neoplasia, and adenocarcinoma: association with gleason score and chromosome 8p deletion. Cancer Res, 2006. 66: 10683-90.
106. Voeller, H.J., M. Augustus, V. Madike, G.S. Bova, K.C. Carter, E.P. Gelmann. Coding region of NKX3.1, a prostate-specific homeobox gene on 8p21, is not mutated in human prostate cancers. Cancer Res, 1997. 57: 4455-9.
107. Ornstein, D.K., M. Cinquanta, S. Weiler, P.H. Duray, M.R. Emmert-Buck, C.D. Vocke, et al. Expression studies and mutational analysis of the androgen regulated homeobox gene NKX3.1 in benign and malignant prostate epithelium. J Urol, 2001. 165: 1329-34.

108. Asatiani, E., W.X. Huang, A. Wang, E. Rodriguez Ortner, L.R. Cavalli, B.R. Haddad, et al. Deletion, methylation, and expression of the NKX3.1 suppressor gene in primary human prostate cancer. Cancer Res, 65: 2005. 1164-73.
109. Bhatia-Gaur, R., A.A. Donjacour, P.J. Sciavolino, M. Kim, N. Desai, P. Young, et al. Roles for Nkx3.1 in prostate development and cancer. Genes Dev, 1999. 13: 966-77.
110. Schneider, A., T. Brand, R. Zweigerdt, H. Arnold. Targeted disruption of the Nkx3.1 gene in mice results in morphogenetic defects of minor salivary glands: parallels to glandular duct morphogenesis in prostate. Mech Dev, 2000. 95: 163-74.
111. Tanaka, M., I. Komuro, H. Inagaki, N.A. Jenkins, N.G. Copeland, S. Izumo. Nkx3.1, a murine homolog of drosophila bagpipe, regulates epithelial ductal branching and proliferation of the prostate and palatine glands. Dev Dyn, 2000. 219: 248-60.
112. Abdulkadir, S.A., J.A. Magee, T.J. Peters, Z. Kaleem, C.K. Naughton, P.A. Humphrey, et al. Conditional loss of Nkx3.1 in adult mice induces prostatic intraepithelial neoplasia. Mol Cell Biol, 2002. 22: 1495-1503.
113. Kim, M.J., R.D. Cardiff, N. Desai, W.A. Banach-Petrosky, R. Parsons, M.M. Shen, et al. Cooperativity of Nkx3.1 and Pten loss of function in a mouse model of prostate carcinogenesis. Proc Natl Acad Sci USA, 2002. 99: 2884-9.
114. Lei, Q., J. Jiao, L. Xin, C.J. Chang, S. Wang, J. Gao, et al. NKX3.1 stabilizes p53, inhibits AKT activation, and blocks prostate cancer initiation caused by PTEN loss. Cancer Cell, 2006. 9: 367-78.
115. Ouyang, X., T.L. DeWeese, W.G. Nelson, C. Abate-Shen. Loss-of-function of Nkx3.1 promotes increased oxidative damage in prostate carcinogenesis. Cancer Res, 2005. 65: 6773-9.
116. Markowski, M.C., C. Bowen, E.P. Gelmann. Inflammatory cytokines induce phosphorylation and ubiquitination of prostate suppressor protein NKX3.1. Cancer Res, 2008. 68: 6896-6901.
117. Bowen, C., E.P. Gelmann. NKX3.1 activates cellular response to DNA damage. Cancer Res, 2010. 70: 3089-97.
118. Bowen, C., L. Bubendorf, H.J. Voeller, R. Slack, N. Willi, G. Sauter, et al. Loss of NKX3.1 expression in human prostate cancers correlates with tumor progression. Cancer Res, 2000. 60: 6111-5.
119. Gurel, B., T.Z. Ali, E.A. Montgomery, S. Begum, J. Hicks, M. Goggins, et al. NKX3.1 as a marker of prostatic origin in metastatic tumors. Am J Surg Pathol, 2010. 34: 1097-1105.
120. Wang, X., M. Kruithof-de Julio, K.D. Economides, D. Walker, H. Yu, M.V. Halili, et al. A luminal epithelial stem cell that is a cell of origin for prostate cancer. Nature, 2009. 461: 495-500.
121. Magee, J.A., S.A. Abdulkadir, J. Milbrandt. Haploinsufficiency at the Nkx3.1 locus. A paradigm for stochastic, dosage-sensitive gene regulation during tumor initiation. Cancer Cell, 2003. 3: 273-83.
122. Muhlbradt, E., E. Asatiani, E. Ortner, A. Wang, E.P. Gelmann. NKX3.1 activates expression of insulin-like growth factor binding protein-3 to mediate insulin-like growth factor-I signaling and cell proliferation. Cancer Res, 2009. 69: 2615-22.
123. Song, H., B. Zhang, M.A. Watson, P.A. Humphrey, H. Lim, J. Milbrandt. Loss of Nkx3.1 leads to the activation of discrete downstream target genes during prostate tumorigenesis. Oncogene, 2009. 28: 3307-19.

124. Gurel, B., T. Iwata, C.M. Koh, R.B. Jenkins, F. Lan, C. Van Dang, et al. Nuclear MYC protein overexpression is an early alteration in human prostate carcinogenesis. Mod Pathol, 2008. 21: 1156-67.
125. Schrecengost, R., K.E. Knudsen. Molecular pathogenesis and progression of prostate cancer. Semin Oncol, 2013. 40: 244-58.
126. Wang, J., J. Kim, M. Roh, O.E. Franco, S.W. Hayward, M.L. Wills, et al., Pim1 kinase synergizes with c-MYC to induce advanced prostate carcinoma. Oncogene, 2010. 29: 2477-87.
127. Gupta, K., S. Gupta. Neuroendocrine differentiation in prostate cancer: key epigenetic players. Transl Cancer Res, 2017. 6: S104-S108.
128. Salmena, L., A. Carracedo, P.P. Pandolfi. Tenets of PTEN tumor suppression. Cell, 2008. 133: 403-14.
129. Wang, S., J. Gao, Q. Lei, N. Rozengurt, C. Pritchard, J. Jiao, et al. Prostate-specific deletion of the murine Pten tumor suppressor gene leads to metastatic prostate cancer. Cancer Cell, 2003. 4: 209-21.
130. Whang, Y.E., X. Wu, H. Suzuki, R.E. Reiter, C. Tran, R.L. Vessella, et al. Inactivation of the tumor suppressor PTEN/MMAC1 in advanced human prostate cancer through loss of expression. Proc Natl Acad Sci USA, 1998. 95: 5246-50.
131. McMenamin, M.E., P. Soung, S. Perera, I. Kaplan, M. Loda, W.R. Sellers. Loss of PTEN expression in paraffin-embedded primary prostate cancer correlates with high Gleason score and advanced stage. Cancer Res, 1999. 59: 4291-6.
132. Dong, J.T., C.L. Li, T.W. Sipe, H.F. Frierson, Jr. Mutations of PTEN/MMAC1 in primary prostate cancers from Chinese patients. Clin Cancer Res, 2001. 7: 304-8.
133. Di Cristofano, A., B. Pesce, C. Cordon-Cardo, P.P. Pandolfi. Pten is essential for embryonic development and tumour suppression. Nat Genet, 1998. 19: 348-55.
134. Podsypanina, K., L.H. Ellenson, A. Nemes, J. Gu, M. Tamura, K.M. Yamada, et al. Mutation of Pten/Mmac1 in mice causes neoplasia in multiple organ systems. Proc Natl Acad Sci USA, 1999. 96: 1563-68.
135. Trotman, L.C., M. Niki, Z.A. Dotan, J.A. Koutcher, A. Di Cristofano, A. Xiao, et al. Pten dose dictates cancer progression in the prostate. PLoS Biol, 2003. 1: E59.
136. Kim, J., I.E. Eltoum, M. Roh, J. Wang, S.A. Abdulkadir. Interactions between cells with distinct mutations in c-MYC and Pten in prostate cancer. PLoS Genet, 2009. 5: e1000542.
137. Carver, B.S., J. Tran, A. Gopalan, Z. Chen, S. Shaikh, A. Carracedo, et al. Aberrant ERG expression cooperates with loss of PTEN to promote cancer progression in the prostate. Nat Genet, 2009. 41: 619-24.
138. King, J.C., J. Xu, J. Wongvipat, H. Hieronymus, B.S. Carver, D.H. Leung, et al. Cooperativity of TMPRSS2-ERG with PI3-kinase pathway activation in prostate oncogenesis. Nat Genet, 2009. 41: 524-6.
139. Di Cristofano, A., M. De Acetis, A. Koff, C. Cordon-Cardo, P.P. Pandolfi. Pten and p27KIP1 cooperate in prostate cancer tumor suppression in the mouse. Nat Genet, 2001. 27: 222-4.
140. Bai, F., X.H. Pei, P.P. Pandolfi, Y. Xiong. p18 Ink4c and Pten constrain a positive regulatory loop between cell growth and cell cycle control. Mol Cell Biol, 2006. 26: 4564-76.
141. Chen, Z., A. Carracedo, H.K. Lin, J.A. Koutcher, N. Behrendt, A. Egia, et al. Differential p53-independent outcomes of p19(Arf) loss in oncogenesis. Sci Signal, 2009. 2: ra44.
142. Ma, L., J. Teruya-Feldstein, N. Behrendt, Z. Chen, T. Noda, O. Hino, et al. Genetic analysis of Pten and Tsc2 functional interactions in the mouse reveals

asymmetrical haploinsufficiency in tumor suppression. Genes Dev, 2005. 19: 1779-86.
143. Nardella, C., Z. Chen, L. Salmena, A. Carracedo, A. Alimonti, A. Egia, et al. Aberrant Rheb-mediated mTORC1 activation and Pten haploinsufficiency are cooperative oncogenic events. Genes Dev, 2008. 22: 2172-7.
144. Guertin, D.A., D.M. Stevens, M. Saitoh, S. Kinkel, K. Crosby, J.H. Sheen, et al. mTOR complex 2 is required for the development of prostate cancer induced by Pten loss in mice. Cancer Cell, 2009. 15: 148-59.
145. Chen, Z., L.C. Trotman, D. Shaffer, H.K. Lin, Z.A. Dotan, M. Niki, et al. Crucial role of p53-dependent cellular senescence in suppression of Pten-deficient tumorigenesis. Nature, 2005. 436: 725-30.
146. Mulholland, D.J., S. Dedhar, H. Wu, C.C. Nelson. PTEN and GSK3beta: key regulators of progression to androgen-independent prostate cancer. Oncogene, 2006. 25: 329-37.
147. Lin, H.K., Y.C. Hu, D.K. Lee, C. Chang. Regulation of androgen receptor signaling by PTEN (phosphatase and tensin homolog deleted on chromosome 10) tumor suppressor through distinct mechanisms in prostate cancer cells. Mol Endocrinol, 2004. 18: 2409-23.
148. Gao, H., X. Ouyang, W.A. Banach-Petrosky, M.M. Shen, C. Abate-Shen. Emergence of androgen independence at early stages of prostate cancer progression in Nkx3.1; Pten mice. Cancer Res, 2006. 66: 7929-33.
149. Wu, Z., M. Conaway, D. Gioeli, M.J. Weber, D. Theodorescu. Conditional expression of PTEN alters the androgen responsiveness of prostate cancer cells. Prostate, 2006. 66: 1114-23.
150. Bertram, J., J.W. Peacock, L. Fazli, A.L. Mui, S.W. Chung, M.E. Cox, et al. Loss of PTEN is associated with progression to androgen independence. Prostate, 2006. 66: 895-902.
151. Cooney, K.A., J.C. Wetzel, S.D. Merajver, J.A. Macoska, T.P. Singleton, K.J. Wojno. Distinct regions of allelic loss on 13q in prostate cancer. Cancer Res, 1996. 56: 1142-5.
152. Melamed, J., J.M. Einhorn, M.M. Ittmann. Allelic loss on chromosome 13q in human prostate carcinoma. Clin Cancer Res, 1997. 3: 1867-72.
153. C. Li, C. Larsson, A. Futreal, J. Lancaster, C. Phelan, U. Aspenblad, et al. Identification of two distinct deleted regions on chromosome 13 in prostate cancer. Oncogene, 1998. 16: 481-7.
154. Bookstein, R., J.Y. Shew, P.L. Chen, P. Scully, W.H. Lee. Suppression of tumorigenicity of human prostate carcinoma cells by replacing a mutated RB gene. Science, 1990. 247: 712-5.
155. Phillips, S.M., C.M. Barton, S.J. Lee, D.G. Morton, D.M. Wallace, N.R. Lemoine, et al. Loss of the retinoblastoma susceptibility gene (RB1) is a frequent and early event in prostatic tumorigenesis. Br J Cancer, 1994. 70: 1252-7.
156. Ittmann, M.M., R. Wieczorek. Alterations of the retinoblastoma gene in clinically localized, stage B prostate adenocarcinomas. Hum Pathol, 1996. 27: 28-34.
157. Wang, Y.Z., Y.C. Wong. Sex hormone-induced prostatic carcinogenesis in the noble rat: the role of insulin-like growth factor-I (IGF-I) and vascular endothelial growth factor (VEGF) in the development of prostate cancer. Prostate, 1998. 35: 165-77.
158. Bookstein, R., P. Rio, S.A. Madreperla, F. Hong, C. Allred, W.E. Grizzle, et al. Promoter deletion and loss of retinoblastoma gene expression in human prostate carcinoma. Proc Natl Acad Sci USA, 1990. 87: 7762-6.

159. Zhao, X., J.E. Gschwend, C.T. Powell, R.G. Foster, K.C. Day, M.L. Day. Retinoblastoma protein-dependent growth signal conflict and caspase activity are required for protein kinase C-signaled apoptosis of prostate epithelial cells. J Biol Chem, 1997. 272: 22751-7.
160. Bowen, C., S. Spiegel, E.P. Gelmann. Radiation-induced apoptosis mediated by retinoblastoma protein. Cancer Res, 1998. 58: 3275-81.
161. Yeh, S., H. Miyamoto, K. Nishimura, H. Kang, J. Ludlow, P. Hsiao, et al. Retinoblastoma, a tumor suppressor, is a coactivator for the androgen receptor in human prostate cancer DU145 cells. Biochem Biophys Res Commun, 1998. 248: 361-7.
162. Macri, E., M. Loda. Role of p27 in prostate carcinogenesis. Cancer Metastasis Rev, 1998. 17: 337-44.
163. Tsihlias, J., L. Kapusta, J. Slingerland. The prognostic significance of altered cyclin-dependent kinase inhibitors in human cancer. Annu Rev Med, 1999. 50: 401-23.
164. Esposito, V., A. Baldi, A. De Luca, A.M. Groger, M. Loda, G.G. Giordano, et al. Prognostic role of the cyclin-dependent kinase inhibitor p27 in non-small cell lung cancer. Cancer Res, 1997. 57: 3381-5.
165. Loda, M., B. Cukor, S.W. Tam, P. Lavin, M. Fiorentino, G.F. Draetta, et al. Increased proteasome-dependent degradation of the cyclin-dependent kinase inhibitor p27 in aggressive colorectal carcinomas. Nat Med, 1997. 3: 231-4.
166. Tan, J., Y. Sharief, K.G. Hamil, C.W. Gregory, D.Y. Zang, M. Sar, et al. Dehydroepiandrosterone activates mutant androgen receptors expressed in the androgen-dependent human prostate cancer xenograft CWR22 and LNCaP cells. Mol Endocrinol, 1997. 11: 450-9.
167. Singh, S.P., J. Lipman, H. Goldman, F.H. Ellis, Jr., L. Aizenman, M.G. Cangi, et al. Loss or altered subcellular localization of p27 in Barrett's associated adenocarcinoma. Cancer Res, 1998. 58: 1730-5.
168. Park, M.S., J. Rosai, H.T. Nguyen, P. Capodieci, C. Cordon-Cardo, A. Koff. p27 and Rb are on overlapping pathways suppressing tumorigenesis in mice. Proc Natl Acad Sci USA, 1999. 96: 6382-7.
169. Otterson, G.A., R.A. Kratzke, A. Coxon, Y.W. Kim, F.J. Kaye. Absence of p16INK4 protein is restricted to the subset of lung cancer lines that retains wildtype RB. Oncogene, 1994. 9: 3375-8.
170. Lukas, J., D. Parry, L. Aagaard, D.J. Mann, J. Bartkova, M. Strauss, et al. Retinoblastoma-protein-dependent cell-cycle inhibition by the tumour suppressor p16. Nature, 1995. 375: 503-6.
171. Ueki, K., Y. Ono, J.W. Henson, J.T. Efird, A. von Deimling, D.N. Louis. CDKN2/p16 or RB alterations occur in the majority of glioblastomas and are inversely correlated. Cancer Res, 56: 1996. 150-3.
172. W. Chen, C.M. Weghorst, C.L. Sabourin, Y. Wang, D. Wang, D.G. Bostwick, et al. Absence of p16/MTS1 gene mutations in human prostate cancer. Carcinogenesis, 1996. 17: 2603-7.
173. Tamimi, Y., P.P. Bringuier, F. Smit, A. van Bokhoven, F.M. Debruyne, J.A. Schalken. p16 mutations/deletions are not frequent events in prostate cancer. Br J Cancer, 1996. 74: 120-2.
174. Gaddipati, J.P., D.G. McLeod, I.A. Sesterhenn, C.J. Hussussian, Y.A. Tong, P. Seth, et al. Mutations of the p16 gene product are rare in prostate cancer. Prostate, 1997. 30: 188-94.

175. Mangold, K.A., H. Takahashi, C. Brandigi, T. Wada, S. Wakui, M. Furusato, et al. p16 (CDKN2/MTS1) gene deletions are rare in prostatic carcinomas in the United States and Japan. J Urol, 1997. 157: 1117-20.
176. Park, D.J., S.P. Wilczynski, E.Y. Pham, C.W. Miller, H.P. Koeffler. Molecular analysis of the INK4 family of genes in prostate carcinomas. J Urol, 1997. 157: 1995-9.
177. Gu, K., A.M. Mes-Masson, J. Gauthier, F. Saad. Analysis of the p16 tumor suppressor gene in early-stage prostate cancer. Mol Carcinog, 1998. 21: 164-70.
178. N. Cancer Genome Atlas Research. The molecular taxonomy of primary prostate cancer. Cell, 2015. 163: 1011-25.
179. Rimar, K.J., P.T. Tran, R.S. Matulewicz, M. Hussain, J.J. Meeks. The emerging role of homologous recombination repair and PARP inhibitors in genitourinary malignancies. Cancer, 2017. 123: 1912-24.
180. Dhawan, M., C.J. Ryan, A. Ashworth. DNA repair deficiency is common in advanced prostate cancer: new therapeutic opportunities. Oncologist, 2016. 21: 940-5.
181. Dhawan, M., C.J. Ryan. BRCAness and prostate cancer: diagnostic and therapeutic considerations. Prostate Cancer Prostatic Dis, 2018. 21: 488-98.
182. Heeke, A.L., M.J. Pishvaian, F. Lynce, J. Xiu, J.R. Brody, W.J. Chen, et al. Prevalence of homologous recombination-related gene mutations across multiple cancer types. JCO Precis Oncol, 2018 (2018).
183. Stover, E.H., P.A. Konstantinopoulos, U.A. Matulonis, E.M. Swisher. Biomarkers of response and resistance to DNA repair targeted therapies. Clin Cancer Res, 2016. 22: 5651-60.
184. Benafif, S., Z. Kote-Jarai, R.A. Eeles, P. Consortium. A review of prostate cancer Genome-Wide Association Studies (GWAS). Cancer Epidemiol Biomarkers Prev, 2018. 27: 845-57.
185. Tan, S.H., G. Petrovics, S. Srivastava. Prostate cancer genomics: recent advances and the prevailing underrepresentation from racial and ethnic minorities. Int J Mol Sci, 2018. 19: 1255.
186. Hoffmann, T.J., M.N. Passarelli, R.E. Graff, N.C. Emami, L.C. Sakoda, E. Jorgenson, et al. Genome-wide association study of prostate-specific antigen levels identifies novel loci independent of prostate cancer. Nature Commun, 2017. 8: 14248.
187. Simard, J., M. Dumont, D. Labuda, D. Sinnett, C. Meloche, M. El-Alfy, et al. Prostate cancer susceptibility genes: lessons learned and challenges posed. Endocr Relat Cancer, 2003. 10: 225-59.
188. Pritchard, C.C., J. Mateo, M.F. Walsh, N. De Sarkar, W. Abida, H. Beltran, et al. Inherited DNA-repair gene mutations in men with metastatic prostate cancer. N Engl J Med, 2016. 375: 443-53.
189. Castro, E., N. Romero-Laorden, A. Del Pozo, R. Lozano, A. Medina, J. Puente, et al. PROREPAIR-B: A prospective cohort study of the impact of germline DNA repair mutations on the outcomes of patients with metastatic castration-resistant prostate cancer. J Clin Oncol, 2019. 37: 490-503.
190. Christenson, E.S., E.S. Antonarakis. PARP inhibitors for homologous recombination-deficient prostate cancer. Expert Opin Emerg Drugs, 2018. 23: 123-33.
191. Faraoni, I., G. Graziani. Role of BRCA mutations in cancer treatment with poly(ADP-ribose) polymerase (PARP) inhibitors. Cancers (Basel), 2018. 10: 487.

192. Brechka, H., R.R. Bhanvadia, C. VanOpstall, D.J. Vander Griend. HOXB13 mutations and binding partners in prostate development and cancer: function, clinical significance, and future directions. Genes Dis, 2017. 4: 75-87.
193. Beebe-Dimmer, J.L., M. Hathcock, C. Yee, L.A. Okoth, C.M. Ewing, W.B. Isaacs, et al. The HOXB13 G84E mutation is associated with an increased risk for prostate cancer and other malignancies. Cancer Epidemiol Biomarkers Prev, 2015. 24: 1366-72.
194. Witte, J.S., J. Mefford, S.J. Plummer, J. Liu, I. Cheng, E.A. Klein, et al. HOXB13 mutation and prostate cancer: studies of siblings and aggressive disease. Cancer Epidemiol Biomarkers Prev, 2013. 22: 675-80.
195. Nupponen, N.N., M.J. Wallen, D. Ponciano, C.M. Robbins, T.L. Tammela, R.L. Vessella, et al. Mutational analysis of susceptibility genes RNASEL/HPC1, ELAC2/HPC2, and MSR1 in sporadic prostate cancer. Genes Chromosomes Cancer, 2004. 39: 119-25.
196. Robbins, C.M., W. Hernandez, C. Ahaghotu, J. Bennett, G. Hoke, T. Mason, et al. Association of HPC2/ELAC2 and RNASEL non-synonymous variants with prostate cancer risk in African American familial and sporadic cases. Prostate, 2008. 68: 1790-7.
197. Malik, S.N., M. Brattain, P.M. Ghosh, D.A. Troyer, T. Prihoda, R. Bedolla, et al. Immunohistochemical demonstration of phospho-Akt in high Gleason grade prostate cancer. Clin Cancer Res, 2002. 8: 1168-71.
198. Thomas, G.V., S. Horvath, B.L. Smith, K. Crosby, L.A. Lebel, M. Schrage, et al. Antibody-based profiling of the phosphoinositide 3-kinase pathway in clinical prostate cancer. Clin Cancer Res, 2004. 10: 8351-6.
199. Chen, M.L., P.Z. Xu, X.D. Peng, W.S. Chen, G. Guzman, X. Yang, et al. The deficiency of Akt1 is sufficient to suppress tumor development in Pten+/- mice. Genes Dev, 2006. 20: 1569-74.
200. Boormans, J.L., K.G. Hermans, G.J. van Leenders, J. Trapman, P.C. Verhagen. An activating mutation in AKT1 in human prostate cancer. Int J Cancer, 2008. 123: 2725-6.
201. Hill, K.M., S. Kalifa, J.R. Das, T. Bhatti, M. Gay, D. Williams, et al. The role of PI 3-kinase p110beta in AKT signally, cell survival, and proliferation in human prostate cancer cells. Prostate, 2010. 70: 755-64.
202. Majumder, P.K., J.J. Yeh, D.J. George, P.G. Febbo, J. Kum, Q. Xue, et al. Prostate intraepithelial neoplasia induced by prostate restricted Akt activation: the MPAKT model. Proc Natl Acad Sci USA, 2003. 100: 7841-6.
203. Uzgare, A.R., J.T. Isaacs. Enhanced redundancy in Akt and mitogen-activated protein kinase-induced survival of malignant versus normal prostate epithelial cells. Cancer Res, 2004. 64: 6190-9.
204. Gao, H., X. Ouyang, W.A. Banach-Petrosky, W.L. Gerald, M.M. Shen, C. Abate-Shen. Combinatorial activities of Akt and B-Raf/Erk signaling in a mouse model of androgen-independent prostate cancer. Proc Natl Acad Sci USA, 2006. 103: 14477-82.
205. Abreu-Martin, M.T., A. Chari, A.A. Palladino, N.A. Craft, C.L. Sawyers. Mitogen-activated protein kinase kinase kinase 1 activates androgen receptor-dependent transcription and apoptosis in prostate cancer. Mol Cell Biol, 1999. 19: 5143-54.
206. Gioeli, D. Signal transduction in prostate cancer progression. Clin Sci (Lond), 2005. 108: 293-308.
207. Paweletz, C.P., L. Charboneau, V.E. Bichsel, N.L. Simone, T. Chen, J.W. Gillespie, et al. Reverse phase protein microarrays which capture disease

progression show activation of pro-survival pathways at the cancer invasion front. Oncogene, 2001. 20: 1981-9.
208. Kinkade, C.W., M. Castillo-Martin, A. Puzio-Kuter, J. Yan, T.H. Foster, H. Gao, et al. Targeting AKT/mTOR and ERK MAPK signaling inhibits hormone-refractory prostate cancer in a preclinical mouse model. J Clin Invest, 2008. 118: 3051-64.
209. Taylor, B.S., N. Schultz, H. Hieronymus, A. Gopalan, Y. Xiao, B.S. Carver, et al. Integrative genomic profiling of human prostate cancer. Cancer Cell, 2010. 18: 11-22.
210. Jeong, J.H., Z. Wang, A.S. Guimaraes, X. Ouyang, J.L. Figueiredo, Z. Ding, et al. BRAF activation initiates but does not maintain invasive prostate adenocarcinoma. PLoS One, 2008. 3: e3949.
211. Pearson, H.B., T.J. Phesse, A.R. Clarke. K-ras and Wnt signaling synergize to accelerate prostate tumorigenesis in the mouse. Cancer Res, 2009. 69: 94-101.
212. Palanisamy, N., B. Ateeq, S. Kalyana-Sundaram, D. Pflueger, K. Ramnarayanan, S. Shankar, et al. Rearrangements of the RAF kinase pathway in prostate cancer, gastric cancer and melanoma. Nat Med, 2010. 16: 793-8.
213. Gschwind, A., O.M. Fischer, A. Ullrich. The discovery of receptor tyrosine kinases: targets for cancer therapy. Nat Rev Cancer, 2004. 4: 361-70.
214. Mellinghoff, I.K., I. Vivanco, A. Kwon, C. Tran, J. Wongvipat, C.L. Sawyers. HER2/neu kinase-dependent modulation of androgen receptor function through effects on DNA binding and stability. Cancer Cell, 2004. 6: 517-27.
215. Fizazi, K. The role of Src in prostate cancer. Ann Oncol, 2007. 18: 1765-73.
216. Migliaccio, A., G. Castoria, M. Di Domenico, A. de Falco, A. Bilancio, M. Lombardi, et al. Steroid-induced androgen receptor-oestradiol receptor beta-Src complex triggers prostate cancer cell proliferation. EMBO J, 2000. 19: 5406-17.
217. Agoulnik, I.U., A. Vaid, W.E. Bingman, 3rd, H. Erdeme, A. Frolov, C.L. Smith, et al. Role of SRC-1 in the promotion of prostate cancer cell growth and tumor progression. Cancer Res, 2005. 65: 7959-67.
218. Kraus, S., D. Gioeli, T. Vomastek, V. Gordon, M.J. Weber. Receptor for activated C kinase 1 (RACK1) and Src regulate the tyrosine phosphorylation and function of the androgen receptor. Cancer Res, 2006. 66: 11047-54.
219. Mani, R.S. The emerging role of speckle-type POZ protein (SPOP) in cancer development. Drug Discov Today, 2014. 19: 1498-1502.
220. Geng, C., B. He, L. Xu, C.E. Barbieri, V.K. Eedunuri, S.A. Chew, et al. Prostate cancer-associated mutations in speckle-type POZ protein (SPOP) regulate steroid receptor coactivator 3 protein turnover. Proc Natl Acad Sci USA, 2013. 110: 6997-7002.
221. X. Gang, L. Xuan, X. Zhao, Y. Lv, F. Li, Y. Wang, et al. Speckle-type POZ protein suppresses lipid accumulation and prostate cancer growth by stabilizing fatty acid synthase. Prostate, 2019. 79: 864-71.
222. Wei, X., J. Fried, Y. Li, L. Hu, M. Gao, S. Zhang, et al. Functional roles of speckle-type Poz (SPOP) protein in genomic stability. J Cancer, 2018. 9: 3257-62.
223. Blattner, M., D. Liu, B.D. Robinson, D. Huang, A. Poliakov, D. Gao, et al. SPOP mutation drives prostate tumorigenesis in vivo through coordinate regulation of PI3K/mTOR and AR signaling. Cancer Cell, 2017. 31: 436-51.
224. An, J., C. Wang, Y. Deng, L. Yu, H. Huang. Destruction of full-length androgen receptor by wild-type SPOP, but not prostate-cancer-associated mutants. Cell Rep, 2014. 6: 657-69.

225. Chen, M.H., C.W. Wilson, Y.J. Li, K.K. Law, C.S. Lu, R. Gacayan, et al. Cilium-independent regulation of Gli protein function by Sufu in Hedgehog signaling is evolutionarily conserved. Genes Dev, 2009. 23: 1910-28.
226. Kim, M.S., E.M. Je, J.E. Oh, N.J. Yoo, S.H. Lee. Mutational and expressional analyses of SPOP, a candidate tumor suppressor gene, in prostate, gastric and colorectal cancers. APMIS, 2013. 121: 626-33.
227. Hernandez-Munoz, I., A.H. Lund, P. van der Stoop, E. Boutsma, I. Muijrers, E. Verhoeven, et al. Stable X chromosome inactivation involves the PRC1 Polycomb complex and requires histone MACROH2A1 and the CULLIN3/SPOP ubiquitin E3 ligase. Proc Natl Acad Sci USA, 2005. 102: 7635-40.
228. Barbieri, C.E., S.C. Baca, M.S. Lawrence, F. Demichelis, M. Blattner, J.P. Theurillat, et al. Exome sequencing identifies recurrent SPOP, FOXA1 and MED12 mutations in prostate cancer. Nat Genet, 2012. 44: 685-9.
229. Tomlins, S.A., D.R. Rhodes, J. Yu, S. Varambally, R. Mehra, S. Perner, et al. The role of SPINK1 in ETS rearrangement - negative prostate cancers. Cancer Cell, 2008. 13: 519-28.
230. Bismar, T.A., M. Yoshimoto, Q. Duan, S. Liu, K. Sircar, J.A. Squire. Interactions and relationships of PTEN, ERG, SPINK1 and AR in castration-resistant prostate cancer. Histopathology, 2012. 60: 645-52.
231. Flavin, R., A. Pettersson, W.K. Hendrickson, M. Fiorentino, S. Finn, L. Kunz, et al. SPINK1 protein expression and prostate cancer progression. Clin Cancer Res, 2014. 20: 4904-11.
232. Ateeq, B., S.A. Tomlins, B. Laxman, I.A. Asangani, Q. Cao, X. Cao, et al. Therapeutic targeting of SPINK1-positive prostate cancer. Sci Transl Med, 2011. 3: 72ra17.
233. Borno, S.T., A. Fischer, M. Kerick, M. Falth, M. Laible, J.C. Brase, et al. Genome-wide DNA methylation events in TMPRSS2-ERG fusion-negative prostate cancers implicate an EZH2-dependent mechanism with miR-26a hypermethylation. Cancer Discov, 2012. 2: 1024-35.
234. Kim, J.H., S.M. Dhanasekaran, J.R. Prensner, X. Cao, D. Robinson, S. Kalyana-Sundaram, et al. Deep sequencing reveals distinct patterns of DNA methylation in prostate cancer. Genome Res, 2011. 21: 1028-41.
235. Liu, W., B.L. Chang, S. Cramer, P.P. Koty, T. Li, J. Sun, et al. Deletion of a small consensus region at 6q15, including the MAP3K7 gene, is significantly associated with high-grade prostate cancers. Clin Cancer Res, 2007. 13: 5028-33.
236. Brand, L.J., S.M. Dehm. Androgen receptor gene rearrangements: new perspectives on prostate cancer progression. Curr Drug Targets, 2013. 14: 441-9.
237. Tan, M.H., J. Li, H.E. Xu, K. Melcher, E.L. Yong. Androgen receptor: structure, role in prostate cancer and drug discovery. Acta Pharmacol Sin, 2015. 36: 3-23.
238. Takayama, K., S. Inoue. Transcriptional network of androgen receptor in prostate cancer progression. Int J Urol, 2013. 20: 756-68.
239. Fujita, K., N. Nonomura. Role of androgen receptor in prostate cancer: a review. World J Mens Health, 2018. 37: 288 -295.
240. Aarnisalo, P., H. Santti, H. Poukka, J.J. Palvimo, O.A. Janne. Transcription activating and repressing functions of the androgen receptor are differentially influenced by mutations in the deoxyribonucleic acid-binding domain. Endocrinology, 1999. 140: 3097-3105.
241. Chen, H., S.J. Libertini, M. George, S. Dandekar, C.G. Tepper, B. Al-Bataina, et al. Genome-wide analysis of androgen receptor binding and gene regulation in two CWR22-derived prostate cancer cell lines. Endocr Relat Cancer, 2010. 17: 857-73.

242. Shen, M.M., C. Abate-Shen. Molecular genetics of prostate cancer: new prospects for old challenges. Genes Dev, 2010. 24: 1967-2000.
243. Shabsigh, A., D.T. Chang, D.F. Heitjan, A. Kiss, C.A. Olsson, P.J. Puchner, et al. Rapid reduction in blood flow to the rat ventral prostate gland after castration: preliminary evidence that androgens influence prostate size by regulating blood flow to the prostate gland and prostatic endothelial cell survival. Prostate, 1998. 36: 201-6.
244. Shabsigh, A., N. Tanji, V. D'Agati, T. Burchardt, M. Burchardt, O. Hayek, et al. Vascular anatomy of the rat ventral prostate. Anat Rec, 1999. 256: 403-11.
245. Yang, S., M. Jiang, M.M. Grabowska, J. Li, Z.M. Connelly, J. Zhang, et al. Androgen receptor differentially regulates the proliferation of prostatic epithelial cells in vitro and in vivo. Oncotarget, 2016. 7: 70404-19.
246. Lonergan, P.E., D.J. Tindall. Androgen receptor signaling in prostate cancer development and progression. J Carcinog, 2011. 10: 20.
247. Balk, S.P., K.E. Knudsen. AR, the cell cycle, and prostate cancer. Nucl Recept Signal, 2008. 6: e001.
248. Schiewer, M.J., M.A. Augello, K.E. Knudsen. The AR dependent cell cycle: mechanisms and cancer relevance. Mol Cell Endocrinol, 2012. 352: 34-45.
249. Cornforth, A.N., J.S. Davis, E. Khanifar, K.L. Nastiuk, J.J. Krolewski. FOXO3a mediates the androgen-dependent regulation of FLIP and contributes to TRAIL-induced apoptosis of LNCaP cells. Oncogene, 2008. 27: 4422-33.
250. Gong, Y., U.D. Chippada-Venkata, W.K. Oh. Roles of matrix metalloproteinases and their natural inhibitors in prostate cancer progression. Cancers (Basel), 2014. 6: 1298-327.
251. Moroz, A., F.K. Delella, R. Almeida, L.M. Lacorte, W.J. Favaro, E. Deffune, et al. Finasteride inhibits human prostate cancer cell invasion through MMP2 and MMP9 downregulation. PLoS One, 2013. 8: e84757.
252. Chuan, Y.C., S.T. Pang, A. Cedazo-Minguez, G. Norstedt, A. Pousette, A. Flores-Morales. Androgen induction of prostate cancer cell invasion is mediated by ezrin. J Biol Chem, 2006. 281: 29938-48.
253. Castoria, G., L. D'Amato, A. Ciociola, P. Giovannelli, T. Giraldi, L. Sepe, et al. Androgen-induced cell migration: role of androgen receptor/filamin A association. PLoS One, 2011. 6: e17218.
254. Harris, W.P., E.A. Mostaghel, P.S. Nelson, B. Montgomery. Androgen deprivation therapy: progress in understanding mechanisms of resistance and optimizing androgen depletion. Nat Clin Pract Urol, 2009. 6: 76-85.
255. Marques, R.B., N.F. Dits, S. Erkens-Schulze, W.M. van Weerden, G. Jenster. Bypass mechanisms of the androgen receptor pathway in therapy-resistant prostate cancer cell models. PLoS One, 2010. 5: e13500.
256. Kita, Y., T. Goto, S. Akamatsu, T. Yamasaki, T. Inoue, O. Ogawa, et al. Castration-resistant prostate cancer refractory to second-generation androgen receptor axis-targeted agents: opportunities and challenges. Cancers (Basel), 2018. 10: 345.
257. Linder, S., H.G. van der Poel, A.M. Bergman, W. Zwart, S. Prekovic. Enzalutamide therapy for advanced prostate cancer: efficacy, resistance and beyond. Endocr Relat Cancer, 2018. 26: R31-R52.
258. Godbole, A.M., V.C. Njar. New insights into the androgen-targeted therapies and epigenetic therapies in prostate cancer. Prostate Cancer, 2011. 2011: 918707.
259. van der Steen, T., D.J. Tindall, H. Huang. Post translational modification of the androgen receptor in prostate cancer. Int J Mol Sci, 2013. 14: 14833-59.

260. Crona, D.J., Y.E. Whang. Androgen receptor-dependent and -independent mechanisms involved in prostate cancer therapy resistance. Cancers (Basel), 2017. 9: 67.
261. Watson, P.A., V.K. Arora, C.L. Sawyers. Emerging mechanisms of resistance to androgen receptor inhibitors in prostate cancer. Nat Rev Cancer, 2015. 15: 701-11.
262. Hoang, D.T., K.A. Iczkowski, D. Kilari, W. See, M.T. Nevalainen. Androgen receptor-dependent and -independent mechanisms driving prostate cancer progression: opportunities for therapeutic targeting from multiple angles. Oncotarget, 2017. 8: 3724-45.
263. Lallous, N., S.V. Volik, S. Awrey, E. Leblanc, R. Tse, J. Murillo, et al. Functional analysis of androgen receptor mutations that confer anti-androgen resistance identified in circulating cell-free DNA from prostate cancer patients. Genome Biol, 2016. 17: 10.
264. Dehm, S.M., D.J. Tindall. Alternatively spliced androgen receptor variants. Endocr Relat Cancer, 2011. 18: R183-96.

CHAPTER 11

Therapeutic Options for Prostate Cancer: A Contemporary Update

Sakthivel Muniyan[1], Jawed A. Siddiqui[1] and Surinder K. Batra[1,2,3]*

[1] Department of Biochemistry and Molecular Biology, University of Nebraska Medical Center, Omaha, NE 68198-5870, USA
[2] Fred and Pamela Buffett Cancer Center, University of Nebraska Medical Center, Omaha, NE 68198-5870, USA
[3] Eppley Institute for Research in Cancer and Allied Diseases, University of Nebraska Medical Center, Omaha, NE 68198-5870, USA

Introduction

The incidence and mortality rates of prostate cancer (PCa) considerably declined in the past few decades due to the early detection and introduction of various novel therapeutic strategies. Still, PCa is the most common type of cancer in US men, which constitutes 19% among all cancers. In 2019, it is projected that 164,690 men will be diagnosed with PCa and out of that 29,430 men will succumb to this cancer [1]. Further, PCa survivors require long-term care to ameliorate the treatment-related adverse effects such as pain, bleeding, and urinary obstruction. Thus, PCa is also a significant cause of decreased QoL and increased health care expenditures.

Classically, the prostate is an endocrine-responsive organ. Steroid hormones such as androgens are essential regulators as well as growth factors for prostate epithelial cell proliferation, survival, and most importantly, for its secretory function [2]. Since Huggins and Hodges [3] have demonstrated that castration can significantly improve the management of PCa patients, androgen deprivation therapy (ADT) remains the mainstay for the treatment of metastatic PCa. Eventually, these tumors will relapse and develop into more aggressive castration-resistant (CR) phenotype. Chemotherapies traditionally managed CR PCa. Since 2010, at least seven newer therapeutic agents were introduced for PCa (Table 1). Still, there is an inevitable tumor progression, and the majority of PCa patients succumb to their disease within three-years after the diagnosis of CRPCa [4]. Despite the fact that there are more therapies for PCa management now than ever, patients may

---

*Corresponding author: sbatra@unmc.edu

not receive all the available therapies during the treatment period. Extensive efforts have been undertaken in the past decade for the possible sequencing of available therapies to increase progression-free survival (PFS) and overall survival (OS). Also, various novel combinational therapies are under clinical trial based on genomic characteristics. In this chapter, we overview current treatment approaches for the management of advanced and relapsed PCa (Fig. 1).

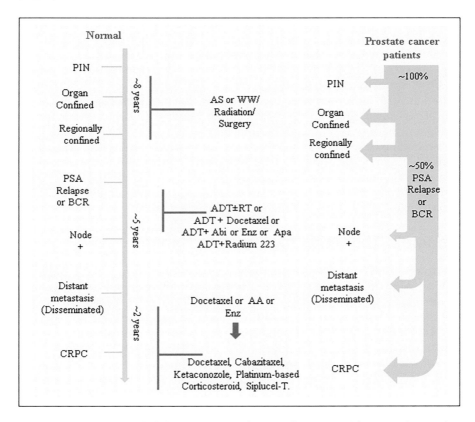

Fig. 1: Prostate cancer (PCa) progression along with stages with approximate time (Left Panel). Primary therapies, including current and FDA approved systemic agents in the management of prostate cancer (Middle Panel). On the right, an approximate percentage of patients diagnosed or developed different stages of PCa in a given time is given.

Prostate Cancer and Current Therapeutic Approaches

PCa is an old men disease, diagnosed by the combination of digital rectal examination (DRE), biopsy, and prostatic acid phosphatase (PAP) [5-7]. The advent of prostate-specific antigen (PSA) has demonstrated a paradigm-shifting impact on the detection, especially the detection of curable disease and management of PCa [8, 9]. Later, the PCa screening combining digital rectal examination (DRE), serum PSA, and ultrasound-guided prostate biopsy demonstrated superior performance of about 70% detection rate [10, 11].

The PCa is a progressive and slow-growing disease which made itself unique in its development and clinical management. There are various therapeutic options available to the PCa patients based on the stage, extent of the disease, possible adverse effects, and patient's preferences and overall health. Majorly, there are seven major treatment options available to PCa patients, and all of the therapy is given either alone or in combination or a sequence based on the disease characteristics. Standard treatment options for PCa include watchful waiting (WW) or active surveillance (AS), surgery, radiation and radiopharmaceutical therapy, hormone therapy (castration, GnRH agonists and antagonists, antiandrogens (AA)), chemotherapy, biological therapy (Sipuleucel-T), and bisphosphonates. At the same time, the management of PCa has become increasingly complex because of available therapeutic options and ever-expanding therapeutic landscape for PCa [12-14]. Though it is heterogeneous, based on the available therapies and its efficacy, the majority of the localized PCa are treatable; even CRPCa is manageable if the tumor is diagnosed early and monitored closely.

Systemic Therapies for Prostate Cancer

There are 16 FDA approved therapeutic agents for PCa [15-31] (Table 1). Among them, majority of the agents target androgen receptor (AR) signaling or are given in combination with ADT. ADT is given classically to the advanced and metastatic PCa and has been shown to extend the survival for 18-24 months [32]. ADT (orchiectomy, gonadotropin-releasing hormones (GnRH) agonists and antagonists, and androgen-receptor blockers) were the primary systemic ADT agents until 2010. However, it invariably results in the development of hormone refractory stage. With the identification of intra-tumoral androgen synthesis and adrenal androgens role in hormone refractory cancer survival, the hormone refractory stage is redefined as castration-resistant PCa [33]. This mechanistic understanding led to the advent of newer second-generation antiandrogens such as abiraterone acetate, enzalutamide, and apalutamide for CRPCa [31, 34]. Even with the next-generation antiandrogens, in almost all cases, the prostate tumor becomes non-responsive to the AR pathway inhibition. Whereas chemotherapies are only palliative and develop resistance in a shorter period, immunotherapy offers hope to advanced PCa patients by significantly improving the survival. However, at present, the immunotherapy clinical trial yielded mixed results in PCa. The bone-targeting radioisotope Radium-223 and bisphosphonates are mainly the treatment choice for metastatic PCa with the involvement of bone.

Androgen Deprivation Therapy

Androgen deprivation is achieved by either surgical (bilateral orchiectomy) or chemical castration (systemic administration). The goal of the ADT is to reduce the circulating androgens to the level of 20-50 ng/dl, which prevents the nourishment of proliferative PCa cells [35, 36]. Orchiectomy was a primary androgen deprivation strategy and was received well in the clinic until the advent of GnRH targeting agents. Castration will bring down the circulating testosterone into the castrate level within 24 hours of the therapy and is permanent. One caveat is that not all the patient population are eligible for surgery (Radical Prostatectomy (RP)) due to other age-related comorbidities. In 1960, Veterans Administration Cooperative Urologic Research Group (VACURG) randomized a trial to compare the efficacy of surgical castration and diethylstilbestrol (DES) [37]. This trial resulted in showing that high-dose diethylstilbestrol (DES) is as effective as ADT, and it opened an alternative option for orchiectomy [37].

Targeting the Gonadal Axis

GnRH agonists are a decapeptide used to suppress the release of the gonadotropins and thus control the testosterone synthesis by testis [38-41]. GnRH agonists (leuprolide, Goserelin, triptorelin, and histrelin) is given either subcutaneous or intramuscularly as a long-term depot. The sustained release of these agents leads to a spike in gonadotropins release and gradually makes the receptors desensitization. In the long term, this leads to a decline in testosterone synthesis and secretion. The initial testosterone surge may lead to undesirable side effects such as spinal cord compression and urinary obstruction due to tumor growth. The initial surge (testosterone flare) is clinically managed by the addition of antiandrogens (combined androgen blockade (CAB)) up to a month [16, 42]. GnRH antagonists share the same pathway (gonadal axis) in controlling the physiological testosterone level. Until recently, the GnRH antagonists were not successful in the clinic due to the instability of the peptides and repeated injection and associated injection-site reactions. Recently, one of the stable antagonists, degarelix, was shown effective in reducing the testosterone level and was made into the clinic by effectively reducing tumor growth which is equal to GnRH agonists [20, 43, 44]. The major advantage of GnRH targeting agents is its reversibility, and it also avoids the psychological belief on surgical castration.

Antiandrogens

Antiandrogens act by binding to AR and inhibiting its signaling, thereby it blocks AR mediated cell survival and reduce PCa growth. However, until recently, in most cases, antiandrogens were not used as a monotherapy in the United States due to its inferiority [45-47]. These agents were added to GnRH agonists to prevent testosterone flare or with any ADT to achieve maximal androgen blockade (MAB). In some cases, these agents were given for the relapsed tumor after ADT. Flutamide, bicalutamide, and nilutamide were the first generation AR antagonists used frequently, with its own merits and demerits. The androgen deprivation-induced mutation favors some of the antagonists into agonists and limits its efficacy in the clinic except bicalutamide (which will act on both wild type and mutant AR) under CAB conditions and as monotherapy in few countries [48-50]. The superiority of the second generation antiandrogen, enzalutamide largely replaced the first generation antiandrogens in castration-resistant conditions except for bicalutamide. Unlike first-generation antiandrogens which block androgen-AR interaction, the newer antiandrogens such as enzalutamide and very recently, apalutamide act on androgen/AR interaction, AR nuclear translocation and Androgen response element (ARE) binding to prevent the action of androgen and its signaling pathway in cancer cells. There are few other AR pathway inhibitors, which show very promising efficacy in preventing the tumor regrowth and are approved for the clinic in the management of PCa (Table 1).

Chemotherapies

Historically, advanced PCa is frequently managed with chemotherapies whenever the patients exhaust other AR-targeting approaches. Taxol-based chemotherapeutic agents were shown to be significantly associated with the beneficial effects of PCa management; for the same reason FDA approved it for CRPCa [22, 24]. Docetaxel is a microtubule stabilizing agent, which will lead to cell cycle arrest at G2/M phase eventually resulting in apoptosis [51, 52]. Microtubules are intracellular filaments that take part in cytoskeleton structural maintenance, vesicular transport, cellular trafficking, and mitochondrial functioning. Taxanes may promote apoptosis by inhibiting the anti-apoptotic function of BCL2 family of proteins [51] and tumor suppressor protein p53 function [52], up-regulating cyclin-dependent kinase inhibitor A (p21, Cip1) [53]. Recently, it has also been proposed that the docetaxel may inhibit AR vesicular transport apart from microtubule formation and thereby induces apoptosis [54, 55]. Cabazitaxel is a next-generation semisynthetic taxane derived from a precursor extracted from yew tree needles. The primary mechanism of action of cabazitaxel is similar to docetaxel in binding and inhibiting β-tubulin, thereby resulting in cell death through G2/M arrest. However, in clinics, it has a limitation of crossing the blood-brain barrier and reduced p-glycoprotein (pgp) binding affinity [56]. *In vitro*, it shows higher efficacy than docetaxel in PCa cells [57]. Mitoxantrone hydrochloride is another chemotherapeutic agent used in the treatment of CR PCa patents along with prednisone, which exhausted all other therapeutic options. Mitoxantrone is a type II topoisomerase inhibitor, which will inhibit cell proliferation by disrupting DNA synthesis and DNA repair by intercalation between the DNA bases [58, 59]. Today, there are various chemotherapeutic agents available but are palliative for CR PCa patients who has exhausted all other therapeutic options.

Table 1: FDA-Approved Prostate Cancer Treatments.

Drug Year of FDA approved	Class/Target	Mechanism of action	Conditions	Phase III study and outcome (Reference)
Bicalutamide (Casodex) 1995	Non-steroidal antiandrogen	Inhibits AR signaling	Advanced prostate cancer (PCa) to achieve CAB	CASODEX Combination Study Group [18]. Bicalutamide plus GnRH agonist increased median time to progression and death
Flutamide (Eulexin) 1996	Non-steroidal antiandrogen	Inhibits AR signaling	Advanced PCa	EORTC study [15]. Flutamide improved the time to progression and survival benefits among metastatic PCa
Nilutamide (Nilandron) 1996	Non-steroidal antiandrogen	Inhibits AR signaling.	Metastatic PCa, which has already undergone surgical castration	Kuhn et al. [16]. Nilutatmide reduced PSA (>50%) in 18 of the 28 patients (64%) who were initially treated with MAB
Goserelin acetate (Zoladex) 1996	GnRH agonists	Continuous inhibition of GnRH receptor signaling	Advanced PCa	ZOLADEX Prostate Study Group [19]. Combined Goserelin and orchiectomy showed similar response in decreasing serum testosterone, acid and alkaline phosphatases, and median times to treatment failure
Leuprolide acetate (Lupron) 1998	GnRH agonists	Continuous inhibition of GnRH receptor signaling	Advanced PCa	Leuprolide Study Group [17]. Comparable efficacy to DES in survival rates
Triptorelin pamoate (Trelstar) 2001	GnRH agonists	Continuous inhibition of GnRH receptor signaling	Advanced PCa	Kuhn et al. [21]. Triptorelin demonstrated significantly higher testosterone reduction than leuprorelin at 12 months

(Contd.)

Table 1: (*Contd.*).

Drug Year of FDA approved	Class/ Target	Mechanism of action	Conditions	Phase III study and outcome (Reference)
Histrelin acetate 2004	GnRH agonists	Continuous inhibition of GnRH receptor signaling	Advanced PCa	Schlegel et al. [23]. Inhibits serum testosterone levels to castrate level as long as the patients were with histrelin implant
Docetaxel (Taxorene) 2004	β-tubulin-binding taxane	Disrupts microtubule assembly	Metastatic PCa	TAX327 [24]. OS: 2.4 months (18.9 versus 16.5 months) SWOG [22]. OS: 1.9 months (17.5 versus 15.6 months)
Degarelix (Firmozon) 2008	GnRH antagonist	GnRH receptor-mediated inhibition of testosterone synthesis	Advanced PCa	Ferring pharma study [20]. Degarelix is as effective as leuprolide in maintaining testosterone levels
Provenge (Sipuleucel-T) 2010	Immunotherapy Targeting prostatic acid phosphatase.	Induces T-cell immunity against prostatic acid phosphatase	Asymptomatic or minimally symptomatic castration-resistant (CR) PCa	IMPACT [25]. OS: 4.1 months (25.8 versus 21.7 months)
Cabazitaxel (Jevtana) 2010	β-tubulin-binding taxane	Disrupts microtubule assembly	Metastatic CR PCa	TROPIC [26]. OS: 2.4 months (15.1 versus 12.7 months)

(Contd.)

Denosumab (Xgeva) 2010	Monoclonal RANKL antibody	Prevents RNAKL mediated RANK receptor activation	Bone metastatic PCa receiving androgen deprivation therapy.	Fizazi et al. [27]. Denosumab had shown extended median time to first SREs than zoledronic acid
Abiraterone Acetate (Zytiga) 2011	17 a-hydroxylase/ C17,20-lyase (CYP17) inhibitor	Inhibits adrenal and intra-tumoral androgen biosynthesis	Advanced PCa	COU-AA-301 [28]. OS: 3.9 months (14.8 versus 10.9 months)
Enzalutamide (Xtandi) 2012	AR antagonist	Blocks androgen binding, inhibits AR nuclear translocation, and DNA binding and transcription	Metastatic CR PCa	AFFIRM [30]. OS: 4.8 months (18.4 versus 13.6 months)
Radium 223 Dichloride (Xofigo) 2013	Radioactive therapeutic agent (Alpha particle-emitter)	Complex formation with bone mineral hydroxyapatite, and induces DNA double-stand breaks in neighboring cells.	PCa with bone metastases	ALSYMPCA [29]. OS: 2.8 months (14.0 versus 11.2 months)
Apalutamide (Erleada) 2018	AR antagonist	Blocks androgen binding, inhibits AR nuclear translocation, and DNA binding and transcription	Non-metastatic CR PCa	Metastasis-free survival: 24.3 months (40.5 vs. 16.2 months) [31].

AA - antiandrogens; AR - androgen receptor; GnRH - Gonadotropin-releasing hormone; OS - overall survival; PCa - prostate cancer.

Immunotherapy

Sipuleucel-T (Provenge) is the only FDA approved biological therapy for advanced PCa. Sipuleucel-T is the first T cell-associated tumor immunotherapeutic vaccine for mCRPCa. Provenge is composed of the PAcP antigen (full-length PA2024 linked to an adjuvant granulocyte-macrophage colony-stimulating factor (GM-CSF)), autologous peripheral blood mononuclear cells (PBMCs) and the activated antigen presenting cells (APCs). Sipuleucel-T works based on the principle of activating one's immune system to kill the foreign (tumor) cells. Sipuleucel-T has been developed with the combination of recombinant tumor antigens (PAcP) infused into the patients' blood with their own activated APCs to induce an immune response against the tumor-associated antigen (TAA).

Management of Localized Therapeutic Approach with a Curative Intention

Primary Prostate Cancer/Clinically Localized Disease (Stage T1 and T2)

Since PSA has become a routine diagnostic test, almost 80% of PCa are diagnosed at early organ confined stage (Stage T1 or Stage T2), and the five-year survival is nearly 100% [60]. This group of patients is well managed with the first-line therapies such as surgery or radiation or by AS. AS (or WW) is one of the treatment options for low-risk and non-aggressive PCa, and for those who have less than five years of survival. These were suggested for the patients whose PSA is between 4 ng/dl and 10 ng/dl, and the primary tumors are organ-confined. AS also considers the treatment-associated risk factors, quality of life (QoL) and cost-associated with the treatment, because overtreatment of low-risk PCa is a significant concern. To date, AS is considered a safe approach to definite treatment and strategic monitoring through periodic PSA level, DRE, imaging, and biopsy, with the intention of moving this population to curative option if needed [12]. In the past decade, there has been an increasing trend towards the selection of AS as an alternative treatment option. Population-based cancer registries identified that patients under AS increased from previous decades (10%) to current period (40% and as high as 70%) [61-63]. Various studies reported that those under AS had better cancer-specific survival under low-risk conditions with median surveillance for up to 11 years. Interestingly, one study reported that 2- and 5-year progression-free survival (PFS) of PCa patients under AS was 80% and 60%, respectively [64]. All these studies suggest that AS to low-risk PCa is still a viable and safer option. Despite its safer choice to low-risk PCa, the majority of the studies disagrees with the selection and monitoring protocols [65].

The RP is the primary type of therapy usually reserved for the men with organ-confined tumor and who are 75 years or younger with at least ten years of predicted survival. RP is a major surgical procedure to remove the whole prostate and seminal vesicles and nearby pelvic lymph nodes. An initial randomized clinical trial demonstrated that RP resulted in significantly higher PFS when compared to radiation therapy (RT) [66]. However, this multicentered cooperative group effort led to various concerns, such as differences in patient recruitment and radiation treatment. Further, the RP has a major drawback of impotence, urinary incontinence, and lymphedema inguinal hernia. Over time, additional and less invasive surgical methods such as nerve-sparing prostatectomy, robotic-assisted laparoscopic RP, and transurethral resection of the prostate (TURP) were developed and shown to reduce some of the risk events and side effects [67, 68]. Since the introduction of PSA, the surgery has significantly improved the PFS due to the early detection. Besides these benefits, it is shown that the primary surgical treatment may avoid the adverse effects of radiation and adjuvant hormonal therapy among those patients who live longer.

Among the patients who were not eligible for surgery due to age-associated comorbidities, RT is the principal treatment option for localized PCa; in the past, its extended and better PFS in PCa patients has been demonstrated [69-74]. External beam radiation (EBRT) and Brachytherapy (BT) are the two main types of radiation therapy used for PCa. EBRT is one of the standard treatment options to treat PCa to destroy the cancer cells by sparing the surrounding healthy cells. With the technological advancement over the past few decades, RT is advanced into intensity-modulated RT (IMRT), image-guided RT (IGRT), and stereotactic body RT (SBRT) which has improved the outcome by escalating the dose and by minimizing the toxicity to adjacent healthy tissue. On the other hand, BT uses small radioactive pellets to deliver its efficacy, either through low-dose BT for prolonged time and high dose BT for short time. There are benefits and pitfalls of each type of treatment for localized PCa [75-77], and despite both RP and RT having similar PFS among low-risk organ-confined tumor, it is also the patient preference towards the selection of RT, RP or AS as treatment strategy [78-80].

Locally Advanced Prostate Cancer

Locally advanced PCa is defined by the tumor which has grown beyond the prostate organ, to seminal vesicle, pelvic lymph node, bladder, but has not spread to distant parts of the body (Stage T3 and Stage T4). This stage of PCa is conventionally managed by conservative management, RP, preoperative radiation, and hormonal therapy and RT [81]. The mono or combinational therapy for locally advanced PCa depends on various factors such as the stage, cancer grade, PSA level, and general health of patients. One of the challenge in treating the locally advanced PCa depends on the extent of clinically undetectable metastatic PCa. The fact is reflected well on its 5-year survival of 30% among metastatic castration-sensitive prostate cancer (mCSPCa) as opposed to 100% of 5-year survival among localized PCa [60]. For the low-grade stage T3 cancers without detectable metastases, local curative therapy is recommended, if eligible (based on age, comorbidities, and life expectancy). There is no definitive treatment regimen for high risk, locally advanced PCa. RP results in 38% to 51% biochemical PFS upon 10-year follow-up. However, the 10-year cancer-specific survival is ~90% among those who underwent adjuvant and salvage procedures [82, 83]. The differential survival benefit is further supported by the fact that out of all the patients primarily treated by RP, more than 50% eventually require adjuvant or salvage RT or hormonal therapy [84]. These studies suggest that there should be a multimodal approach for the management of advanced PCa based on the extent of the tumor, PSA value, tumor stage, and lymph node involvement. Further, it is also suggested that the age, life span, and other comorbidities of the patients should be taken into consideration.

Management of Castration-sensitive (Advanced) Prostate Cancer

In most cases, castration-sensitive (CS) PCa (CSPCa) is a relapsed tumor after first-line curative therapy and might be detected either by biochemical recurrence or clinical (radiological) recurrence before the hormonal intervention. There is no consensus for the treatment of CS non-metastatic PCa. It is suggested that these high-risk localized PCa must be managed by multimodal approach due to the likelihood of positive surgical margins (33.5–66% of patients) and the presence of node-positive tumor (7.9–49% of patients) [84-87]. The immediate therapy for this recurrent tumor is the salvage or with adjuvant hormonal therapy [84, 85]. It was further shown that asymptomatic T3-T4 tumor might benefit with adjuvant ADT only in patients with higher PSA (>50 ng/ml), less PSA doubling time (<12 months) and longer life expectancy [87, 88]. Recently, taxel based chemotherapy is proposed in combination with standard of care ADT for the high-risk localized PCa patients. Currently, the evidence from any phase III randomized trials to study the benefit of adding chemotherapy to the standard of care for the management of non-metastatic CSPCa patients is not mature enough to come to any clinical decision. However, the analyses of phase II clinical studies have yielded mixed results on the clinical improvement of CSPCa [89-92]. Some of the suggested reasons are that PCa is quite heterogeneous and in many cases the standard of care depends on multiple factors including, tumor stage, patient's age, kind of primary therapy and expected survival [12-14].

Node-positive Prostate Cancer

Nodal positive is the most common manifestation during the prostatectomy. Among the 164,690 expected new cases, only 12% of patients have lymph node metastasis at the time of diagnosis [60], and generally, these PCa patients with regional spread have close to 100% 5-year survival. However, the lymph node involvement (LNI) represents an unfavorable prognosis when compared with the absence of lymph node metastases [93]. Several institutional and population-based studies suggest that the combination of RT and ADT or RP and ADT might be associated with favorable outcomes in node-positive tumor compared to ADT alone. Notably, Lin and colleagues showed that patients who were treated with RT were associated with favorable survival when compared to standard of care [94]. Using SEER database, Rusthoven et al. [95] observed that local therapy (RP 57%, EBRT 10%, or both 11%) had significantly better 10-year OS (65% vs. 42%; P<.001) and prostate cancer-specific survival (PCSS) (78% vs. 56% P<.001). Similarly, Tward and colleagues [96] observed that RT is significantly associated with reduced all-cause mortality and PCSS. Recently, Seisen and colleagues [97] compared OS among the node-positive metastatic PCa patients who received ADT with and without local therapy. These analyses suggest that any form of local therapy to standard ADT may give significant survival benefit. Local therapy, in general, represents a valuable option for lymph node positive tumors; these individuals show a significant risk of recurrence and progression as compared to those diagnosed with advanced PCa without LNI. However, studies which analyzed the tumor extension suggest that the number of positive lymph nodules might be helpful in predicting the tumor prognosis and recurrence [98-101].

Castration-sensitive Prostate Cancer

Despite the decline in PCa incidence, recently, there has been a 72% higher incidence of metastatic PCa (castration-sensitive) [102, 103]. Metastatic PCa includes stage IVB cancers with cancer metastasized to lymph nodes (nearby and distant), vertebrae, and other visceral organs. There are approximately 4% of PCa patients diagnosed as metastatic mCSPC at the time of diagnosis. However, this *de nova* mCSPC is usually fatal and has a shorter time to develop CR phenotype than the patients who develop CRPCa after curative treatments [104]. Further, the relative overall 5-year survival rate for distant stage PCa is about 30% [60]. Advanced-stage PCa often develops bone metastasis, which occurs in approximately 90% of metastatic PCa patients [105].

The treatment landscape of mCSPC has significantly transformed over the past decade. Historically, metastatic PCa was commonly managed with ADT. However, this paradigm is shifting in favor of adding additional agents to the standard of care (ADT) for the management of mCSPC. After the superior effect of docetaxel in extending the survival of castration-resistant (CR) PCa [22, 24], the docetaxel-based chemotherapy has been combined with the various combinational agent for PCa. Consequently, the docetaxel is aimed to combine with the standard of care in CSPCa conditions. Chemotherapy was even more successful when given in the hormone-sensitive setting. A decade later, the large RCTs, CHAARTED, STAMPEDE, and GETUG-AFU

15 present the evidence for first-line docetaxel chemotherapy alongside ADT which has shown the benefit of combining docetaxel with androgen deprivation therapy [106-110]. GETUG-AFU 15 trial showed that nine-three week cycles of docetaxel plus ADT increased overall survival (OS) when combined with ADT (median survival 62.1 versus 48.6 months), but did not achieve statistical significance (HR 0.88, (0.68-1.14; p = 0.3). However, the biochemical and radiographic PFS were significantly higher in the docetaxel plus ADT arm [(HR = 0.73 (0.56-0.94); p = 0.014 and (HR = 0.75 (0.58-0.97); p = 0.030, respectively] [111]. The results of CHAARTED trial [106] clearly showed that the median OS in the chemohormonal arm was 57.6 months against 47.2 months in ADT arm (hazard ratio [HR], 0.72 (0.59 - 0.89); p = 0.0018). The long term survival analyses of the CHAARTED trial [106] demonstrated that high volume disease (defined as having the visceral disease and more than four bone lesions) had benefited well from the combination of docetaxel and the standard of care ADT. The high-volume disease has higher median OS among the combinational arm than ADT arm [51.2 vs. 34.4 months, HR = 0.63 (0.50 to 0.79); P < .001], whereas for those with low-volume disease, the OS benefit was not statistically significant (HR, 1.04 (0.70 to 1.55); P = 0.86) [110]. Similarly, the STAMPEDE [107] trial showed that the upfront treatment of docetaxel with ADT significantly increased the overall survival (81 months versus 71 months) among the standard of care treatment. Later, the two phase III clinical studies showed similar survival improvement when the ADT is combined with another hormonal agent, abiraterone plus prednisone [112, 113]. The STAMPEDE showed that upfront treatment of abiraterone and prednisolone was significantly associated with OS and disease-free survival (DFS) when compared to the standard of care [112]. Similarly, in LATITUDE trial, the addition of abiraterone plus prednisone to ADT significantly increased the median OS than the placebo group (HR = 0.62 (0.51 to 0.76); P<0.001) [113]. Further, the median radiographic PFS in the abiraterone plus prednisone vs. placebo group was 33.0 versus 14.8 months (HR = 0.47 (0.39 to 0.55); P<0.001) [113].

A novel second-generation antiandrogen, enzalutamide which has shown significant efficacy in metastatic CRPC under chemotherapy naïve and chemotherapy exposed conditions, was also queried under CSPC conditions. An initial efficacy was assessed in a single arm phase II clinical study with a hormone-sensitive disease that would generally be treated with ADT [114]. From 67 men at baseline, 62 patients had a decline in PSA of 80% or higher at week 25 (92.5%, 95% CI 86.2-98.8). The most common side effects were gynecomastia, fatigue, nipple pain, and hot flashes (36, 34, 19, and 18 percent, respectively). However, the activity duration and the efficacy relative to standard ADT will require comparative clinical trials and longer follow-up. Apart from this, the use of enzalutamide in men with hormone-sensitive PCa remains experimental.

A recent phase III clinical trial showed that ADT plus Apalutamide (Erleada) was found to significantly associate with radiographic PFS (rPFS) and OS versus placebo in metastatic CSPCa settings. SPARTAN phase III study which randomized 1207 patients in a 2:1 ratio already revealed that addition of ADT led reduced risk of metastasis (72%) and death in nonmetastatic CRPC patients. This trial shows that apalutamide significantly increases median metastasis-free survival [40.5 months vs. 16.2 months (HR = 0.28, 0.23 to 0.35; P<0.001)]. Similarly, the apalutamide treatment significantly extends the time to symptomatic progression (HR = 0.45; 0.32 to 0.63; P<0.001) [31]. Similarly, unpublished Phase 3 TITAN study from Janssen Pharmaceuticals describes that apalutamide achieved primary endpoint in metastatic CSPCa. [115].

All these studies have shown that early intervention with either chemo or hormonal agent significantly improved the prognosis of mCSPC patients for the first time; however, they also added further challenges in patient selection for treatment and in sequencing the best agents. Further, the PCa is highly heterogeneous, and the treatment is varied according to tumor stages and progression and response rate. To date, there is no direct comparison between different treatment combinations with the standard of care (ADT). However, the meta-analyses compared the overlapped treatment arms (ADT+docetaxel vs ADT+abiraterone acetate) of STAMPEDE trial to give the evidence of no significant difference in OS and PCSS or symptomatic skeletal events between these two (new standard) treatment approaches [116]. However, given that localized CSPC patients may undergo androgen deprivation for a longer time than those with metastatic disease, additional therapeutic agents may lead to unwanted treatment-related adverse events. This creates the debate on the timing of hormonal therapy and suggests further studies.

Castration-resistant Prostate Cancer

Most PCa patients will initially respond to androgen ablation (hormonal treatment), but the vast majority of castration-sensitive disease develops the castration-resistant phenotype. These group of patients frequently develop metastases, particularly to lymph nodes, bone and, to a lesser extent, to visceral organs such as the liver and lungs. Men with CRPC often have pain and other comorbidities leading to decreased QOL. Despite various approved therapeutic agents, CRPCa is always fatal.

Classically, CRPC is managed by chemotherapy as a palliative agent. Until 2004, mitoxantrone was used as a major therapeutic agent. In 2004, docetaxel largely replaced mitoxantrone as the first-line chemotherapeutic agent and showed ~2-4 months of the extended OS. However, since 2010, various new agents have been approved for metastatic castrate-resistant PCa (mCRPC) and these agents changed the treatment landscape significantly. Currently, CRPCa patients may be eligible to receive docetaxel, cabazitaxel, siplucel-T, abiraterone acetate, enzalutamide, apalutamdie, and various experimental agents under clinical trials.

Chemotherapies for Advanced Prostate Cancer

Chemotherapeutic agents were developed to improve the QoL and to substantially increase the time to progression (PFS) among mCRPCa. Initial agents, such as cyclophosphamide, vinorelbine, and mitoxantrone plus prednisone were poorly received in the clinic to demonstrate its effects on palliation and improvement in OS [117-119]. Early in 2004, two phase III clinical trials demonstrated that docetaxel-based trials have been shown to improve both the QoL and median overall survival (OS) of 2.4 and 1.9 months with a 3-weekly six cycle schedule [22, 24]. For TAX327 study, 1006 men with mCRPC patients were randomized to receive docetaxel (every three weeks), weekly docetaxel, and mitoxantrone with prednisone as an anti-inflammatory agent. Among them, patients who received 3-week once docetaxel for six cycles demonstrated 2.4 months extended OS. A later study which analyzed the benefit of docetaxel with extended follow-up confirms that docetaxel regimen has significantly higher median survival compared to the mitoxantrone group among the men with metastatic CRPC [120]. Further analyses also found that docetaxel (every three weeks) had a significantly better pain response rate, PSA, and QoL versus mitoxantrone arm. The Southwest Oncology Group (SWOG) 9916 study compared docetaxel (60 mg/m^2) plus estramustine with mitoxantrone plus estramustine and found significantly better median survival of 1.9 months (17.5 vs. 15.6 months p=0·02) among mCRPCa patients. However, the docetaxel-efficacy was limited to the rapid development of resistance, was modest (a few months) in improving survival and had a high toxicity profile. Subsequently, a second generation taxel, cabazitaxel has been developed and compared for its efficacy in mCRPC patients who have progressed after docetaxel chemotherapy. In the TROPIC trial [26], 755 patients were randomized to receive either cabazitaxel plus prednisone or mitoxantrone and prednisone. This study found that patients who received cabazitaxel have significantly longer median OS of 2.4 months versus mitoxantrone group (HR 0.70 (0.59-0.83); P < 0.0001). Cabazitaxel also had significantly longer median PFS (2.8 months vs. 1.4 months), and reduced PSA and supportively decrease the tumor lesions size than mitoxantrone group. Cabazitaxel is also shown to retain its activity after abiraterone and enzalutamide [121, 122]. However, as an inherent property of any chemotherapy, cabazitaxel is also limited to higher toxicity and modest survival benefit. Even a later phase III trial, FIRSTANA [123] randomly assigned mCRPCa patients to receive 20 mg/m^2 and 25 mg/m^2 cabazitaxel and 75 mg/m^2 docetaxel and found that both cabazitaxel doses did not demonstrate a superior survival benefit over docetaxel.

Second Generation Antiandrogens for Castration-resistant Prostate Cancer

Despite the efficacy of hormonal therapy in containing tumor progression, hormonal therapy becomes ineffective over a period. Various preclinical and clinical evidence suggest PCa develops multiple ways to feed itself. This understanding led to the identification and development of numerous AR pathway targeting agents such as abiraterone acetate, enzalutamide, apalautamide, etc. The later studies suggest that AR pathway is still critical in CRPCa, and it is shown that the androgen receptor antagonists to some extent inhibit the CRPCa or it provides palliative symptoms in metastatic CRPCa conditions.

In the COU-AA-301 trial [124], 1195 eligible patients were randomized to receive either abiraterone acetate plus prednisone or receive placebo plus prednisone in 2:1 ratio. With 20.2 months median follow-up, this study found that patients who received abiraterone acetate have significantly longer median OS of 15.8 months versus 11.2 months among the placebo group (HR = 0.74 (64–0.86) p<0.0001). Abiraterone acetate is also shown to significantly improve the median time to PSA progression and median radiologic PFS versus the placebo group [124]. Further, the interim analyses of COU-AA-302 trial found that abiraterone acetate delayed disease progression, reduced pain, and functional deterioration and delayed time to initiate chemotherapy among the patients with mCRPC in chemo naïve conditions [125].

A potent and oral second-generation antiandrogen, enzalutamide, was evaluated in a randomized phase 3 trial (AFFIRM) [30] in 2:1 ratio. This study demonstrated that enzalutamide intake significantly prolonged survival (HR = 0.63 (0.53 to 0.75); P<0.001), PSA level reduced by 50% or more (54% vs. 2%, P<0.001), and time to biochemical and radiographic progression in CRPC patients. Further, this study revealed that in the majority of the patients, tumor has biochemically progressed while receiving enzalutamide [30], which suggests that the tumors remained driven by the low level of circulating androgens possibly through the low androgenic synthesis by the adrenal gland. Later, the PREVAIL study on chemo naïve patients found that enzalutamide significantly decreased the risk of radiographic progression and death. Importantly, enzalutamide treatment improved the time to subsequent chemotherapy and the time to first skeletal-related event (SREs) among the men with mPCa [126]. Recently, enzalutamide was also approved for men with nonmetastatic CRPC based on the PROSPER clinical study [127]. This study compared enzalutamide and placebo with ADT which were biochemically progressed and are highly risky for metastasis. The results showed that patients who received enzalutamide had a significantly lower risk of metastasis or death (HR = 0.29 (0.24-0.35); p<0.001). Further, the study also showed enzalutamide treatment significantly prolonged the time for subsequent antineoplastic therapy (15% vs. 48%) and PSA progression (37.2 vs. 3.9 months) compared to placebo [127].

Around the same time, another next-generation AR antagonist, apalutamide (also known as ARN-509) is introduced into the clinic. The initial preclinical and phase II studies suggest that promising clinical efficacy and lower central nervous system (CNS) accumulation in mCRPC patients. With the promising efficacy of apalutamide in mCRPC patients (phase II studies) [128] and phase III nonmetastatic CRPCa [31], several clinical studies are ongoing [129]. Though the initial findings from these phase II studies are encouraging, the mature findings are yet to publish.

Immunotherapy

Originally, Sipuleucel-T was approved for patients who developed metastatic CRPCa and progressed on chemotherapy. In phase III randomized trial, Sipuleucel-T treatment significantly reduced the death compared to placebo (HR = 0.78; 95 % CI, 0.61–0.98; p = 0.03). This study demonstrated additional 4.1-month median survival compared to the placebo group; however, the time for disease progression is the same, and no decline in PSA was observed between the arms. Currently, there are various other immunotherapies against several targets and are at different clinical stages such as PROSTVAC-VF, GVAX, Ipilimumab (checkpoint inhibitor) and pembrolizumab (PD-1/PD-L1 inhibitor), either as monotherapy or in combination.

Majority of the above-said drugs has demonstrated significant clinical benefit. Still, a large number of patients are not responding after a brief period. In such a condition, when the patients have exhausted all other therapeutic options, platinum-based therapies are given. In most cases, these therapies are only palliative. There are various targeted therapies still in the experimental and clinical developmental stage but they are yet to mature to demonstrate their beneficial effect.

Prostate Cancer-bone Metastasis

Advanced-stage PCa often develops bone metastasis, which occurs in approximately 90% of metastatic patients [60]. The metastatic castration-resistant PCa (mCRPC) leads to skeletal-related events (commonly known as SREs) that are associated with bone pain, spinal cord compression, the risk of multiple pathological fractures, decreased quality of life (QoL) and increased mortality.

Effect of Androgen Deprivation Therapy on Bone Metastasis of Prostate Cancer

ADT is the most frequently used primary treatment option for metastatic PCa, which includes either surgical or medical castration. Furthermore, in the case of non-metastatic PCa, the combination of ADT with other treatment option is a frequently used strategy to combat PCa [130]. ADT, in combination with radiotherapy, is commonly used in patients with rising PSA levels after failed primary treatment.

Bone metastasis is the characteristic clinical symptom for patients with CRPCa and has a negative impact on prognosis. Most of the chemotherapy and RT to treat PCa and bone metastasis are mainly palliative. It has been reported that approved standard localized PCa therapies, abiraterone (CYP17 inhibitor) and enzalutamide (AR antagonist) both have demonstrated to significantly delay the SREs in patients with mCRPC [28, 131]. The outcome of a recent clinical trial suggests that cabozantinib (a tyrosine kinase inhibitor) improves the skeletal health in mCRPC [132]. The exact mechanism by which cabozantinib reduces the SREs is not explored, but it may target both cancer cells and osteoblasts [132]. However, direct bone-targeted therapies have a significant advantage for patients with PCa bone metastasis. In past years, advances in the understanding of the molecular mechanism of bone metastasis and cancer cell-bone microenvironment interactions have allowed the development of new targets and therapeutic agents to treat bone metastasis (Table 1).

Apart from the therapeutic benefit, patients on ADT have significantly lower bone mineral density (BMD) or deleterious effect at several skeletal sites compare to naïve or healthy age-matched individuals. Various studies have shown that ADT is associated with loss of bone density and the higher risk of bone fracture [133-136]. ADT-associated bone loss has such a prominent effect on skeletal integrity that is not reversed with calcium or Vitamin D supplementation. Low levels of testosterone due to ADT lead to a significant reduction in the serum estradiol level [137]. Estrogen receptor (ER) is the well-known receptor for estradiol, which is expressed by both osteoblasts and osteoclasts and is a primary regulator of bone remodeling, suggesting that the loss of estradiol is responsible for ADT-mediated decrease in BMD [138]. The outcome of a recent clinical trial indicated that adding external beam radiation therapy (EBRT) to the primary prostatic tumor with the continuation of ADT did not improve overall survival [139].

Bone-targeted Therapies

Several bone-targeted therapies have been approved for bone pain management and to prevent SREs. The CRPCa with bone metastasis is primarily managed with the addition of denosumab and zoledronic acid, which mainly targets osteoclasts and prevents bone deformities.

Bisphosphonates

Bisphosphonates are derivatives of pyrophosphate that bind to hydroxyapatite crystals and impede osteoclast-mediated resorption of bone matrix. Third-generation nitrogen-containing bisphosphonates such as zoledronic acid, ibandronate, neridronate, and risedronate are the most potent bisphosphonates. While bisphosphonates are commonly used to treat osteoporosis, they are also used to reduce the incidence of SREs associated with PCa-bone metastasis. Nitrogen-containing bisphosphonates (Zoledronic acid) are a potent inhibitor of mevalonate pathway enzymes (Farnesyl diphosphate synthase), which directly suppresses osteoclast function. The serum bone resorption markers significantly decreased with initial treatment of zoledronic acid which further suggests its role in inhibition of osteoclasts functions. In addition to inhibiting the osteoclast-mediated bone resorption, zoledronic acid also increased PCa cells' apoptosis, and inhibited cancer cell adhesion to bone matrix and tumor's growth. Further, patients treated with zoledronic acid have significantly reduced PCa-induced bone pain. It has been shown that bisphosphonates have a beneficial effect on patients with non-metastatic PCa treated with ADT in terms of better bone health rather than high-risk localized PCa [140]. However, bisphosphonate withdrawal leads to loss of BMD in patients on ADT. In several clinical trials, zoledronic acid has been shown to improve BMD [141-143].

Denosumab

Receptor activator of nuclear factor kappa-B (RANK) ligand RANKL is the crucial molecule for osteoclast formation and their activation. Several preclinical studies suggest that osteoclast inhibition might prevent bone metastases. The data from the preclinical study indicate that targeting RANKL might inhibit PCa-bone metastasis [144, 145]. A randomized study shows that targeting the bone microenvironment can delay bone metastasis in men with PCa. A humanized monoclonal antibody, denosumab binds and neutralizes the RANKL and controls osteoclastogenesis and restores the bone resorption. The outcome of a clinical study has shown that Denosumab significantly increased bone-metastasis-free survival. Denosumab had better efficacy in terms of reducing bone resorption marker and incidence of SREs in patients not adequately responding to bisphosphonates therapy [146]. Denosumab has been shown to improve the BMD at all skeletal sites and in all subgroups. The improvement of BMD was better in patients with high baseline bone turnover markers [147]. However, incidence of cataracts have been reported in some patients who received denosumab. The outcome of Hormone Ablation Bone Loss Trial (HALT 138) study demonstrated that denosumab increases the BMD (mostly at lumbar spine) and decreases the bone turnover markers in patients with non-metastatic PCa on ADT compared to placebo [147]. Osteonecrosis of the jaw (ONJ) is reported in nearly 5% of patients treated with denosumab. In another clinical trial, compared to zoledronic acid, denosumab showed better efficacy in the suppression of bone turnover markers [27]. Due to lack of progression-free survival and high incidence of ONJ, denosumab did not receive FDA approval for earlier intervention among the patients with non-metastatic CRPC.

Radiopharmaceuticals

Currently, several radiopharmaceuticals such as strontium-89, samarium-153, and radium-223 are approved for the treatment of SREs in mCRPC. These radiopharmaceutical agents are preferentially deposited in the area of bone with an increased bone turnover due to cancer metastases. Strontium-89 and samarium-153 are mainly beta-particle-emitter and have been approved for the bone pain management arising due to PCa bone metastasis but unfortunately have not been shown to extend the survival [148]. Being beta-particle-emitters, both particles have high tissue penetration, which raises concern over bone marrow toxicity or myelosuppression.

A preclinical study showed that uptake of calcium-mimetic radioisotope, radium-223 (alpha-particle emitter) was increased in osteoblastic lesions compared to healthy bone. Patient's bioavailability data suggests that total uptake of radium-223 was 40-60% followed by the intestine. Phase II studies showed that radium-223 reduced the SREs related pain and also improved the bone metastasis biomarker [29, 149]. In phase III study, the Alpharadin in Symptomatic PCa Patients (ALSYMPCA) trial, which was randomized, double-blind, placebo-controlled, in which 921 patients with CRPC and symptomatic bone metastases were randomly assigned to receive intravenous treatments of radium-223 at 50 kBq/kg IV monthly for six months, or placebo showed the benefit of radium-223 over placebo in median overall survival [29, 150]. Alpha-emitting radioisotopes are safer than beta emitters in terms of toxicity while having greater potency to irradiate cancer cells. The incidence of myelosuppression, which is common in beta emitters, was very low in patients treated with radium-223. Therefore, radium-223 has received regulatory approval for use in CRPC with symptomatic bone metastases.

Summary and Conclusion

Historically, PCa was managed well with ADT with local therapies. However, a subset of patients still develops resistance and does not respond to available treatments, which is evident from the fact that ~29,000 men are dying because of cancer through various therapeutic options. Ongoing studies and clinical evidence deepen our knowledge on prostate biology, novel therapeutics, and resistance mechanisms. Since 2010, the introduction of the second generation of antiandrogens and second-line chemotherapies led to a significant decline in PCa death. Equally, even after eight decades of the initiation of first ADT, the development of metastatic CRPCa remains a deadly disease which frequently adjusts or evades AR signaling. Of note, there is an increased incidence of NEPC subtype after the second generation AA treatments.

Interestingly, the recent success of chemo-hormonal therapy also highlighted the significance of the right sequence and combination of available therapies for successful management. With the availability of multiple AR pathway inhibitors at various stages, varied survival benefits and adverse events, the current goal is the best sequential use of these agents in the management of mCRPCa patients to prolong survival, minimize adverse events and maintain QoL.

Acknowledgments

The authors are, in part, supported by the grants from National Institutes of Health UO1 CA185148 and Department of Defense - Idea Award (PC170891).

References

1. Siegel, R.L., K.D. Miller, A. Jemal. Cancer statistics, 2019. CA Cancer J Clin, 2019. 69: 7-34.
2. Isaacs, J.T. The biology of hormone refractory prostate cancer. Why does it develop? Urol Clin North Am, 1999. 26: 263-73.
3. Huggins, C., C.V. Hodges. Studies on prostatic cancer: I. The effect of castration, of estrogen and of androgen on serum phosphatases in metastatic carcinoma of the prostate. Cancer Research, 1941. 1: 293-7.
4. A.U. Association. Castration-Resistant Prostate Cancer (Published 2013; Amended 2018, Acessed 04 February 2019).
5. Nadji, M., Z. Tabei, A. Castro, A.R. Morales. Immunohistological demonstration of prostatic origin of malignant neoplasms. Lancet, 1979. 1: 671-2.
6. Jobsis, A.C., G.P. De Vries, R.R. Anholt, G.T. Sanders. Demonstration of the prostatic origin of metastases: an immunohistochemical method for formalin-fixed embedded tissue. Cancer, 1978. 41: 1788-93.
7. Nadji, M., S.Z. Tabei, A. Castro, T.M. Chu, A.R. Morales. Prostatic origin of tumors: an immunohistochemical study. Am J Clin Pathol, 1980. 73: 735-9.
8. Nadji, M., S.Z. Tabei, A. Castro, T.M. Chu, G.P. Murphy, M.C. Wang, et al. Prostatic-specific antigen: an immunohistologic marker for prostatic neoplasms. Cancer, 1981. 48: 1229-32.
9. Papsidero, L.D., M.C. Wang, L.A. Valenzuela, G.P. Murphy, T.M. Chu. A prostate antigen in sera of prostatic cancer patients. Cancer Res, 1980. 40: 2428-32.
10. Brawer, M.K., M.P. Chetner, J. Beatie, D.M. Buchner, R.L. Vessella, P.H. Lange. Screening for prostatic carcinoma with prostate specific antigen. J Urol, 1992. 147: 841-5.
11. Catalona, W.J., D.S. Smith, T.L. Ratliff, J.W. Basler. Detection of organ-confined prostate cancer is increased through prostate-specific antigen-based screening. JAMA, 1993. 270: 948-54.
12. Heidenreich, A., P.J. Bastian, J. Bellmunt, M. Bolla, S. Joniau, T. van der Kwast, et al. U. European Association of, EAU guidelines on prostate cancer. Part 1: Screening, diagnosis, and local treatment with curative intent-update 2013. Eur Urol, 2014. 65: 124-37.
13. Mottet, N., J. Bellmunt, M. Bolla, E. Briers, M.G. Cumberbatch, M. De Santis, et al. EAU-ESTRO-SIOG guidelines on prostate cancer. Part 1: Screening, diagnosis, and local treatment with curative intent. Eur Urol, 2017. 71: 618-29.
14. Pignot, G., D. Maillet, E. Gross, P. Barthelemy, J.B. Beauval, F. Constans-Schlurmann, et al. Systemic treatments for high-risk localized prostate cancer. Nat Rev Urol, 2018. 15: 498-510.
15. Denis, L.J., J.L. Carnelro de Moura, A. Bono, R. Sylvester, P. Whelan, D. Newling, et al. Goserelin acetate and flutamide versus bilateral orchiectomy: a phase III EORTC trial (30853). EORTC GU Group and EORTC Data Center. Urology, 1993. 42: 119-29; discussion 129-30.
16. Kuhn, J.M., T. Billebaud, H. Navratil, A. Moulonguet, J. Fiet, P. Grise, et al. Prevention of the transient adverse effects of a gonadotropin-releasing hormone analogue (buserelin) in metastatic prostatic carcinoma by administration of an antiandrogen (nilutamide). N Engl J Med, 1989. 321: 413-8.
17. G. Leuprolide Study. Leuprolide versus diethylstilbestrol for metastatic prostate cancer. N Engl J Med, 1984. 311: 1281-6.

18. Schellhammer, P.F., R. Sharifi, N.L. Block, M.S. Soloway, P.M. Venner, A.L. Patterson, et al. Clinical benefits of bicalutamide compared with flutamide in combined androgen blockade for patients with advanced prostatic carcinoma: final report of a double-blind, randomized, multicenter trial. Casodex Combination Study Group. Urology, 1997. 50: 330-6.
19. Soloway, M.S., G. Chodak, N.J. Vogelzang, N.L. Block, P.F. Schellhammer, J.A. Smith Jr., et al. Zoladex versus orchiectomy in treatment of advanced prostate cancer: a randomized trial. Zoladex Prostate Study Group. Urology, 1991. 37: 46-51.
20. Klotz, L., L. Boccon-Gibod, N.D. Shore, C. Andreou, B.E. Persson, P. Cantor, et al. The efficacy and safety of degarelix: a 12-month, comparative, randomized, open-label, parallel-group phase III study in patients with prostate cancer. BJU Int, 2008. 102: 1531-8.
21. Kuhn, J.M., H. Abourachid, P. Brucher, J.C. Doutres, J. Fretin, A. Jaupitre, et al. A randomized comparison of the clinical and hormonal effects of two GnRH agonists in patients with prostate cancer. Eur Urol, 1997. 32: 397-403.
22. Petrylak, D.P., C.M. Tangen, M.H. Hussain, P.N. Lara Jr., J.A. Jones, M.E. Taplin, et al. Docetaxel and estramustine compared with mitoxantrone and prednisone for advanced refractory prostate cancer. N Engl J Med, 2004. 351: 1513-20.
23. Schlegel, P.N., P. Kuzma, J. Frick, A. Farkas, A. Gomahr, I. Spitz, et al. Effective long-term androgen suppression in men with prostate cancer using a hydrogel implant with the GnRH agonist histrelin. Urology, 2001. 58: 578-82.
24. Tannock, I.F., R. de Wit, W.R. Berry, J. Horti, A. Pluzanska, K.N. Chi, et al. T.A.X. Investigators, Docetaxel plus prednisone or mitoxantrone plus prednisone for advanced prostate cancer. N Engl J Med, 2004. 351: 1502-12.
25. Small, E.J., P.F. Schellhammer, C.S. Higano, C.H. Redfern, J.J. Nemunaitis, F.H. Valone, et al. Placebo-controlled phase III trial of immunologic therapy with sipuleucel-T (APC8015) in patients with metastatic, asymptomatic hormone refractory prostate cancer. J Clin Oncol, 2006. 24: 3089-94.
26. de Bono, J.S., S. Oudard, M. Ozguroglu, S. Hansen, J.P. Machiels, I. Kocak, et al. Prednisone plus cabazitaxel or mitoxantrone for metastatic castration-resistant prostate cancer progressing after docetaxel treatment: a randomised open-label trial. Lancet, 2010. 376: 1147-54.
27. Fizazi, K., M. Carducci, M. Smith, R. Damiao, J. Brown, L. Karsh, et al. Denosumab versus zoledronic acid for treatment of bone metastases in men with castration-resistant prostate cancer: a randomised, double-blind study. Lancet, 2011. 377: 813-22.
28. de Bono, J.S., C.J. Logothetis, A. Molina, K. Fizazi, S. North, L. Chu, et al. Abiraterone and increased survival in metastatic prostate cancer. N Engl J Med, 2011. 364: 1995-2005.
29. Parker, C., S. Nilsson, D. Heinrich, S.I. Helle, J.M. O'Sullivan, S.D. Fossa, et al. Alpha emitter radium-223 and survival in metastatic prostate cancer. N Engl J Med, 2013. 369: 213-23.
30. Scher, H.I., K. Fizazi, F. Saad, M.E. Taplin, C.N. Sternberg, K. Miller, et al. Increased survival with enzalutamide in prostate cancer after chemotherapy. N Engl J Med, 2012. 367: 1187-97.
31. Smith, M.R., F. Saad, S. Chowdhury, S. Oudard, B.A. Hadaschik, J.N. Graff, et al. Apalutamide treatment and metastasis-free survival in prostate cancer. N Engl J Med, 2018. 378: 1408-18.

32. Sharifi, N., J.L. Gulley, W.L. Dahut. An update on androgen deprivation therapy for prostate cancer. Endocr Relat Cancer, 2010. 17: R305-15.
33. Hoimes, C.J., W.K. Kelly. Redefining hormone resistance in prostate cancer. Ther Adv Med Oncol, 2010. 2: 107-23.
34. D'Amico, A.V. US Food and Drug Administration approval of drugs for the treatment of prostate cancer: a new era has begun. J Clin Oncol, 2013. 32: 362-4.
35. Klotz, L., C. O'Callaghan, K. Ding, P. Toren, D. Dearnaley, C.S. Higano, et al. Nadir testosterone within first year of androgen-deprivation therapy (ADT) predicts for time to castration-resistant progression: a secondary analysis of the PR-7 trial of intermittent versus continuous ADT. J Clin Oncol, 2015. 33: 1151-6.
36. Nishiyama, T. Serum testosterone levels after medical or surgical androgen deprivation: a comprehensive review of the literature. Urol Oncol, 2014. 32: 38 e17-28.
37. Byar, D.P. Proceedings: The Veterans Administration Cooperative Urological Research Group's studies of cancer of the prostate. Cancer, 1973. 32: 1126-30.
38. Amoss, M., R. Burgus, R. Blackwell, W. Vale, R. Fellows, R. Guillemin. Purification, amino acid composition and N-terminus of the hypothalamic luteinizing hormone releasing factor (LRF) of ovine origin. Biochem Biophys Res Commun, 1971. 44: 205-10.
39. Matsuo, H.. Y. Baba, R.M. Nair, A. Arimura, A.V. Schally. Structure of the porcine LH- and FSH-releasing hormone. I: The proposed amino acid sequence. Biochem Biophys Res Commun, 1971. 43: 1334-9.
40. Plant, T.M. 60 Years of neuroendocrinology: the hypothalamo-pituitary-gonadal axis. J Endocrinol, 2015. 226: T41-54.
41. Warner, B., T.J. Worgul, J. Drago, L. Demers, M. Dufau, D. Max, et al. Effect of very high dose D-leucine6-gonadotropin-releasing hormone proethylamide on the hypothalamic-pituitary testicular axis in patients with prostatic cancer. J Clin Invest, 1983. 71: 1842-53.
42. Labrie, F., A. Dupont, A. Belanger, J. Emond, G. Monfette. Simultaneous administration of pure antiandrogens, a combination necessary for the use of luteinizing hormone-releasing hormone agonists in the treatment of prostate cancer. Proc Natl Acad Sci USA, 1984. 81: 3861-3.
43. Van Poppel, H., B. Tombal, J.J. de la Rosette, B.E. Persson, J.K. Jensen, T. Kold Olesen. Degarelix: a novel gonadotropin-releasing hormone (GnRH) receptor blocker—results from a 1-yr, multicentre, randomised, phase 2 dosage-finding study in the treatment of prostate cancer. Eur Urol, 2008. 54: 805-13.
44. Gittelman, M., P.J. Pommerville, B.E. Persson, J.K. Jensen, T.K. Olesen. G. Degarelix study: a 1-year, open label, randomized phase II dose finding study of degarelix for the treatment of prostate cancer in North America. J Urol, 2008. 180: 1986-92.
45. Iversen, P., C.J. Tyrrell, A.V. Kaisary, J.B. Anderson, H. Van Poppel, T.L. Tammela, et al. Bicalutamide monotherapy compared with castration in patients with nonmetastatic locally advanced prostate cancer: 6.3 years of followup. J Urol, 2000. 164: 1579-82.
46. Bales, G.T., G.W. Chodak. A controlled trial of bicalutamide versus castration in patients with advanced prostate cancer. Urology, 1996. 47: 38-43; discussion 48-53.
47. Tyrrell, C.J., A.V. Kaisary, P. Iversen, J.B. Anderson, L. Baert, T. Tammela, et al. A randomised comparison of 'Casodex' (bicalutamide) 150 mg monotherapy versus castration in the treatment of metastatic and locally advanced prostate cancer. Eur Urol, 1998. 33: 447-56.

48. Anderson, J. The role of antiandrogen monotherapy in the treatment of prostate cancer. BJU Int, 2003. 91: 455-61.
49. Wirth, M., C. Tyrrell, M. Wallace, K.P. Delaere, M. Sanchez-Chapado, J. Ramon, et al. Bicalutamide (Casodex) 150 mg as immediate therapy in patients with localized or locally advanced prostate cancer significantly reduces the risk of disease progression. Urology, 2001. 58: 146-51.
50. Urushibara, M., J. Ishioka, N. Hyochi, K. Kihara, S. Hara, P. Singh, et al. Effects of steroidal and non-steroidal antiandrogens on wild-type and mutant androgen receptors. Prostate, 2007. 67: 799-807.
51. Moos, P.J., F.A. Fitzpatrick. Taxanes propagate apoptosis via two cell populations with distinctive cytological and molecular traits. Cell Growth Differ, 1998. 9: 687-97.
52. Caraglia, M., G. Giuberti, M. Marra, E. Di Gennaro, G. Facchini, F. Caponigro, et al. Docetaxel induces p53-dependent apoptosis and synergizes with farnesyl transferase inhibitor r115777 in human epithelial cancer cells. Front Biosci, 2005. 10: 2566-75.
53. Liu, C., Y. Zhu, W. Lou, N. Nadiminty, X. Chen, Q. Zhou, et al. Functional p53 determines docetaxel sensitivity in prostate cancer cells. Prostate, 2013. 73: 418-27.
54. Komlodi-Pasztor, E., D. Sackett, J. Wilkerson, T. Fojo. Mitosis is not a key target of microtubule agents in patient tumors. Nat Rev Clin Oncol, 2011. 8: 244-50.
55. Thadani-Mulero, M., D.M. Nanus, P. Giannakakou. Androgen receptor on the move: boarding the microtubule expressway to the nucleus. Cancer Res, 2012. 72: 4611-5.
56. Mita, A.C., L.J. Denis, E.K. Rowinsky, J.S. Debono, A.D. Goetz, L. Ochoa, et al. Phase I and pharmacokinetic study of XRP6258 (RPR 116258A), a novel taxane, administered as a 1-hour infusion every 3 weeks in patients with advanced solid tumors. Clin Cancer Res, 2009. 15: 723-30.
57. de Morrée, E., R. van Soest, A. Aghai, C. de Ridder, P. de Bruijn, I. Ghobadi Moghaddam-Helmantel, et al. Understanding taxanes in prostate cancer; importance of intratumoral drug accumulation. The Prostate, 2016. 76: 927-36.
58. Mazerski, J., S. Martelli, E. Borowski. The geometry of intercalation complex of antitumor mitoxantrone and ametantrone with DNA: molecular dynamics simulations. Acta Biochim Pol, 1998. 45: 1-11.
59. Kapuscinski, J., Z. Darzynkiewicz. Interactions of antitumor agents Ametantrone and Mitoxantrone (Novatrone) with double-stranded DNA. Biochem Pharmacol, 1985. 34: 4203-13.
60. SEER, Epidemiology and End Results Program. Cancer Stat Facts. Prostate Cancer. https://seer.cancer.gov/statfacts/html/prost.html, Assessed on Jan 20th 2019, National Cancer Institute.
61. Cooperberg, M.R., P.R. Carroll. Trends in management for patients with localized prostate cancer, 1990-2013. JAMA, 2015. 314: 80-82.
62. Loeb, S., Y. Folkvaljon, C. Curnyn, D. Robinson, O. Bratt, P. Stattin. Uptake of active surveillance for very-low-risk prostate cancer in Sweden. JAMA Oncol, 2017. 3: 1393-8.
63. Weerakoon, M., N. Papa, N. Lawrentschuk, S. Evans, J. Millar, M. Frydenberg, et al. The current use of active surveillance in an Australian cohort of men: a pattern of care analysis from the Victorian Prostate Cancer Registry. BJU Int, 2015. 115 (Suppl 5): 50-56.
64. Adamy, A., D.S. Yee, K. Matsushita, A. Maschino, A. Cronin, A. Vickers, et al. Role of prostate specific antigen and immediate confirmatory biopsy in predicting progression during active surveillance for low risk prostate cancer. J Urol, 2011. 185: 477-82.

65. Kinsella, N., J. Helleman, S. Bruinsma, S. Carlsson, D. Cahill, C. Brown, et al. Active surveillance for prostate cancer: a systematic review of contemporary worldwide practices. Transl Androl Urol, 2018. 7: 83-97.
66. Paulson, D.F. Randomized series of treatment with surgery versus radiation for prostate adenocarcinoma. NCI Monogr, 1988. 127-31.
67. Quinlan, D.M., J.I. Epstein, B.S. Carter, P.C. Walsh. Sexual function following radical prostatectomy: influence of preservation of neurovascular bundles. J Urol, 1991. 145: 998-1002.
68. Catalona, W.J., J.W. Basler. Return of erections and urinary continence following nerve sparing radical retropubic prostatectomy. J Urol, 1993. 150: 905-7.
69. Bagshaw, M.A., R.S. Cox, G.R. Ray. Status of radiation treatment of prostate cancer at Stanford University. NCI monographs: a publication of the National Cancer Institute. 1988. 47-60.
70. Kuban, D.A., S.L. Tucker, L. Dong, G. Starkschall, E.H. Huang, M.R. Cheung, et al. Long-term results of the M.D. Anderson randomized dose-escalation trial for prostate cancer. Int J Radiat Oncol Biol Phys, 2008. 70: 67-74.
71. Zietman, A.L., K. Bae, J.D. Slater, W.U. Shipley, J.A. Efstathiou, J.J. Coen, et al. Randomized trial comparing conventional-dose with high-dose conformal radiation therapy in early-stage adenocarcinoma of the prostate: long-term results from Proton Radiation Oncology Group/American College of Radiology 95-09. J Clin Oncol, 2010. 28: 1106-11.
72. Dearnaley, D.P., M.R. Sydes, J.D. Graham, E.G. Aird, D. Bottomley, R.A. Cowan, et al. Escalated-dose versus standard-dose conformal radiotherapy in prostate cancer: first results from the MRC RT01 randomised controlled trial. Lancet Oncol, 2007. 8: 475-87.
73. Peeters, S.T., W.D. Heemsbergen, P.C. Koper, W.L. van Putten, A. Slot, M.F. Dielwart, et al. Dose-response in radiotherapy for localized prostate cancer: results of the Dutch multicenter randomized phase III trial comparing 68 Gy of radiotherapy with 78 Gy. J Clin Oncol, 2006. 24: 1990-6.
74. Beckendorf, V., S. Guerif, E. Le Prise, J.M. Cosset, A. Bougnoux, B. Chauvet, et al. 70 Gy versus 80 Gy in localized prostate cancer: 5-year results of GETUG 06 randomized trial. Int J Radiat Oncol Biol Phys, 2011. 80: 1056-63.
75. Ferrer, M., F. Guedea, J.F. Suarez, B. de Paula, V. Macias, A. Marino, et al. Clinically localized prostate, quality of life impact of treatments for localized prostate cancer: cohort study with a 5-year follow-up. Radiother Oncol, 2013. 108: 306-13.
76. Kotecha, R., T. Djemil, R.D. Tendulkar, C.A. Reddy, R.A. Thousand, A. Vassil, et al. Dose-escalated stereotactic body radiation therapy for patients with intermediate- and high-risk prostate cancer: initial dosimetry analysis and patient outcomes. Int J Radiat Oncol Biol Phys, 2016. 95: 960-4.
77. Lardas, M., M. Liew, R.C. van den Bergh, M. De Santis, J. Bellmunt, T. Van den Broeck, et al. Quality of life outcomes after primary treatment for clinically localised prostate cancer: a systematic review. Eur Urol, 2017. 72: 869-85.
78. Chen, R.C., R. Basak, A.M. Meyer, T.M. Kuo, W.R. Carpenter, R.P. Agans, et al. Association between choice of radical prostatectomy, external beam radiotherapy, brachytherapy, or active surveillance and patient-reported quality of life among men with localized prostate cancer. JAMA, 2017. 317: 1141-50.
79. Hamdy, F.C., J.L. Donovan, J.A. Lane, M. Mason, C. Metcalfe, P. Holding, et al. 10-year outcomes after monitoring, surgery, or radiotherapy for localized prostate cancer. N Engl J Med, 2016. 375: 1415-24.

80. Lamers, R.E., M. Cuypers, M. de Vries, L.V. van de Poll-Franse, J.L. Ruud Bosch, P.J. Kil. How do patients choose between active surveillance, radical prostatectomy, and radiotherapy? The effect of a preference-sensitive decision aid on treatment decision making for localized prostate cancer. Urol Oncol, 2017. 35: 37 e39-37 e17.
81. Catalona, W.J. Management of cancer of the prostate. N Engl J Med, 1994. 331: 996-1004.
82. Spahn, M., S. Joniau, P. Gontero, S. Fieuws, G. Marchioro, B. Tombal, et al. Outcome predictors of radical prostatectomy in patients with prostate-specific antigen greater than 20 ng/ml: a European multi-institutional study of 712 patients. Eur Urol, 2010. 58: 1-7.
83. Gontero, P., M. Spahn, B. Tombal, P. Bader, C.Y. Hsu, G. Marchioro, et al. Is there a prostate-specific antigen upper limit for radical prostatectomy? BJU International, 2011. 108: 1093-1100.
84. Ward, J.F., J.M. Slezak, M.L. Blute, E.J. Bergstralh, H. Zincke. Radical prostatectomy for clinically advanced (cT3) prostate cancer since the advent of prostate-specific antigen testing: 15iyear outcome. BJU International, 2005. 95: 751-6.
85. Hsu, C.-Y., S. Joniau, R. Oyen, T. Roskams, H. Van Poppel. Outcome of surgery for clinical unilateral T3a prostate cancer: a single-institution experience. Eur Urol, 2007. 51: 121-9.
86. Pierorazio, P.M., T.J. Guzzo, M. Han, T.J. Bivalacqua, J.I. Epstein, E.M. Schaeffer, et al. Long-term survival after radical prostatectomy for men with high Gleason sum in pathologic specimen. Urology, 2010. 76: 715-21.
87. van den Bergh, R.C., N.J. van Casteren, T. van den Broeck, E.R. Fordyce, W.K. Gietzmann, F. Stewart, et al. Role of hormonal treatment in prostate cancer patients with nonmetastatic disease recurrence after local curative treatment: a systematic review. Eur Urol, 2016. 69: 802-20.
88. Studer, U.E., L. Collette, P. Whelan, W. Albrecht, J. Casselman, T. de Reijke, et al. Using PSA to guide timing of androgen deprivation in patients with T0-4 N0-2 M0 prostate cancer not suitable for local curative treatment (EORTC 30891). Eur Urol, 2008. 53: 941-9.
89. Cha, E.K., J.A. Eastham. Chemotherapy and novel therapeutics before radical prostatectomy for high-risk clinically localized prostate cancer. Urol Oncol, 2015. 33: 217-25.
90. Lou, D.Y., L. Fong. Neoadjuvant therapy for localized prostate cancer: examining mechanism of action and efficacy within the tumor. Urol Oncol, 2016. 34: 182-92.
91. McKay, R.R., T.K. Choueiri, M.E. Taplin. Rationale for and review of neoadjuvant therapy prior to radical prostatectomy for patients with high-risk prostate cancer. Drugs, 2013. 73: 1417-30.
92. Pietzak, E.J., J.A. Eastham. Neoadjuvant treatment of high-risk, clinically localized prostate cancer prior to radical prostatectomy. Curr Urol Rep, 2016. 17: 37.
93. Moschini, M., V. Sharma, F. Zattoni, J.F. Quevedo, B.J. Davis, E. Kwon, et al. Natural history of clinical recurrence patterns of lymph node-positive prostate cancer after radical prostatectomy. Eur Urol, 2016. 69: 135-42.
94. Lin, C.C., P.J. Gray, A. Jemal, J.A. Efstathiou. Androgen deprivation with or without radiation therapy for clinically node-positive prostate cancer. J Natl Cancer Inst, 2015. 107: 122-3.
95. Rusthoven, C.G., J.A. Carlson, T.V. Waxweiler, D. Raben, P.E. Dewitt, E.D. Crawford, et al. The impact of definitive local therapy for lymph node-positive

prostate cancer: a population-based study. Int J Radiat Oncol Biol Phys, 2014. 88: 1064-73.
96. Tward, J.D., K.E. Kokeny, D.C. Shrieve. Radiation therapy for clinically node-positive prostate adenocarcinoma is correlated with improved overall and prostate cancer-specific survival. Pract Radiat Oncol, 2013. 3: 234-40.
97. Seisen, T., M.W. Vetterlein, P. Karabon, T. Jindal, A. Sood, L. Nocera, et al. Efficacy of local treatment in prostate cancer patients with clinically pelvic lymph node-positive disease at initial diagnosis. Eur Urol, 2017. 73: 452-61.
98. Boorjian, S.A., R.H. Thompson, S. Siddiqui, S. Bagniewski, E.J. Bergstralh, R.J. Karnes, et al. Long-term outcome after radical prostatectomy for patients with lymph node-positive prostate cancer in the prostate specific antigen era. J Urol, 2007. 178: 864-71.
99. Briganti, A., J.R. Karnes, L.F. Da Pozzo, C. Cozzarini, A. Gallina, N. Suardi, et al. Two positive nodes represent a significant cut-off value for cancer specific survival in patients with node positive prostate cancer. A new proposal based on a two-institution experience on 703 consecutive N+ patients treated with radical prostatectomy, extended pelvic lymph node dissection and adjuvant therapy. Eur Urol, 2009. 55: 261-70.
100. Preisser, F., M. Marchioni, S. Nazzani, M. Bandini, Z. Tian, R.S. Pompe, et al. The impact of lymph node metastases burden at radical prostatectomy. Eur Urol Focus, 2018. 5: 399-406.
101. Schumacher, M.C., F.C. Burkhard, G.N. Thalmann, A. Fleischmann, U.E. Studer. Good outcome for patients with few lymph node metastases after radical retropubic prostatectomy. Eur Urol, 2008. 54: 344-52.
102. Weiner, A.B., R.S. Matulewicz, S.E. Eggener, E.M. Schaeffer. Increasing incidence of metastatic prostate cancer in the United States (2004-2013). Prostate Cancer Prostatic Dis, 2016. 19: 395-7.
103. Dalela, D., M. Sun, M. Diaz, P. Karabon, T. Seisen, Q.-D. Trinh, et al. Contemporary trends in the incidence of metastatic prostate cancer among US men: results from nationwide analyses. Eur Urol Focus, 2017. 5: 77-80.
104. Finianos, A., K. Gupta, B. Clark, S.J. Simmens, J.B. Aragon-Ching. Characterization of differences between prostate cancer patients presenting with de novo versus primary progressive metastatic disease. Clin Genitourin Cancer, 2018. 16: 85-9.
105. Scher, H.I., M.J. Morris, W.K. Kelly, L.H. Schwartz, G. Heller. Prostate cancer clinical trial end points: "RECIST"ing a step backwards. Clin Cancer Res, 2005. 11: 5223-32.
106. Sweeney, C.J., Y.H. Chen, M. Carducci, G. Liu, D.F. Jarrard, M. Eisenberger, et al. Chemohormonal therapy in metastatic hormone-sensitive prostate cancer. N Engl J Med, 2015. 373: 737-46.
107. James, N.D., M.R. Sydes, N.W. Clarke, M.D. Mason, D.P. Dearnaley, M.R. Spears, et al. Addition of docetaxel, zoledronic acid, or both to first-line long-term hormone therapy in prostate cancer (STAMPEDE): survival results from an adaptive, multiarm, multistage, platform randomised controlled trial. Lancet, 2016. 387: 1163-77.
108. Gravis, G., K. Fizazi, F. Joly, S. Oudard, F. Priou, B. Esterni, et al. Androgen-deprivation therapy alone or with docetaxel in non-castrate metastatic prostate cancer (GETUG-AFU 15): a randomised, open-label, phase 3 trial. Lancet Oncol, 2013. 14: 149-58.
109. Oudard, S., I. Latorzeff, A. Caty, L. Miglianico, E. Sevin, A.C. Hardy-Bessard, et al. Effect of adding docetaxel to androgen-deprivation therapy in patients

with high-risk prostate cancer with rising prostate-specific antigen levels after primary local therapy: a randomized clinical trial. JAMA Oncol, 2019. 5: 623-32.
110. Kyriakopoulos, C.E., Y.H. Chen, M.A. Carducci, G. Liu, D.F. Jarrard, N.M. Hahn, et al. Chemohormonal therapy in metastatic hormone-sensitive prostate cancer: long-term survival analysis of the randomized phase III E3805 CHAARTED trial. J Clin Oncol, 2018. 36: 1080-7.
111. Gravis, G., J.M. Boher, F. Joly, M. Soulie, L. Albiges, F. Priou, et al. Androgen deprivation therapy (adt) plus docetaxel versus adt alone in metastatic non castrate prostate cancer: impact of metastatic burden and long-term survival analysis of the randomized phase 3 GETUG-AFU15 trial. Eur Urol, 2016. 70: 256-62.
112. James, N.D., J.S. de Bono, M.R. Spears, N.W. Clarke, M.D. Mason, D.P. Dearnaley, et al. Abiraterone for prostate cancer not previously treated with hormone therapy. N Engl J Med, 2017. 377: 338-51.
113. Fizazi, K., N. Tran, L. Fein, N. Matsubara, A. Rodriguez-Antolin, B.Y. Alekseev, et al. Abiraterone plus prednisone in metastatic, castration-sensitive prostate cancer. N Engl J Med, 2017. 377: 352-60.
114. Tombal, B., M. Borre, P. Rathenborg, P. Werbrouck, H. Van Poppel, A. Heidenreich, et al., Enzalutamide monotherapy in hormone-naive prostate cancer: primary analysis of an open-label, single-arm, phase 2 study. Lancet Oncol, 2014. 15: 592-600.
115. Janssen. A Study of ERLEADA® (apalutamide) Phase 3 TITAN Study Unblinded as Dual Primary Endpoints Achieved in Clinical Program Evaluating Treatment of Patients with Metastatic Castration-Sensitive Prostate Cancer. https://www.janssen.com/janssen-announces-erleada-apalutamide-phase-3-titan-study-unblinded-dual-primary-endpoints-achieved.
116. Sydes, M.R., M.R. Spears, M.D. Mason, N.W. Clarke, D.P. Dearnaley, J.S. de Bono, et al. Adding abiraterone or docetaxel to long-term hormone therapy for prostate cancer: directly randomised data from the STAMPEDE multi-arm, multi-stage platform protocol. Ann Oncol, 2018. 29: 1235-48.
117. Wozniak, A.J., B.A. Blumenstein, E.D. Crawford, M. Boileau, S.E. Rivkin, W.S. Fletcher. Cyclophosphamide, methotrexate, and 5-fluorouracil in the treatment of metastatic prostate cancer. A Southwest Oncology Group study. Cancer, 1993. 71: 3975-8.
118. Abratt, R.P., D. Brune, M.A. Dimopoulos, J. Kliment, J. Breza, F.P. Selvaggi, et al. Randomised phase III study of intravenous vinorelbine plus hormone therapy versus hormone therapy alone in hormone-refractory prostate cancer. Ann Oncol, 2004. 15: 1613-21.
119. Tannock, I.F., D. Osoba, M.R. Stockler, D.S. Ernst, A.J. Neville, M.J. Moore, et al. Chemotherapy with mitoxantrone plus prednisone or prednisone alone for symptomatic hormone-resistant prostate cancer: a Canadian randomized trial with palliative end points. J Clin Oncol, 1996. 14: 1756-64.
120. Berthold, D.R., G.R. Pond, F. Soban, R. de Wit, M. Eisenberger, I.F. Tannock. Docetaxel plus prednisone or mitoxantrone plus prednisone for advanced prostate cancer: updated survival in the TAX 327 study. J Clin Oncol, 2008. 26: 242-5.
121. van Soest, R.J., E.S. de Morree, C.F. Kweldam, C.M.A. de Ridder, E.A.C. Wiemer, R.H.J. Mathijssen, et al. Targeting the androgen receptor confers in vivo cross-resistance between enzalutamide and docetaxel, but not cabazitaxel, in castration-resistant prostate cancer. Eur Urol, 2015. 67: 981-5.
122. Sella, A., T. Sella, A. Peer, R. Berger, S.J. Frank, E. Gez, et al. Activity of

cabazitaxel after docetaxel and abiraterone acetate therapy in patients with castration-resistant prostate cancer. Clin Genitourin Cancer, 2014. 12: 428-32.
123. Oudard, S., K. Fizazi, L. Sengelov, G. Daugaard, F. Saad, S. Hansen, et al. Cabazitaxel versus docetaxel as first-line therapy for patients with metastatic castration-resistant prostate cancer: a randomized phase III trial-FIRSTANA. J Clin Oncol, 2017. 35: 3189-97.
124. Fizazi, K., H.I. Scher, A. Molina, C.J. Logothetis, K.N. Chi, R.J. Jones, et al. Abiraterone acetate for treatment of metastatic castration-resistant prostate cancer: final overall survival analysis of the COU-AA-301 randomised, double-blind, placebo-controlled phase 3 study. Lancet Oncol, 2012. 13: 983-92.
125. Rathkopf, D.E., M.R. Smith, J.S. de Bono, C.J. Logothetis, N.D. Shore, P. de Souza, et al. Updated interim efficacy analysis and long-term safety of abiraterone acetate in metastatic castration-resistant prostate cancer patients without prior chemotherapy (COU-AA-302). Eur Urol, 2014. 66: 815-25.
126. Beer, T.M., A.J. Armstrong, D.E. Rathkopf, Y. Loriot, C.N. Sternberg, C.S. Higano, et al. Enzalutamide in metastatic prostate cancer before chemotherapy. N Engl J Med, 2014. 371: 424-33.
127. Hussain, M., K. Fizazi, F. Saad, P. Rathenborg, N. Shore, U. Ferreira, et al. Enzalutamide in men with nonmetastatic, castration-resistant prostate cancer. N Engl J Med, 2018. 378: 2465-74.
128. Rathkopf, D.E., E.S. Antonarakis, N.D. Shore, R.F. Tutrone, J.J. Alumkal, C.J. Ryan, et al. Safety and antitumor activity of apalutamide (ARN-509) in metastatic castration-resistant prostate cancer with and without prior abiraterone acetate and prednisone. Clin Cancer Res, 2017. 23: 3544-51.
129. Koshkin, V.S., E.J. Small. Apalutamide in the treatment of castrate-resistant prostate cancer: evidence from clinical trials. Ther Adv Urol, 2018. 10: 445-54.
130. Owen, P.J., R.M. Daly, P.M. Livingston, S.F. Fraser. Lifestyle guidelines for managing adverse effects on bone health and body composition in men treated with androgen deprivation therapy for prostate cancer: an update. Prostate Cancer Prostatic Dis, 2017. 20: 137-45.
131. Rizzo, S., A. Galvano, F. Pantano, M. Iuliani, B. Vincenzi, F. Passiglia, et al. The effects of enzalutamide and abiraterone on skeletal related events and bone radiological progression free survival in castration resistant prostate cancer patients: an indirect comparison of randomized controlled trials. Crit Rev Oncol Hematol, 2017. 120: 227-33.
132. Dai, J., H. Zhang, A. Karatsinides, J.M. Keller, K.M. Kozloff, D.T. Aftab, et al. Cabozantinib inhibits prostate cancer growth and prevents tumor-induced bone lesions. Clin Cancer Res, 20: 2014. 617-30.
133. B.J. Kiratli, S. Srinivas, I. Perkash, M.K. Terris. Progressive decrease in bone density over 10 years of androgen deprivation therapy in patients with prostate cancer. Urology, 2001. 57: 127-32.
134. Stoch, S.A., R.A. Parker, L. Chen, G. Bubley, Y.J. Ko, A. Vincelette, et al. Bone loss in men with prostate cancer treated with gonadotropin-releasing hormone agonists. J Clin Endocrinol Metab, 2001. 86: 2787-91.
135. Wang, A., Z. Obertova, C. Brown, N. Karunasinghe, K. Bishop, L. Ferguson, et al. Risk of fracture in men with prostate cancer on androgen deprivation therapy: a population-based cohort study in New Zealand. BMC Cancer, 2015. 15: 837.
136. Yamada, Y., S. Takahashi, T. Fujimura, H. Nishimatsu, A. Ishikawa, H. Kume, et al. The effect of combined androgen blockade on bone turnover and bone mineral density in men with prostate cancer. Osteoporos Int, 2008. 19: 321-7.

137. Novara, G., A. Galfano, S. Secco, V. Ficarra, W. Artibani. Impact of surgical and medical castration on serum testosterone level in prostate cancer patients. Urol Int, 2009. 82: 249-55.
138. Vidal, O., L.G. Kindblom, C. Ohlsson. Expression and localization of estrogen receptor-beta in murine and human bone. J Bone Miner Res, 1999. 14: 923-9.
139. Boeve, L.M.S., M. Hulshof, A.N. Vis, A.H. Zwinderman, J.W.R. Twisk, W.P.J. Witjes, et al. Effect on survival of androgen deprivation therapy alone compared to androgen deprivation therapy combined with concurrent radiation therapy to the prostate in patients with primary bone metastatic prostate cancer in a prospective randomised clinical trial: data from the HORRAD trial. Eur Urol, 2019. 75: 410-418.
140. Wirth, M., T. Tammela, V. Cicalese, F. Gomez Veiga, K. Delaere, K. Miller, et al. Prevention of bone metastases in patients with high-risk nonmetastatic prostate cancer treated with zoledronic acid: efficacy and safety results of the Zometa European Study (ZEUS). Eur Urol, 2015. 67: 482-91.
141. Saad, F. Treatment of bone complications in advanced prostate cancer: rationale for bisphosphonate use and results of a phase III trial with zoledronic acid. Semin Oncol, 2002. 29: 19-27.
142. Saad, F., D.M. Gleason, R. Murray, S. Tchekmedyian, P. Venner, L. Lacombe, et al. G. Zoledronic acid prostate cancer study: a randomized, placebo-controlled trial of zoledronic acid in patients with hormone-refractory metastatic prostate carcinoma. J Natl Cancer Inst, 2002. 94: 1458-68.
143. Saad, F., D.M. Gleason, R. Murray, S. Tchekmedyian, P. Venner, L. Lacombe, et al. G. Zoledronic acid prostate cancer study: long-term efficacy of zoledronic acid for the prevention of skeletal complications in patients with metastatic hormone-refractory prostate cancer. J Natl Cancer Inst, 2004. 96: 879-82.
144. Chu, G.C., H.E. Zhau, R. Wang, A. Rogatko, X. Feng, M. Zayzafoon, et al. RANK- and c-Met-mediated signal network promotes prostate cancer metastatic colonization. Endocr Relat Cancer, 2014. 21: 311-26.
145. Jones, D.H., T. Nakashima, O.H. Sanchez, I. Kozieradzki, S.V. Komarova, I. Sarosi, et al. Regulation of cancer cell migration and bone metastasis by RANKL. Nature, 2006. 440: 692-6.
146. Fizazi, K., A. Lipton, X. Mariette, J.J. Body, Y. Rahim, J.R. Gralow, et al. Randomized phase II trial of denosumab in patients with bone metastases from prostate cancer, breast cancer, or other neoplasms after intravenous bisphosphonates. J Clin Oncol, 2009. 27: 1564-71.
147. Smith, M.R., F. Saad, B. Egerdie, M. Szwedowski, T.L. Tammela, C. Ke, et al. Effects of denosumab on bone mineral density in men receiving androgen deprivation therapy for prostate cancer. J Urol, 2009. 182: 2670-5.
148. Bauman, G., M. Charette, R. Reid, J. Sathya. Radiopharmaceuticals for the palliation of painful bone metastasis: a systemic review. Radiother Oncol, 2005. 75: 258-70.
149. Parker, C.C., S. Pascoe, A. Chodacki, J.M. O'Sullivan, J.R. Germa, C.G. O'Bryan-Tear, et al. A randomized, double-blind, dose-finding, multicenter, phase 2 study of radium chloride (Ra 223) in patients with bone metastases and castration-resistant prostate cancer. Eur Urol, 2013. 63: 189-97.
150. Hoskin, P., O. Sartor, J.M. O'Sullivan, D.C. Johannessen, S.I. Helle, J. Logue, et al. Efficacy and safety of radium-223 dichloride in patients with castration-resistant prostate cancer and symptomatic bone metastases, with or without previous docetaxel use: a prespecified subgroup analysis from the randomised, double-blind, phase 3 ALSYMPCA trial. Lancet Oncol, 2014. 15: 1397-1406.

CHAPTER 12

Regulation and Targeting of MUCINS in Pancreatic Cancer

Shailendra K. Gautam[1], Abhijit Aithal[1], Grish C. Varshney[1] and Parthasarathy Seshacharyulu[1,2*]

[1] Department of Biochemistry and Molecular Biology, University of Nebraska Medical Center
[2] Fred and Pamela Buffett Cancer Center, Eppley Institute for Research in Cancer and Allied Diseases, University of Nebraska Medical Center, Omaha, Nebraska, USA

Introduction

Pancreatic cancer (PC) is the 4th leading cause of cancer-related mortalities in the United States with an estimated 56,770 new cases and more than 45,750 deaths in 2019 [1]. Despite the clinical advances and ongoing efforts in therapeutic management, PC is predicted to be the second leading cause of cancer-related deaths in the United States by 2030 [2]. Its asymptomatic nature attributed to the worst five-year survival rate in PC patients, late diagnosis, early metastasis, un-resectability, high recurrence rate, drug resistance, and immunosuppressive tumor microenvironment (TME) [3-6]. Only less than 20% PC patients who were eligible for resection, showed improved survival rates [7, 8]. In all other cases, survival did not improve significantly, emphasizing the need for novel therapeutic approaches in PC management. A better understanding of PC pathogenesis associated genetic and epigenetic regulations, molecular mechanisms and altered signaling pathways, metabolic alterations and mechanisms of drug resistance, tumor microenvironment, and immunosuppression, is essential to improve the ongoing diagnostic and therapeutic approaches [4, 5]. In addition, efforts are also needed to be directed towards the biomarker development for early detection and targeted therapies in a combination of immunotherapy and anti-stromal therapies for better therapeutic response in PC patients.

The malignant cells undergo a change in their genetic and molecular signatures to support the growing structural and metabolic needs during cancer progression and metastasis. These genetic and epigenetic changes

*Corresponding author: p.seshacharyulu@unmc.edu

drive the initiation and progression of PC from early PanIN to advanced PDAC. As an initial event, PC starts with K-Ras mutation(s) followed by CDKN2A, TP53, SMAD4 and BRCA1/2 mutations that further contribute, in parts or altogether, to the aggressiveness of the disease [9-11]. Aberrant glycosylation and increased mucin expression are the two other important characteristics of PC [12-14]. In recent years, studies in PC have been aimed to exploit gene-specific mutations and altered structural and functional proteins as a biomarker as well as for therapeutic target. In addition, the genetic and molecular alterations contributing to PC progression and metastasis have also been associated with a critical role in drug resistance [15-18].

Mucins are family of high molecular weight glycoproteins that have been reported to play a structural and functional role in PC pathogenesis. Both secretory mucins e.g. MUC2, MUC5AC, MUC5B, MUC6-9, and MUC19 and membrane-bound mucins e.g. MUC1, MUC3A/B, MUC4, MUC11-13, MUC15-17, and MUC20-22 have clinical significance as predictive biomarkers for disease progression and as putative therapeutic targets [14, 19]. Importantly, MUC1, MUC4, MUC5AC, and MUC16 have been thoroughly investigated for their role in PC progression [14, 20]. However, more efforts need to be directed towards the optimization of mucin-based therapeutic approaches in PC. Strategically, mucin-based diagnostic and therapeutic approaches can be divided based on their secretory and membrane-bound nature. For instance, the membrane-bound mucins like MUC1, MUC4, and MUC16 are more suitable for targeted therapies including small molecule inhibitors, antibody-based therapy, and other vaccine-based immunotherapy. In contrast, the secretory mucins MUC2 and MUC5AC are more likely to have potential as biomarkers. This chapter will highlight the significant findings related to transmembrane mucins such as MUC1, MUC 4 and MUC16 due to their well-characterized structure and biological functions. Further, the mucin-based targeting approach will emphasize the global perspective to delineate the therapeutic potentials of different membrane bound mucins in PC.

Mucins Regulations in Pancreatic Cancer

Mucins are glycoproteins regulated by pathogenic agents like bacteria, through their products such as lipopolysaccharides (LPS), and viruses e.g. respiratory syncytial virus regulation of MUC1 and MUC5AC in the epithelial cells [21, 22]. Other agents that could influence mucins regulation or expression include differentiation factors (e.g. Retinoic acid) [23, 24], ligands of growth factors (EGF) [25, 26], inflammatory cytokines (e.g. interferon-γ (IFN-γ)) [27, 28], and environmental or carcinogenic agents (e.g. cigarette smoke, chemicals (phorbol esters or pollutants) [29-32]. Other than infectious diseases, mucins have also been found to be abnormally over expressed in various cancers, while its expression is required for physiological functions of normal intestine and lung cells [33, 34]. For instance, biosynthesis of intestinal mucins such as MUC2 and MUC3 plays a vital role in protecting the normal intestinal cells by decreasing the adherence capacity of pathogenic microbes and tissue insulting agents (bile acids) present in the gastrointestinal (GI) tract. Similarly, MUCIN 4 (MUC4) which is a transmembrane mucin is aberrantly expressed in vast majority of cancers such as lung [33], stomach [35], breast [36], prostate [37], ovarian [38], biliary tract [39], esophageal [40] and pancreatic cancer [14, 27, 41, 42]. In contrast, MUC4 and MUC 5B were shown to be expressed at basal level in human colon and the moderate level of expression in normal nasopharyngeal and bronchial epithelial cells of [43]. MUC1, MUC6, and MUC5AC are expressed in the glandular or luminal epithelial cells of GI tract, pancreas, mammary gland, uterus, prostate, and lungs [44-46]. Another transmembrane mucin MUC16 was shown to be associated with the normal physiological function of eye and uterus while its overexpression was found to be associated with breast, ovarian and pancreatic cancer [47-49].

Mucin Domains in Regulating Pancreatic Cancer Signaling

Mucins are mainly thought to play a vital role in protecting the normal/cancer cell from the injury by extensive modification through glycosylation or mucin domain interaction with membrane-anchored receptors (EGFR, HER2, and HER3) activation. One such example can be provided by describing the domain structure, signaling of MUC4 domains in PC and MUC4 mucin interactions with EGFR family proteins. MUC4 is present in chromosome 3 and confined to the region q29 concomitantly with MUC20 mucin. MUC4 contains 26 exons out of which 24 exons code for MUC4 apoprotein function, while exon 1 and 2 code for the 5'-untranslated region (UTR), leader peptide and tandem repeat (TR) region/domain. The serine and threonine amino acid residues present in the TR domain undergoes extensive N and O-linked glycosylation, as a result of which MUC4 (mucins in general) attains high molecular weight. These TR regions present in MUC4 vary and provide unique biophysical property to each mucin. The TR also varies in numbers due to polymorphism, which is commonly known as the variable number of tandem repeats (VNTRs). Due to the presence of VNTRs, MUC4 protein size varies from 550 to 930 kDa. Human MUC4 gene was predicted to possess >60% sequence homology with rat Muc4. MUC4 harbors a glycine-asparagine-proline-histidine rich sequence which dissects MUC4 functional domains into MUC4 α (cell adhesion) and MUC4 β (signaling). MUC4 α domain is predicted to contain large and centrally located TR regions, nidogen-like domain (NIDO) and cysteine amino acid-rich adhesion-associated domain (AMOP). MUC4β domain contains von Willebrand factor (vWD; type D domain) and three epidermal growth factor (EGF) like domains. Thus, in the process of tumorogenesis and metastasis, these functional domains are predicted to play a vital role. In addition to these domains, a short 22 amino acid sequence corresponds to the cytoplasmic tail (CT) of MUC4. Among all these domains, the three EGF-like domains present in the MUC4 were shown to interact and stabilize HER2 in the transmembrane region of pancreatic and ovarian cancer cells [14, 38, 50]. In this context, recently, apart from HER2-MUC4 association, HER3 was also shown to interact with MUC4, compensating for the loss or HER2 deficiency in pancreatic cancer. The authors had shown that stable knockdown of HER2 leads to increase in HER3 and MUC4, which further interact with each other, resulting in PC cells proliferation and growth via activation of PI3K and cMYC signaling [51]. More importantly, the unique domains such as NIDO, AMOP, and vWD are present only in MUC4 mucin but not in other membrane-bound mucins i.e. MUC1, MUC13, MUC16 and MUC20 [14, 52]. Notably, the structural analysis revealed that critical amino acids in each of the domains of NIDO, AMOP, vWD, transmembrane domain (TD), EGF and CT were highly conserved among humans, rat, mice, dogs, and chickens [53]. Senapati et al. had demonstrated that the nidogen-like domain (NIDO) present in the MUC4 promotes metastasis of PC cells by interacting with the basement membrane protein fibulin-2 [54]. Another recent study had documented that deletion of AMOP domain from a splice variant of MUC4 (MUC4/Y) that lacks the exon2 coding for tandem repeat region of MUC4 could reverse the angiogenesis and aggressiveness of PC cells that are typically promoted by MUC4/Y [55]. The researchers also found that upon exogenous introduction of MUC4/Y-AMOP engineered construct in PC cells, the NOTCH pathway and its corresponding target genes VEGF, MMP-9 and angiopoietin-2 (ANG-2) were activated [55]. Thus, AMOP domain present in the MUC4 has a critical role in tumor angiogenesis and metastasis.

MUC4 participates in the intracellular signaling through its short CT consisting of 22 amino acid with three serine and one tyrosine phosphorylation sites [14, 52]. Another alternative spice variant of MUC4, MUC4/X that lacks exon2 and exon3 of wild-type MUC4 protein, was also shown to be associated with aggressive properties of PC cells. Stable introduction of MUC4/X construct in wild-type MUC4 expressing cells leads to increased proliferation, invasion and metastasis by facilitating integrin-β1/FAK/ERK signaling [56]. On the other hand, using the specific domain deletion models of MUC4 (MUC4/Y), the synergistic roles of NIDO, AMOP and vWD on MUC4/Y in promoting oncogenic phenotype, intracellular signaling and promoting tumor-related positive feedback loops were identified and validated against single domain alone [57]. These findings confirm the unique role of each domain of MUC4 mucin, regulating and facilitating the aggressive nature of the PC.

The domain structure of major transmembrane mucins such as MUC1, MUC3, MUC4, MUC12, MUC13, MUC16 and MUC17 have been discussed in detail [58, 59]. Specifically, the unique difference in the structure of the domains of mucins is the existence of sea urchin sperm protein, enterokinase, and agrin (SEA) domain, epidermal growth factor (EGF)-like domain and CT. The membrane-bound mucins such as MUC1, MUC3, MUC12, MUC16, and MUC17 contain SEA domain with a conserved sequence of GSVVV motif as an autoproteolytic cleavage site. This cleavage in the SEA domain was predicted to undergo further processing in the endoplasmic reticulum resulting in larger N-terminal α-subunit and C-terminal β-subunits of mucins. Both α and β subunits are attached by hydrogen bonding [58, 60]. In contrast to other transmembrane mucins, MUC4 lacks SEA domains but possesses three EGF-like domains. Other transmembrane mucins such as MUC2, MUC3, MUC13, and MUC17 harbor EGF-like domains flanking the SEA domain [58]. The intracellular signaling of mucin function through the cytoplasmic domain was well-studied in MUC1 and MUC16. MUC1 CT contains 72 amino acid with highly conserved serine and tyrosine amino acids that are later phosphorylated by epidermal growth factor receptor or kinases. MUC16 contains 32 amino acid residues with serine and threonine as the phosphorylation sites along with proline and leucine residues [59, 60]. Hence, it is critical to annotate that these domains are not only structurally important but also essential to elucidate the biological role of mucins.

Tissue Specific Regulation/Overexpression of Mucins in Pancreatic Cancer

Mucin overexpression is common in PC, and therefore, several efforts have been directed towards developing mucins as targets for both disease prognosis and therapy. Elevated expressions of structurally and functionally important mucins in tissues and/or serum of PC patients correlate with disease progression and metastasis; therefore, these high molecular weight glycoproteins have high clinical significance in PC management. Previously, the membrane-bound mucins like MUC1, MUC4, MUC16, and secretory mucin MUC5AC have been reported for their potentials as biomarkers and therapeutic targets in different cancers including PC. Mucins are differentially regulated in different organ cancers. For example, MUC4 is not present in normal pancreas, but progressively increase in PanIN lesions and is overexpressed in 70-90% of pancreatic tumors [19, 42]. Similarly, antibody-based (8G7) detection of MUC4 in normal ovary showed faint, or undetectable level of MUC4 expression whereas, its expression was very high/strong in ovarian tumor tissues [61]. In the case of breast cancer, MUC4 is a well-studied mucin among all mucin family members and is well known for its trastuzumab resistance mechanism [62-64]. Approximately, 92-100% of normal breast tissue exhibit positive expression for MUC4 and it was further reported to be overexpressed in breast cancer (BC)specimens (64-66). Further, MUC4 overexpression is associated with the aggressive Triple-negative breast cancer (TNBC) [36]. Recently, secretory carcinoma of the breast had shown a unique MUC4 expression pattern (13 positive out of 16 cases) holding promise as a therapeutic target in breast cancer [67]. In head and neck cancer, MUC4 was overexpressed in 78% of tumors, while a faint to basal level expression was observed in 10% of benign tissues [68]. In contrast, prostate, lung and colon cancer MUC4 expression was downregulated/decreased, which is further shown to be associated with increased tumor grade and protective in function as compared to respective normal/benign counterparts [33, 37, 69]. Thus, we need to investigate mucin expression and its potential as a therapeutic target in the context of cancer and cell type-specific manner.

MiRNAs-mediated Regulation of Mucin Expression in Pancreatic Cancer

Direct regulation of mucins has been successfully demonstrated through gene-silencing *in vitro* as well as *in vivo*. In case of MUC1, miR-29a and miR-330-5p have been shown to directly target 3'-UTR, which consequently reduced the proliferation, invasion, and migration (*in vitro*) as well as tumor size after intratumoral injection of PC cells in a xenograft mouse model. In addition, the MUC1 targeting strategy through these miRs sensitized the PC cells to gemcitabine *in vitro* suggesting the role of mucins in drug sensitization [70]. Since downregulation of mucins expression increases the drug uptake and sensitivity, it is possible to implicate mucin mediated sensitization of PC cells for current USFDA approved chemotherapies. Other mucins like MUC4 have also been demonstrated as a direct target for miRs, selected by *in silico* screening. Importantly, miR-Let-7b, miR-150, miR-200c, and miR-219-1-3p have been shown to target MUC4, leading to reduction in growth and proliferation in PC [71-74]. Specifically, Let-7b downregulation is associated with PC progression as well as poor prognosis, and restoration of its expression leads to downregulation of K-Ras, MUC4, and NCOA3 [75]. Among others, the miR-150 and miR-219-1-3p have been reported to target 3'-UTR of MUC4 whereas, miR-200c has been demonstrated to target the coding mRNA sequence of MUC4 and MUC16 mucins [72-74]. Functionally, direct targeting of MUC4 using miR-150 has been shown to reduce proliferation and metastasis *in vitro* as well as *in vivo* in PC by downregulation of its interacting partner HER2 and subsequent oncogenic downstream signaling [72] whereas, miR-219-3p overexpression in PC cell lines has been shown to decrease the expression of cyclinD1, AKT, and ERK, suggesting its role in survival, proliferation, and metastasis [74]. Recently, Khan et al. demonstrated miR-145 targeting 3'-UTR of MUC13, which has been reported to play a role in PC progression. The miR-145 showed downregulation of MUC13 and its interacting partner HER2 and subsequent downstream signaling molecules when transfected in Capan1 and HPAF PC cell lines. In *in vivo* xenograft mouse model, miR-145 inhibited the tumor growth, which further validated its effect observed *in vitro* in PC cell lines. Furthermore, the miR-145 increased the drug-sensitivity to gemcitabine in ASPC1 cells, suggesting the potential role of miR-145 as a sensitizer to the existing chemotherapy [76]. Overall, the miRNA-regulated mucin will provide a platform for direct targeting of mucins; therefore, it might be a useful approach in PC therapy optimization [9]. However, their efficacy and specificity need to be optimized before promoting in clinical settings.

Transcriptional Regulation of Mucin Expression

Mucin expression was majorly linked with cancer pathogenesis, and it is mainly regulated by the transcriptional and translational mechanism(s). It has been shown that regulation of mucin expression appears to be a complex process, and an improved understanding of the underlying molecular mechanisms may help to identify the biochemical events and biologically relevant factors that are responsible for the aberrant up-regulation of this mucin gene which helps PC cell growth, proliferation, invasion, and metastasis. Recently, a systemic integrated analysis of miRNA and mRNA from a set of normal and PC specimens identified MUC4 as one of the top differentially expressed gene (DEG) in the mucin family to differentiate low and high-risk PC groups [77]. Earlier studies using mammary epithelial cells were shown to upregulate MUC4 mucin upon IGF treatment. Extensive studies have been performed on characterizing MUC4 promoter and identified that MUC4 expression is driven by active transcriptional factors in the proximal and distal promoter and TATA box is present in the distal promoter region. Previous work from Andrianifahanana et al. had demonstrated that the MUC4 promoter has STAT1 binding sites [27]. In continuation of this STAT1 regulation of MUC4, a recent study also documented that transient knockdown of EGFR specific siRNA resulted in decreased STAT1 phosphorylation which correlates with decreased MUC4 expression of PC cells [26]. Previously, MUC4 expression was known to stabilize HER2/ErbB2, thereby enhancing its downstream signaling to promote pancreatic and ovarian cancer cell growth and proliferation [38, 50]. On the other hand, a recent study by Jonckheere et al. have demonstrated that MUC4 suppression leads to impairment of JNK signaling whereas, HER2/ErbB2 knockdown alters MAPK signaling. Thus, MUC4 and HER2/ErbB2 signaling regulate different intracellular signaling to control the biological properties of PC cells [78]. MUC4 mucin can be differentially upregulated by nuclear receptor coactivator 3 (NCOA3) in PC. When chromatin remodeling enzymes responsible for MUC4 expression in PC were analyzed in MUC4-expressing (CAPAN1) and non-expressing cells (Panc1) by chromatin factor PCR array (consisting of 84 key genes involved in chromatin remodeling), the authors observed that NCOA3 was differentially expressed in the MUC4 expressing cell lines, as compared to the non-expressing cell line, both at transcript and translational levels. Further, RNA interference-mediated suppression of NCOA3 in the PC cells significantly downregulated MUC4 expression, establishing the direct role played by NCOA3 on MUC4 expression. Besides these biochemical roles, NCOA3 knockdown also resulted in decreased MUC1 expression levels [79]. In last two decades, multiple transcription factors such as AP-1, SP-1, Growth response element (GRE), Nuclear Factor kappa B (NFκB), were shown to regulate MUC4 by binding to the proximal region of MUC4 promoter [80]. In addition, other transcription factors like HNF, CDX, FOXA, GATA, and HNF1α have also been shown to regulate MUC4 expression [81]. Recently, *in silico* analysis revealed three putative binding sites for TCF/LEF in the promoter region of MUC4. Thus, MUC4 can be regulated by betacatenin in PC [82]. Apart from promoter binding mediated regulation, mucins can also be transcriptionally activated by the treatment of cytokines (IFNγ), retinoic acid, bile acids and TGFβ [23, 29, 83]. MUC4 also regulates acute phase protein lipocalin 2 to support HER2/AKT/NF-κB axis in PC [84]. Similarly, the MUC1 and MUC16 promoter also contain binding sites for some common transcription factors such as AP-1, ERα, EBP-β and SP-1 [58]. Through our *in silico* analysis, we have identified few putative binding sites for transcription factors such as SP-1, ERα, SF-1, PPARα, EBP-α, β and δ, RARβ, AP-1, and POU in the promoter region of MUC1 and MUC16 [85, 86]. Briefly, we have summarized the transcription factors binding to the promoter region of MUC1, MUC4 and MUC 16 (Fig. 1). Apart from the discussed MUC4 mucin, MUC1 and MUC16 mucins have been elaborately discussed in several review articles [59, 60, 87].

Mucins Regulation as a Perspective Biomarker

Biomarkers are molecules that, in response to any pathological insults including oncogenic transformation, progression, and metastasis, exhibit altered expression profile and can differentiate the pathological condition to the normal in the patient-derived biopsies with a high clinical significance. These biomarkers are analyzed on any biological material derived from the patient including, tissues, blood, urine, sputum, saliva, etc. while in cancers the most common specimens include tissue biopsies either derived from the specific organ or the blood samples derived from the patients. Biomarker analysis in the patient serum is correlative but easier to perform. In addition, the blood samples derived from the patients can be used for an array of analyses including, circulating nucleic acids, exosomes, proteins, tumor cells, stem cells, enzymes, and other metabolic products, etc., which can be correlated with cancer progression as well as with the treatment response in cancer patients. In contrast, the tissue biopsies followed by specific staining provide a closer and more accurate examination of any molecular or cellular abnormality but sometimes undergo the technically difficult procedure.

The poor prognosis and lack of predictive biomarkers attribute unfavorable survival rate in PC patients; therefore, more efforts have been directed in recent years for biomarker development. Thus far, both tissues and serum biomarkers have been investigated in PC patients, alone or in combination. Importantly, the carbohydrate antigens have been broadly investigated in PC include CA19.9, CEA, and other carbohydrate antigens like CA125 (MUC16), CA242, CA50, CA195, CA-72.4, etc. [88-90]. However, most of these biomarkers have low specificity and therefore, need to be used in combination of other antigens. Mucin antigens MUC1, MUC4, MUC5AC, and MUC16 (CA125) are either low or negative in the normal pancreas, but with disease progression, the expression elevates and is observed high in PDAC patients with advanced stages [14, 91]. Some of these mucins have been investigated as biomarkers in PC [92, 93]. Earlier studies suggest that

PAM4 monoclonal antibody (mAb) reactive MUC1 is a potential biomarker for early detection of PC [94, 95]. However, recent reports suggest that PAM4 is reactive to MUC5AC epitope and therefore, MUC5AC might be a more potential predictive and prognostic biomarker in PC [96]. In addition, the secretory glycoproteins like MUC1, MUC2, and MUC5AC, besides the staining in tissue samples, have been investigated for their elevated serum levels, which suggest their potential as biomarkers for PC patients [97-99]. Earlier, our group and others had demonstrated high MUC5AC in PC patient serum. In addition, MUC5AC in combination with CA19.9 antigen has been demonstrated to show high specificity and sensitivity as a biomarker in PC patients [93]. Among other mucins, MUC16 (CA125) and MUC4 have demonstrated their potential as biomarkers [91, 100, 101]. In fine needle aspiration (FNA) samples of pancreas, MUC1, MUC2, MUC5AC, and MUC16 (CA125) expression have been investigated and directly correlated with disease progression and metastasis and therefore, might be used as a prognostic marker in PC patients [99, 102, 103]. Similarly, MUC4 alone or in combination with other established biomarkers has been investigated in tissue samples, fine needle aspirates, and serum samples of PC patients [104]. Results suggest that MUC4 is a potential biomarker when analyzed in tissue samples. While in PC serum samples, MUC4 was detectable in surface-enhanced Raman scattering (SRS) method, not in conventional ELISA and RIA based techniques [105], emphasizing the need of improving the detection techniques. Interestingly, MUC4 and MUC16 (CA125) analysis in FNA samples from PC patients showed high specificity and therefore, seems important adjunct for cytomorphological analysis of FNA samples [102]. Overall, mucins due to their overexpression and secretory nature demonstrate this biomarker potential in PC. In addition, high expression

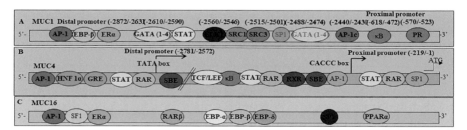

Fig. 1: Schematic representation of binding sites for various transcription factors in promoter regions of MUC1, MUC4, and MUC16. MUC1 (A) and MUC4 (B) expressions are driven by active transcriptional factor in proximal and distal promoter and TATA box is present in distal promoter region. Through *in silico* analysis, putative binding sites of proteins/transcription factors are identified in the MUC 16 promoter region (C) GRE: Growth response element, HNF: Hepatocyte nuclear factor, RAP: Retinoic acid receptor, NFκB: Nuclear factor Kappa B, RAR: Retinoid X receptor, SP-1: Specificity protein-1, AP1: Activator protein 1, SF1: Steroidogenic factor 1, SBE: SMAD binding elements, ERα: Estrogen receptor alpha, EBP: Enhancer binding protein, PR: Progesterone response region, SRC: Steroid receptor co-activator.

of MUC1 in cancer stem cells (CSCs) in circulation as well as in the tumor tissues show pathological significance in progression and metastasis, which correlates with disease pathogenesis [106]. Therefore, analysis of MUC1 expression in circulating CSCs can be used for predicting PC metastasis. However, caution must be taken in selecting the reagents and detection methods to achieve high specificity. For example, lots of variabilities exist in the immunohistochemical analysis of tissues. Different antibodies for the same target have been published in the literature showing difference in staining, which hinders the biomarker development process in cancer.

Overall, mucins alone or in combination with other potential biomarkers might be useful as predictive and prognostic biomarkers in PC. However, to improve the current survival rates in PC patients, PC research is facing the unmet needs of a successful biomarker that can predict the disease at an early stage and can also serve as a good prognostic marker in clinical settings. The challenges still exist in bringing the specificity and sensitivity of these biomarkers in the clinically acceptable range. Therefore, more combinations must be evaluated with other pipeline molecules to improve clinical outcomes.

Mucins Based Targeted Therapies in Pancreatic Cancer

Mucin overexpression is one of the characteristics of PC and reports suggest that many of these mucins contribute to oncogenic progression and metastasis [14, 91, 101]. Essentially, MUC1 and MUC4 have been reported for their role in survival, proliferation, angiogenesis, hypoxia, and metastasis of PC cells through different signaling pathways *in vitro* as well as *in vivo* [26, 51, 83, 84, 107-113]. These tumor-associated mucins are different from the mucins expressed in normal physiological conditions in their expression, glycosylation, and genetic splicing and therefore could be used for targeted therapies [114-117]. Targeting key signaling molecules in PC showed downregulation of related mucins and vice versa. For instance, gene silencing of MUC4 by siRNA, miRNAs, and shRNA attenuated different signaling molecules important for proliferation and metastasis on PC cells [20, 71, 78, 118]. Furthermore, the pharmacological inhibition of key signaling molecules correlates with downregulation of mucins and decrease in survival and proliferation *in vitro* and *in vivo* [20, 119-121]. Interestingly, mucins, in particular MUC1 and MUC4, have been observed to be associated with EGFR-regulated signaling pathways in PC. Earlier studies showed that inhibition of signaling axes regulated through EGFR showed decreased expression in mucins MUC1 and MUC4 and thereafter, decreased proliferation and metastasis in PC [26, 122]. The consequences of both direct and indirect mucin targeting cellular proliferation and metastasis have been demonstrated both *in vitro* as well as *in vivo* in PC models. Overall, in addition to their implication in biomarker development, mucins are suitable molecules for designing different therapeutic approaches in PC including antibody-based targeting, differentially glycosylated and immunogenic fragment for developing vaccine, molecular and pharmacological inhibition of functional domains, etc. Overall, targeting mucins might be a useful strategy in PC diagnosis and therapy. Here, we highlight the recent developments related to mucins in PC diagnosis and treatment (Fig. 2a; Table 1).

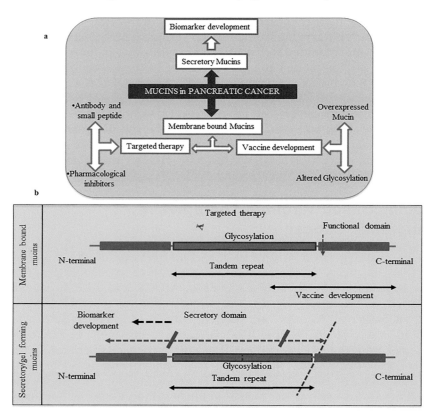

Fig. 2: Dissecting the mucin domains for therapeutic optimizations for pancreatic cancer treatment. (a) Schematic flowchart dissecting the important membrane bound and secretory mucins having different therapeutic implications in PC therapy. (b) Upper panel shows domains present in membrane bound mucins that have different therapeutic significances including targeted therapies and vaccine development, whereas lower panel depicts the secretory mucins and their significant association in biomarker development.

Therapeutic Importance of Structural and Functional Domains of Mucins

Despite high similarities in basic structure and function under normal physiological conditions, the mucins under pathological conditions exhibit

Table 1: Clinically relevant mucins[‡] in pancreatic cancer

S.No.	Mucin	Gene location	Size[#] (kDa)	Pathological function and clinical significance
				A. Membrane bound mucins
1	MUC1	Chr 1; q22	300-600	Binds to EGFR, β-catenin, and NF-KB and promotes proliferation via MAPK/ AKT/WNT-β-catenin signaling, upregulate ERK signaling. Role in drug chemoresistance and angiogenesis. Potent vaccine candidate and important as biomarker.
2	MUC4	Chr 3; q29	600-900	Interacts and activates ERBB-family member(s). Regulate EGFR-STAT1 signaling, MAPK/AP1, JAK/STAT, and Src/FAK signaling. Role in proliferation and metastasis. Promotes chemoresistance. Potential for vaccine development, biomarker, and targeted therapies.
3	MUC16	Chr 19; p13.2	300-500	Involved in WNT/β-catenin signaling, FAK-signaling. Promotes PC progression and metastasis. Immunologically important mucin.
4	MUC17	Chr 7; q22.1	400-500	Upregulated in PDAC. Involved in HIF-1α mediated signaling. Role in PDAC progression.
				B. Secretory mucins
5	MUC2	Chr 11; p15.5	540-550	Upregulated in mucinous neoplasm Potential as a biomarker
6	MUC5AC	Chr 11; p15.5	641.00	Role in inflammation Upregulated in pancreatic cancer Potential as a biomarker
7	MUC6	Chr 11; p15.5	257.05	Potential as a biomarker

[‡] Clinical relevance is based on the current information available for the particular mucin in PC.
[#] Molecular weights may vary depending on the extent of post translational modifications.

differences in the number of splice variants, variable numbers of tandem repeats (VNTRs), glycosylation patterns, and oncogenic signaling, which provide exclusive avenues for mucin targeting. Genetic polymorphism and difference in exon-size and number, tandem-repeats and post-translational modifications, lead to the alteration in structural and functional properties of mucins in pathological conditions [10, 14, 22]. Structurally, the membrane-bound mucins are composed of extracellular and intracellular polypeptides with a hydrophobic transmembrane region tethering them to the cell membrane. In contrast, the secretory mucins lack the transmembrane domain but, are generally gel-forming and undergo polymerization due to disulfide linkages (Fig. 2b) [21, 52, 91]. For example, the variable sizes of tandem repeat regions (TRRs) and extent of glycosylation are not only important determinants in the size, structure, and function of mucins, but are also crucial for antibody-based targeted therapies and other immunotherapeutic approaches. For example, antibodies generated against tandem repeats show high-affinity due to multiple accessible epitopes. Monoclonal antibody 8G7 (mAb8G7) generated by Batra et al. is one of the example of high-affinity antibodies against MUC4 TRR [123]. In addition, the C-terminal of membrane-tethered mucins including MUC1 and MUC4 have been reported to have putative cleavage sites and the interacting and functional domains [14, 52, 91]; therefore, targeted therapies against these domains might be useful to block the mucin-associated functions in PC. The functional domains, tandem repeat regions (TRRs) and altered N- and O-glycosylation in these PC pathogenesis associated over-expressed mucins are thought to be hot spots for designing mucin-based biomarker and therapeutics [20]. In addition, the altered glycosylation in over-expressed mucins makes them an immunologically good target for vaccine design. Therefore, glycoepitopes along with novel peptide epitopes are being considered important for pathogenesis specific immune response in PC patients. Overall, the PC pathogenesis associated structural and functional attributes of mucins have significant therapeutic values for future therapeutic developments in PC.

Mucins Based Vaccines in Targeting Pancreatic Cancer

Despite several similarities in structure and functions, each mucin has unique characteristics attributed to its specific role in normal as well as a pathological condition. The membrane-bound mucins are more important for vaccine development because the cancer cells expressing these mucins are targetable by the specific immune response generated against surface accessible epitopes present in them. In addition, other targeted approaches like neutralizing antibodies and antibody-drug conjugates are also feasible with membrane-bound mucins. Due to their large molecular sizes, the purification of recombinant mucins has been a challenge so far; therefore, the immune response against mucins has been investigated mostly by using the peptide epitopes predicted though different *in silico* prediction methods [124]. Different regions of mega-dalton mucins from signal peptide to N-terminal region, Sea urchin sperm protein, Enterokinase, and Agrin (SEA) domain, tandem repeat regions (TRRs), and C-terminal domain have different characteristics that contribute to structure and function of mucins. Immunologically, these mucin-domains have not been fully explored for vaccine development in PC. Except for MUC1 and partly MUC4, none of the other mucins have been evaluated for PC specific immune response [20, 124, 125]. Most of the studies related to mucin vaccines evaluated the TRR-peptides of the signal peptide sequence.

MUC1 is the top studied mucin for its immunological potential as a vaccine candidate. The cancer-specific aberrant O-glycosylation in overexpressed MUC1 generates novel peptide and glycopeptides epitopes that have been used for anticancer vaccine development in different epithelial malignancies in both preclinical and clinical settings [126-131]. Importantly, MUC1 in combination with radiation or other components like immunogenic epitopes, adjuvants, checkpoint inhibitors, and cytokines, etc., elicited both humoral and cell-mediated immune responses, which led to prolonged progression-free survival (PFS) in a transgenic mouse model [132-137]. Studies from clinical trials suggest that MUC1 vaccination is well-tolerated in PC patients without any significant toxicity, and elicited specific CTL response, high IFN-γ release and elevated humoral response [138-141]. Importantly, not only glycosylated but also non-glycosylated small peptides derived from MUC1 protein have been shown to enrich IFN-γ producing CD4 and CD8 cells in PC MUC1 transgenic mouse model [133]. Indeed, the change in glycosylation pattern generates not only novel glycoepitopes but also exposes unique immunodominant peptide epitopes from the native protein. Thus, the mucins associated with PC pathogenesis provide an important template for vaccine development. Studies on cancer vaccine development so far suggest that a successful vaccine development depends on numerous parameters including dose, formulation, regimen, adjuvant system, *in vivo* stability, release kinetics, etc. [142]. Therefore, the optimization of each parameter is important for better therapeutic response in cancer vaccine studies. Similar approaches have been implemented for MUC4 vaccine optimization, mostly *in vitro* and in preclinical settings. Theoretically, MUC4 seems to be a potential candidate for PC vaccine design due to its exclusive expression in PC. Like MUC1, MUC4 *in silico* derived peptides in combination with different adjuvants and cytokines demonstrated TH1 type immune response *in vitro* as well as *in vivo*. Earlier, the dendritic cells (DC) vaccines having MUC4 along with survivin or universal T-helper epitope PADRE demonstrated activation of DC surface markers and antigen presentation [143, 144]. In another study, when MUC4-glycopeptide conjugated to tetanus toxoid used for vaccination in mice, the antigen-specific strong immune response was observed [145]. The study also suggested that glycosylation is important for a robust immune response. However, lack of transgenic mouse model and insufficiency of reagents impeded the MUC4 vaccine development in PC.

Overall, the vaccine development against PC is challenging in many ways when compared to other malignancies. Immunosuppressive and highly desmoplastic tumor microenvironment are among the major challenges. Nevertheless, the combination of various immunotherapeutic approaches needs to be implemented in a way so that it can improve the survival rate in patients. Importantly, the use of checkpoint inhibitors, cytokine therapy, stroma depletion therapy, in combination with vaccines might be useful additions to conventional PC therapies to improve the therapeutic outcome effectively.

Small Molecule Inhibitors and Natural Products to Target Aberrantly Expressed Mucins in Pancreatic Cancer

Despite their structural and functional significance in oncogenic transformation and signaling during PC progression, mucin-targeted therapies are in infancy. The targeted therapies are more focused on membrane-bound mucins rather than the secretory or gel-forming mucins because targeting the latter could leach the inhibitor or mAbs [98, 146]. However, the gene-silencing based approaches could be implicated to both membrane-tethered as well as secretory mucins. There is no report for targeting secretory mucins for therapeutic implications in PC; therefore, we will emphasize more on membrane-bound mucins except the gene silencing. Although the targeting strategies like gene silencing, mAb mediated targeting, indirect targeting by pharmacological inhibitors, and micro-RNA mediated targeting have been implicated in targeting mucins, it has been challenging to target bulky mucin glycoproteins in a way that their structural and functional roles in PC progression and metastasis could be specifically disrupted. Gene silencing mediated through si-RNA, Sh-RNA, and micro-RNAs (miRs) has been implicated in establishing the functional significance of mucins in oncogenic progression rather than focusing on targeted approach for therapeutic intervention in PC [20, 26, 70, 71, 118]. The membrane-bound mucins have a large extracellular accessible domain and functional intracellular domains that could be utilized for targeted therapies. In addition, exclusive expression of the target molecule in malignant cells, accessibility of target site, epitope multiplicity in case of Ab-mediated targeting, the specificity of targeting agents with no off-target effect are important parameters to be considered for designing the targeted therapies. Based on these parameters, mucins qualify as targets for targeted therapies in PC. Importantly, MUC4, MUC5AC, and MUC16 are overexpressed exclusively under PC pathological condition, and not in the normal pancreas [91, 92]. Besides, the epitope multiplicity in TRRs and altered glycosylation provide more exclusivity to these overexpressed mucins including MUC1 for targeted therapy in PC. On the other hand, several other targeted approaches have been reported where inhibition of PC growth and metastasis is regulated through mucin downregulation, suggesting that though mucins are not the direct target, they are functionally important in PC pathogenesis. Among these, the clinically acceptable chemotherapies, natural products, and other antibiotics and small molecule inhibitors, etc. have been demonstrated to downregulate mucins and, therefore, shown to inhibit PC progression. Among natural products, guggulsterone (GS) [147], thymoquinone (TQ) [148], and graviola extracts [149] significantly inhibited the PC growth and proliferation [20]. Interestingly, the effect was MUC4 mediated suggesting that MUC4 is functionally important and, therefore, has potential as a therapeutic target in PC. The crude extracts are indicative of their potential as anticancer agents but need to be dosed at a higher concentration for effective results. Therefore, more efforts are required to isolate the compounds that specifically inhibit PC growth followed by a delineation of mechanisms of their action. In another report, we have demonstrated that inhibition of EGFR-STAT1 axis using Pan-EGFR small molecule inhibitors afatinib and canertinib is mediated through MUC4 where inhibition of this axis significantly downregulated MUC4 expression, leading to a decreased proliferation *in vitro* and metastasis *in vivo* [26]. Several pieces of evidence support the mucin-mediated effect of therapeutic agents on PC growth and metastasis, but few reports describe the direct targeting approach. Therefore, more efforts are needed to optimize the screening and development of therapeutic agents directly targeting the functional domains of mucins to prove the druggability of mucins in PC.

Conclusions and Future Directions

Poor survival in PC patients is mainly attributed to lack of available biomarkers for early detection, early metastasis, high relapse rate, and resistance to therapies. Keeping in view the structural and functional significance of mucins in PC pathogenesis, they have been projected as potential targets for both the development of biomarker and therapeutic strategies against PC. Though their high molecular size, heavy glycosylation, genetic polymorphism, and reported splice variants pose challenges, significant advancements have been made while optimizing mucins as diagnostic and therapeutic targets in PC. First, the pathogenesis associated secretory and membrane-bound mucins have been shown as potential biomarkers in PC patients' sera as well as tumor specimens [93, 99, 102-105, 150]. Second, the immunologically active epitopes from overexpressed and aberrantly glycosylated mucins, especially MUC1 [126, 134, 139], MUC4 [124, 144, 151], and MUC16 [146], have been reported to have potential against PC. Third, the mAbs against different structural and functional domains of mucins have been demonstrated to target mucins with high specificity, suggesting their futuristic role as targeted agents or vehicles for delivering the cytotoxic agents in pancreatic tumors [123, 152-154]. An important aspect of mucins is their functional role in oncogenic signaling through multiple domains that promote PC progression and metastasis. In particular, AMOP, NIDO, EGF-like domain, and vWD have been reported to contribute to PC pathogenesis. Therefore, domain specific neutralizing antibodies may serve as effective anti-cancer agents in PC. Fourth, a significant inhibition in proliferation and metastasis has been demonstrated in PC by downregulating mucins using various biological and pharmacological inhibitors. At genetic level, different mucin-specific miRs, shRNAs, and siRNAs have been shown to inhibit proliferation and metastasis in PC. Therefore, gene-silencing approaches and pharmacological inhibitors derived from synthetic or natural sources could be developed for targeting mucins. Overall, mucins are structurally and functionally important for PC progression and metastasis and these mucin targeting approaches are essential for designing the future diagnostics and therapeutics in PC.

References

1. Siegel, R.L., K.D. Miller, A. Jemal. Cancer statistics, 2019. CA Cancer J Clin, 2019. 69(1): 7-34.
2. Siegel, R.L., K.D. Miller, A. Jemal. Cancer statistics, 2018. CA Cancer J Clin, 2018. 68(1): 7-30.
3. Kleeff, J., M. Korc, M. Apte, C. La Vecchia, C.D. Johnson, A.V. Biankin, et al. Pancreatic cancer. Nature Reviews Disease Primers, 2016. 2: 16022.
4. Hidalgo, M., S. Cascinu, J. Kleeff, R. Labianca, J.M. Lohr, J. Neoptolemos, et al. Addressing the challenges of pancreatic cancer: future directions for improving outcomes. Pancreatology, 2015. 15(1): 8-18.
5. Al Haddad, A.H., T.E. Adrian. Challenges and future directions in therapeutics for pancreatic ductal adenocarcinoma. Expert Opin Investig Drugs, 2014. 23(11): 1499-1515.
6. Kaur, S., M.J. Baine, M. Jain, A.R. Sasson, S.K. Batra. Early diagnosis of pancreatic cancer: challenges and new developments. Biomark Med, 2012. 6(5): 597-612.
7. Neoptolemos, J.P., J. Kleeff, P. Michl, E. Costello, W. Greenhalf, D.H. Palmer. Therapeutic developments in pancreatic cancer: current and future perspectives. Nature Reviews Gastroenterology & Hepatology, 2018. 15(6): 333-48.
8. Werner, J., S.E. Combs, C. Springfeld, W. Hartwig, T. Hackert, M.W. Büchler. Advanced-stage pancreatic cancer: therapy options. Nature Reviews Clinical Oncology, 2013. 10: 323.
9. Khan, M.A., S. Azim, H. Zubair, A. Bhardwaj, G.K. Patel, Khushman Md, et al. Molecular drivers of pancreatic cancer pathogenesis: looking inward to move forward. International Journal of Molecular Sciences, 2017. 18(4): 779.
10. Di Domenico, A., T. Wiedmer, I. Marinoni, A. Perren. Genetic and epigenetic drivers of neuroendocrine tumours (NET). Endocr Relat Cancer, 2017. 24(9): R315-R334.
11. Tatarian, T., J.M. Winter. Genetics of pancreatic cancer and its implications on therapy. Surg Clin North Am, 2016. 96(6): 1207-21.
12. Krishnan, S., H.J. Whitwell, J. Cuenco, A. Gentry-Maharaj, U. Menon, S.P. Pereira, et al. Evidence of altered glycosylation of serum proteins prior to pancreatic cancer diagnosis. Int J Mol Sci, 2017. 18(12).

13. Pan, S., T.A. Brentnall, R. Chen. Glycoproteins and glycoproteomics in pancreatic cancer. World J Gastroenterol, 2016. 22(42): 9288-99.
14. Kaur, S., S. Kumar, N. Momi, A.R. Sasson, S.K. Batra. Mucins in pancreatic cancer and its microenvironment. Nat Rev Gastroenterol Hepatol, 2013. 10(10): 607-20.
15. Keleg, S., Büchler, R. Ludwig, M.W. Büchler, H. Friess. Invasion and metastasis in pancreatic cancer. Molecular Cancer 2003. 2: 14.
16. Nath, S., K. Daneshvar, L.D. Roy, P. Grover, A. Kidiyoor, L. Mosley, et al. MUC1 induces drug resistance in pancreatic cancer cells via upregulation of multidrug resistance genes. Oncogenesis, 2013. 2: e51.
17. Mezencev, R., L.V. Matyunina, G.T. Wagner, J.F. McDonald. Acquired resistance of pancreatic cancer cells to cisplatin is multifactorial with cell context-dependent involvement of resistance genes. Cancer Gene Therapy, 2016. 23(12): 446-53.
18. Chand, S., K. O'Hayer, F.F. Blanco, J.M. Winter, J.R. Brody. The landscape of pancreatic cancer therapeutic resistance mechanisms. International Journal of Biological Sciences, 2016. 12(3): 273-82.
19. Moniaux, N., M. Andrianifahanana, R.E. Brand, S.K. Batra. Multiple roles of mucins in pancreatic cancer, a lethal and challenging malignancy. Br J Cancer, 2004. 91(9): 1633-8.
20. Gautam, S.K., S. Kumar, A. Cannon, B. Hall, R. Bhatia, M.W. Nasser, et al. MUC4 mucin – a therapeutic target for pancreatic ductal adenocarcinoma. Expert Opin Ther Targets, 2017. 21(7): 657-69.
21. Devine, P.L., I.F. McKenzie. Mucins: structure, function, and associations with malignancy. Bioessays, 1992. 14(9): 619-25.
22. Bansil, R., B.S. Turner. Mucin structure, aggregation, physiological functions and biomedical applications. Current Opinion in Colloid & Interface Science, 2006. 11(2): 164-70.
23. Choudhury, A., R.K. Singh, N. Moniaux, T.H. El-Metwally, J.P. Aubert, S.K. Batra. Retinoic acid-dependent transforming growth factor-beta 2-mediated induction of MUC4 mucin expression in human pancreatic tumor cells follows retinoic acid receptor-alpha signaling pathway. J Biol Chem, 2000. 275(43): 33929-36.**
24. Andrianifahanana, M., A. Agrawal, A.P. Singh, N. Moniaux, I. van Seuningen, J.P. Aubert, et al. Synergistic induction of the MUC4 mucin gene by interferon-gamma and retinoic acid in human pancreatic tumour cells involves a reprogramming of signalling pathways. Oncogene, 2005. 24(40): 6143-54.
25. Neeraja, D., B.J. Engel, D.D. Carson. Activated EGFR stimulates MUC1 expression in human uterine and pancreatic cancer cell lines. J Cell Biochem, 2013. 114(10): 2314-22.
26. Seshacharyulu, P., M.P. Ponnusamy, S. Rachagani, I. Lakshmanan, D. Haridas, Y. Yan, et al. Targeting EGF-receptor(s) - STAT1 axis attenuates tumor growth and metastasis through downregulation of MUC4 mucin in human pancreatic cancer. Oncotarget, 2015. 6(7): 5164-81.
27. Andrianifahanana, M., A.P. Singh, C. Nemos, M.P. Ponnusamy, N. Moniaux, P.P. Mehta, et al. IFN-gamma-induced expression of MUC4 in pancreatic cancer cells is mediated by STAT-1 upregulation: a novel mechanism for IFN-gamma response. Oncogene, 2007. 26(51): 7251-61.
28. Andrianifahanana, M., S.C. Chauhan, A. Choudhury, N. Moniaux, R.E. Brand, A.A. Sasson, et al. MUC4-expressing pancreatic adenocarcinomas show elevated

levels of both T1 and T2 cytokines: potential pathobiologic implications. Am J Gastroenterol, 2006. 101(10): 2319-29.
29. Kunigal, S., M.P. Ponnusamy, N. Momi, S.K. Batra, S.P. Chellappan. Nicotine, IFN-gamma and retinoic acid mediated induction of MUC4 in pancreatic cancer requires E2F1 and STAT-1 transcription factors and utilize different signaling cascades. Mol Cancer, 2012. 11: 24.
30. Kim, S.W., J.S. Hong, S.H. Ryu, W.C. Chung, J.H. Yoon, J.S. Koo. Regulation of mucin gene expression by CREB via a nonclassical retinoic acid signaling pathway. Mol Cell Biol, 2007. 27(19): 6933-47.
31. Momi, N., M.P. Ponnusamy, S. Kaur, S. Rachagani, S.S. Kunigal, S. Chellappan, et al. Nicotine/cigarette smoke promotes metastasis of pancreatic cancer through alpha7nAChR-mediated MUC4 upregulation. Oncogene, 2013. 32(11): 1384-95.
32. Poachanukoon, O., S. Koontongkaew, P. Monthanapisut, N. Pattanacharoenchai. Macrolides attenuate phorbol ester-induced tumor necrosis factor-alpha and mucin production from human airway epithelial cells. Pharmacology, 2014. 93(1-2): 92-9.
33. Majhi, P.D., I. Lakshmanan, M.P. Ponnusamy, M. Jain, S. Das, S. Kaur, et al. Pathobiological implications of MUC4 in non-small-cell lung cancer. J Thorac Oncol, 2013. 8(4): 398-407.
34. Weiss, A.A., M.W. Babyatsky, S. Ogata, A. Chen, S.H. Itzkowitz. Expression of MUC2 and MUC3 mRNA in human normal, malignant, and inflammatory intestinal tissues. J Histochem Cytochem, 1996. 44(10): 1161-6.
35. Senapati, S., P. Chaturvedi, P. Sharma, G. Venkatraman, J.L. Meza, W. El-Rifai, et al. Deregulation of MUC4 in gastric adenocarcinoma: potential pathobiological implication in poorly differentiated non-signet ring cell type gastric cancer. Br J Cancer, 2008. 99(6): 949-56.
36. Mukhopadhyay, P., I. Lakshmanan, M.P. Ponnusamy, S. Chakraborty, M. Jain, P. Pai, et al. MUC4 overexpression augments cell migration and metastasis through EGFR family proteins in triple negative breast cancer cells. PLoS One, 2013. 8(2): e54455.
37. Singh, A.P., S.C. Chauhan, S. Bafna, S.L. Johansson, L.M. Smith, N. Moniaux, et al. Aberrant expression of transmembrane mucins, MUC1 and MUC4, in human prostate carcinomas. Prostate, 2006. 66(4): 421-9.
38. Ponnusamy, M.P., I. Lakshmanan, M. Jain, S. Das, S. Chakraborty, P. Dey, et al. MUC4 mucin-induced epithelial to mesenchymal transition: a novel mechanism for metastasis of human ovarian cancer cells. Oncogene, 2010. 29(42): 5741-54.
39. Matull, W.R., F. Andreola, A. Loh, Z. Adiguzel, M. Deheragoda, U. Qureshi, et al. MUC4 and MUC5AC are highly specific tumour-associated mucins in biliary tract cancer. Br J Cancer, 2008. 98(10): 1675-81.
40. Bruyere, E., N. Jonckheere, F. Frenois, C. Mariette, I. Van Seuningen. The MUC4 membrane-bound mucin regulates esophageal cancer cell proliferation and migration properties: implication for S100A4 protein. Biochem Biophys Res Commun, 2011. 413(2): 325-9.
41. Chaturvedi, P., A.P. Singh, N. Moniaux, S. Senapati, S. Chakraborty, J.L. Meza, et al. MUC4 mucin potentiates pancreatic tumor cell proliferation, survival, and invasive properties and interferes with its interaction to extracellular matrix proteins. Mol Cancer Res, 2007. 5(4): 309-20.
42. Swartz, M.J., S.K. Batra, G.C. Varshney, M.A. Hollingsworth, C.J. Yeo, J.L. Cameron, et al. MUC4 expression increases progressively in pancreatic intraepithelial neoplasia. Am J Clin Pathol, 2002. 117(5): 791-6.

43. Fischer, B.M., J.G. Cuellar, M.L. Diehl, A.M. deFreytas, J. Zhang, K.L. Carraway, et al. Neutrophil elastase increases MUC4 expression in normal human bronchial epithelial cells. Am J Physiol Lung Cell Mol Physiol, 2003. 284(4): L671-9.
44. Gendler, S.J. MUC1, the renaissance molecule. J Mammary Gland Biol Neoplasia, 2001. 6(3): 339-53.
45. Larsson, J.M., H. Karlsson, J.G. Crespo, M.E. Johansson, L. Eklund, H. Sjovall, et al. Altered O-glycosylation profile of MUC2 mucin occurs in active ulcerative colitis and is associated with increased inflammation. Inflamm Bowel Dis, 2011. 17(11): 2299-307.
46. Hasnain, S.Z., A.L. Gallagher, R.K. Grencis, D.J. Thornton. A new role for mucins in immunity: insights from gastrointestinal nematode infection. Int J Biochem Cell Biol, 2013. 45(2): 364-74.
47. Felder, M., A. Kapur, J. Gonzalez-Bosquet, S. Horibata, J. Heintz, R. Albrecht, et al. MUC16 (CA125): tumor biomarker to cancer therapy, a work in progress. Mol Cancer, 2014. 13: 129.
48. Haridas, D., S. Chakraborty, M.P. Ponnusamy, I. Lakshmanan, S. Rachagani, E. Cruz, et al. Pathobiological implications of MUC16 expression in pancreatic cancer. PLoS One, 2011. 6(10): e26839.
49. Lakshmanan, I., M.P. Ponnusamy, S. Das, S. Chakraborty, D. Haridas, P. Mukhopadhyay, et al. MUC16 induced rapid G2/M transition via interactions with JAK2 for increased proliferation and anti-apoptosis in breast cancer cells. Oncogene, 2012. 31(7): 805-17.
50. Chaturvedi, P., A.P. Singh, S. Chakraborty, S.C. Chauhan, S. Bafna, J.L. Meza, et al. MUC4 mucin interacts with and stabilizes the HER2 oncoprotein in human pancreatic cancer cells. Cancer Res, 2008. 68(7): 2065-70.
51. Lakshmanan, I., P. Seshacharyulu, D. Haridas, S. Rachagani, S. Gupta, S. Joshi, et al. Novel HER3/MUC4 oncogenic signaling aggravates the tumorigenic phenotypes of pancreatic cancer cells. Oncotarget, 2015. 6(25): 21085-99.
52. Desseyn, J.L., D. Tetaert, V. Gouyer. Architecture of the large membrane-bound mucins. Gene, 2008. 410(2): 215-22.
53. Chaturvedi, P., A.P. Singh, S.K. Batra. Structure, evolution, and biology of the MUC4 mucin. Faseb J, 2008. 22(4): 966-81.
54. Senapati, S., V.S. Gnanapragassam, N. Moniaux, N. Momi, S.K. Batra. Role of MUC4-NIDO domain in the MUC4-mediated metastasis of pancreatic cancer cells. Oncogene, 2012. 31(28): 3346-56.
55. Tang, J., Y. Zhu, K. Xie, X. Zhang, X. Zhi, W. Wang, et al. The role of the AMOP domain in MUC4/Y-promoted tumour angiogenesis and metastasis in pancreatic cancer. J Exp Clin Cancer Res, 2016. 35(1): 91.
56. Jahan, R., M.A. Macha, S. Rachagani, S. Das, L.M. Smith, S. Kaur, et al. Axed MUC4 (MUC4/X) aggravates pancreatic malignant phenotype by activating integrin-beta1/FAK/ERK pathway. Biochim Biophys Acta Mol Basis Dis, 2018. 1864(8): 2538-49.
57. Zhu, Y., J.J. Zhang, Y.P. Peng, X. Liu, K.L. Xie, J. Tang, et al. NIDO, AMOP and vWD domains of MUC4 play synergic role in MUC4 mediated signaling. Oncotarget, 2017. 8(6): 10385-99.
58. van Putten, J.P.M., K. Strijbis. Transmembrane mucins: signaling receptors at the intersection of inflammation and cancer. J Innate Immun, 2017. 9(3): 281-99.
59. Haridas, D., M.P. Ponnusamy, S. Chugh, I. Lakshmanan, P. Seshacharyulu, S.K. Batra. MUC16: molecular analysis and its functional implications in benign and malignant conditions. Faseb J, 2014. 28(10): 4183-99.

60. Nath, S., P. Mukherjee. MUC1: a multifaceted oncoprotein with a key role in cancer progression. Trends Mol Med, 2014. 20(6): 332-42.
61. Chauhan, S.C., A.P. Singh, F. Ruiz, S.L. Johansson, M. Jain, L.M. Smith, et al. Aberrant expression of MUC4 in ovarian carcinoma: diagnostic significance alone and in combination with MUC1 and MUC16 (CA125). Mod Pathol, 2006. 19(10): 1386-94.
62. Mercogliano, M.F., M. De Martino, L. Venturutti, M.A. Rivas, C.J. Proietti, G. Inurrigarro, et al. TNFalpha-induced mucin 4 expression elicits trastuzumab resistance in HER2-positive breast cancer. Clin Cancer Res, 2017. 23(3): 636-48.
63. Mukohara, T. Mechanisms of resistance to anti-human epidermal growth factor receptor 2 agents in breast cancer. Cancer Sci, 2011. 102(1): 1-8.
64. Mukhopadhyay, P., S. Chakraborty, M.P. Ponnusamy, I. Lakshmanan, M. Jain, S.K. Batra. Mucins in the pathogenesis of breast cancer: implications in diagnosis, prognosis and therapy. Biochim Biophys Acta, 2011. 1815(2): 224-40.
65. Rakha, E.A., R.W. Boyce, D. Abd El-Rehim, T. Kurien, A.R. Green, E.C. Paish, et al. Expression of mucins (MUC1, MUC2, MUC3, MUC4, MUC5AC and MUC6) and their prognostic significance in human breast cancer. Mod Pathol, 2005. 18(10): 1295-1304.
66. Workman, H.C., J.K. Miller, E.Q. Ingalla, R.P. Kaur, D.I. Yamamoto, L.A. Beckett, et al. The membrane mucin MUC4 is elevated in breast tumor lymph node metastases relative to matched primary tumors and confers aggressive properties to breast cancer cells. Breast Cancer Res, 2009. 11(5): R70.
67. Shet, T., S. Valsangar, S. Dhende. Secretory carcinoma of breast: pattern of MUC 2/MUC 4/MUC 6 expression. Breast J, 2013. 19(2): 222-4.
68. Macha, M.A., S. Rachagani, P. Pai, S. Gupta, W.M. Lydiatt, R.B. Smith, et al. MUC4 regulates cellular senescence in head and neck squamous cell carcinoma through p16/Rb pathway. Oncogene, 2015. 34(13): 1698-708.
69. Krishn, S.R., S. Kaur, L.M. Smith, S.L. Johansson, M. Jain, A. Patel, et al. Mucins and associated glycan signatures in colon adenoma-carcinoma sequence: prospective pathological implication(s) for early diagnosis of colon cancer. Cancer Lett, 2016. 374(2): 304-14.
70. Trehoux, S., F. Lahdaoui, Y. Delpu, F. Renaud, E. Leteurtre, J. Torrisani, et al. Micro-RNAs miR-29a and miR-330-5p function as tumor suppressors by targeting the MUC1 mucin in pancreatic cancer cells. Biochim Biophys Acta, 2015. 1853(10 Pt A): 2392-403.
71. Yang, K., M. He, Z. Cai, C. Ni, J. Deng, N. Ta, et al. A decrease in miR-150 regulates the malignancy of pancreatic cancer by targeting c-Myb and MUC4. Pancreas, 2015. 44(3): 370-9.
72. Srivastava, S.K., A. Bhardwaj, S. Singh, S. Arora, B. Wang, W.E. Grizzle, et al. MicroRNA-150 directly targets MUC4 and suppresses growth and malignant behavior of pancreatic cancer cells. Carcinogenesis, 2011. 32(12): 1832-9.
73. Radhakrishnan, P., A.M. Mohr, P.M. Grandgenett, M.M. Steele, S.K. Batra, M.A. Hollingsworth. MicroRNA-200c modulates the expression of MUC4 and MUC16 by directly targeting their coding sequences in human pancreatic cancer. PLoS One, 2013. 8(10): e73356.
74. Lahdaoui, F., Y. Delpu, A. Vincent, F. Renaud, M. Messager, B. Duchene, et al. miR-219-1-3p is a negative regulator of the mucin MUC4 expression and is a tumor suppressor in pancreatic cancer. Oncogene, 2015. 34(6): 780-8.
75. Rachagani, S., M.A. Macha, M.S. Menning, P. Dey, P. Pai, L.M. Smith, et al. Changes in microRNA (miRNA) expression during pancreatic cancer

development and progression in a genetically engineered KrasG12D; Pdx1-Cre mouse (KC) model. Oncotarget, 2015. 6(37): 40295-309.
76. Khan, S., M.C. Ebeling, M.S. Zaman, M. Sikander, M.M. Yallapu, N. Chauhan, et al. MicroRNA-145 targets MUC13 and suppresses growth and invasion of pancreatic cancer. Oncotarget, 2014. 5(17): 7599-609.
77. Sun, H., L. Zhao, K. Pan, Z. Zhang, M. Zhou, G. Cao. Integrated analysis of mRNA and miRNA expression profiles in pancreatic ductal adenocarcinoma. Oncol Rep, 2017. 37(5): 2779-86.
78. Jonckheere, N., N. Skrypek, J. Merlin, A.F. Dessein, P. Dumont, E. Leteurtre, et al. The mucin MUC4 and its membrane partner ErbB2 regulate biological properties of human CAPAN-2 pancreatic cancer cells via different signalling pathways. PLoS One, 2012. 7(2): e32232.
79. Kumar, S., S. Das, S. Rachagani, S. Kaur, S. Joshi, S.L. Johansson, et al. NCOA3-mediated upregulation of mucin expression via transcriptional and post-translational changes during the development of pancreatic cancer. Oncogene, 2015. 34(37): 4879-89.
80. Perrais, M., P. Pigny, M.P. Ducourouble, D. Petitprez, N. Porchet, J.P. Aubert, et al. Characterization of human mucin gene MUC4 promoter: importance of growth factors and proinflammatory cytokines for its regulation in pancreatic cancer cells. J Biol Chem, 2001. 276(33): 30923-33.
81. Jonckheere, N., A. Vincent, M. Perrais, M.P. Ducourouble, A.K. Male, J.P. Aubert, et al. The human mucin MUC4 is transcriptionally regulated by caudal-related homeobox, hepatocyte nuclear factors, forkhead box A, and GATA endodermal transcription factors in epithelial cancer cells. J Biol Chem, 2007. 282(31): 22638-50.
82. Pai, P., S. Rachagani, I. Lakshmanan, M.A. Macha, Y. Sheinin, L.M. Smith, et al. The canonical Wnt pathway regulates the metastasis-promoting mucin MUC4 in pancreatic ductal adenocarcinoma. Mol Oncol, 2016. 10(2): 224-39.
83. Joshi, S., E. Cruz, S. Rachagani, S. Guha, R.E. Brand, M.P. Ponnusamy, et al. Bile acids-mediated overexpression of MUC4 via FAK-dependent c-Jun activation in pancreatic cancer. Mol Oncol, 2016. 10(7): 1063-77.
84. Kaur, S., N. Sharma, S.R. Krishn, I. Lakshmanan, S. Rachagani, M.J. Baine, et al. MUC4-mediated regulation of acute phase protein lipocalin 2 through HER2/AKT/NF-kappaB signaling in pancreatic cancer. Clin Cancer Res, 2014. 20(3): 688-700.
85. Zaretsky, J.Z., R. Sarid, Y. Aylon, L.A. Mittelman, D.H. Wreschner, I. Keydar. Analysis of the promoter of the MUC1 gene overexpressed in breast cancer. FEBS Lett, 1999. 461(3): 189-95.
86. Rao, T.D., H. Tian, X. Ma, X. Yan, S. Thapi, N. Schultz, et al. Expression of the carboxy-terminal portion of MUC16/CA125 induces transformation and tumor invasion. PLoS One, 2015. 10(5): e0126633.
87. Das, S., S.K. Batra. Understanding the unique attributes of MUC16 (CA125): potential implications in targeted therapy. Cancer Res, 2015. 75(22): 4669-74.
88. Scara, S., P. Bottoni, R. Scatena. CA 19-9: Biochemical and clinical aspects. Adv Exp Med Biol, 2015. 867: 247-60.
89. Loosen, S.H., U.P. Neumann, C. Trautwein, C. Roderburg, T. Luedde. Current and future biomarkers for pancreatic adenocarcinoma. Tumour Biol, 2017. 39(6): 1010428317692231.
90. Lei, X.F., S.Z. Jia, J. Ye, Y.L. Qiao, G.M. Zhao, X.H. Li, et al. Application values of detection of serum CA199, CA242 and CA50 in the diagnosis of pancreatic cancer. J Biol Regul Homeost Agents, 2017. 31(2): 383-8.

91. Jonckheere, N., N. Skrypek, I. Van Seuningen. Mucins and pancreatic cancer. Cancers, 2010. 2(4): 1794-812.
92. Rachagani, S., M.P. Torres, S. Kumar, D. Haridas, M. Baine, M.A. Macha, et al. Mucin (Muc) expression during pancreatic cancer progression in spontaneous mouse model: potential implications for diagnosis and therapy. J Hematol Oncol, 2012. 5: 68.
93. Kaur, S., L.M. Smith, A. Patel, M. Menning, D.C. Watley, S.S. Malik, et al. A combination of MUC5AC and CA19-9 improves the diagnosis of pancreatic cancer: a multicenter study. Am J Gastroenterol, 2017. 112(1): 172-83.
94. Han, S., G. Jin, L. Wang, M. Li, C. He, X. Guo, et al. The role of PAM4 in the management of pancreatic cancer: diagnosis, radioimmunodetection, and radioimmunotherapy. J Immunol Res, 2014. 2014: 268479.
95. Gold, D.V., Z. Karanjawala, D.E. Modrak, D.M. Goldenberg, R.H. Hruban. PAM4-reactive MUC1 is a biomarker for early pancreatic adenocarcinoma. Clin Cancer Res, 2007. 13(24): 7380-7.
96. Liu, D., C.H. Chang, D.V. Gold, D.M. Goldenberg. Identification of PAM4 (clivatuzumab)-reactive epitope on MUC5AC: a promising biomarker and therapeutic target for pancreatic cancer. Oncotarget, 2015. 6(6): 4274-85.
97. Yamasaki, H., S. Ikeda, M. Okajima, Y. Miura, T. Asahara, N. Kohno, et al. Expression and localization of MUC1, MUC2, MUC5AC and small intestinal mucin antigen in pancreatic tumors. Int J Oncol, 2004. 24(1): 107-13.
98. Kufe, D.W. Mucins in cancer: function, prognosis and therapy. Nature Reviews Cancer, 2009. 9: 874.
99. Wang, Y., J. Gao, Z. Li, Z. Jin, Y. Gong, X. Man. Diagnostic value of mucins (MUC1, MUC2 and MUC5AC) expression profile in endoscopic ultrasound-guided fine-needle aspiration specimens of the pancreas. Int J Cancer, 2007. 121(12): 2716-22.
100. Rachagani, S., M.P. Torres, S. Kumar, D. Haridas, M. Baine, M.A. Macha, et al. Mucin (Muc) expression during pancreatic cancer progression in spontaneous mouse model: potential implications for diagnosis and therapy. Journal of Hematology & Oncology, 2012. 5(1): 68.
101. Nagata, K., M. Horinouchi, M. Saitou, M. Higashi, M. Nomoto, M. Goto, et al. Mucin expression profile in pancreatic cancer and the precursor lesions. 2007. 14(3): 243-54.
102. Horn, A., S. Chakraborty, P. Dey, D. Haridas, J. Souchek, S.K. Batra, et al. Immunocytochemistry for MUC4 and MUC16 is a useful adjunct in the diagnosis of pancreatic adenocarcinoma on fine-needle aspiration cytology. Arch Pathol Lab Med, 2013. 137(4): 546-51.
103. Higashi, M., S. Yokoyama, T. Yamamoto, Y. Goto, I. Kitazono, T. Hiraki, et al. Mucin expression in endoscopic ultrasound-guided fine-needle aspiration specimens is a useful prognostic factor in pancreatic ductal adenocarcinoma. Pancreas, 2015. 44(5): 728-34.
104. Jhala, N., D. Jhala, S.M. Vickers, I. Eltoum, S.K. Batra, U. Manne, et al. Biomarkers in diagnosis of pancreatic carcinoma in fine-needle aspirates. Am J Clin Pathol, 2006. 126(4): 572-9.
105. Wang, G., R.J. Lipert, M. Jain, S. Kaur, S. Chakraboty, M.P. Torres, et al. Detection of the potential pancreatic cancer marker MUC4 in serum using surface-enhanced Raman scattering. Anal Chem, 2011. 83(7): 2554-61.
106. Curry, J.M., K.J. Thompson, S.G. Rao, D.M. Besmer, A.M. Murphy, V.Z. Grdzelishvili, et al. The use of a novel MUC1 antibody to identify cancer stem

cells and circulating MUC1 in mice and patients with pancreatic cancer. J Surg Oncol, 2013. 107(7): 713-22.
107. Zhou, R., J.M. Curry, L.D. Roy, P. Grover, J. Haider, L.J. Moore, et al. A novel association of neuropilin-1 and MUC1 in pancreatic ductal adenocarcinoma: role in induction of VEGF signaling and angiogenesis. Oncogene, 2016. 35(43): 5608-18.
108. Zhao, P., M. Meng, B. Xu, A. Dong, G. Ni, L. Lu. Decreased expression of MUC1 induces apoptosis and inhibits migration in pancreatic cancer PANC-1 cells via regulation of Slug pathway. Cancer Biomark, 2017. 20(4): 469-76.
109. Shukla, S.K., V. Purohit, K. Mehla, V. Gunda, N.V. Chaika, E. Vernucci, et al. MUC1 and HIF-1alpha signaling crosstalk induces anabolic glucose metabolism to impart gemcitabine resistance to pancreatic cancer. Cancer Cell, 2017. 32(1): 71-87.e7.
110. Murthy, D., K.S. Attri, P.K. Singh. Phosphoinositide 3-kinase signaling pathway in pancreatic ductal adenocarcinoma progression, pathogenesis, and therapeutics. Front Physiol, 2018. 9: 335.
111. Gunda, V., J. Souchek, J. Abrego, S.K. Shukla, G.D. Goode, E. Vernucci, et al. MUC1-mediated metabolic alterations regulate response to radiotherapy in pancreatic cancer. Clin Cancer Res, 2017. 23(19): 5881-91.
112. Zhi, X., J. Tao, K. Xie, Y. Zhu, Z. Li, J. Tang, et al. MUC4-induced nuclear translocation of beta-catenin: a novel mechanism for growth, metastasis and angiogenesis in pancreatic cancer. Cancer Lett, 2014. 346(1): 104-13.
113. Vasseur, R., N. Skrypek, B. Duchene, F. Renaud, D. Martinez-Maqueda, A. Vincent, et al. The mucin MUC4 is a transcriptional and post-transcriptional target of K-ras oncogene in pancreatic cancer. Implication of MAPK/AP-1, NF-kappaB and RalB signaling pathways. Biochim Biophys Acta, 2015. 1849(12): 1375-84.
114. Choudhury, A., N. Moniaux, J.P. Winpenny, M.A. Hollingsworth, J.P. Aubert, S.K. Batra. Human MUC4 mucin cDNA and its variants in pancreatic carcinoma. J Biochem, 2000. 128(2): 233-43.
115. Choudhury, A., N. Moniaux, J. Ringel, J. King, E. Moore, J.P. Aubert, et al. Alternate splicing at the 3'-end of the human pancreatic tumor-associated mucin MUC4 cDNA. Teratog Carcinog Mutagen, 2001. 21(1): 83-96.
116. Wu, Y.M., D.D. Nowack, G.S. Omenn, B.B. Haab. Mucin glycosylation is altered by pro-inflammatory signaling in pancreatic-cancer cells. J Proteome Res, 2009. 8(4): 1876-86.
117. Remmers, N., J.M. Anderson, E.M. Linde, D.J. DiMaio, A.J. Lazenby, H.H. Wandall, et al. Aberrant expression of mucin core proteins and o-linked glycans associated with progression of pancreatic cancer. Clin Cancer Res, 2013. 19(8): 1981-93.
118. Li, Y., C. Wu, T. Chen, J. Zhang, G. Liu, Y. Pu, et al. Effects of RNAi-mediated MUC4 gene silencing on the proliferation and migration of human pancreatic carcinoma BxPC-3 cells. Oncol Rep, 2016. 36(6): 3449-55.
119. Xu, H.L., X. Zhao, K.M. Zhang, W. Tang, N. Kokudo. Inhibition of KL-6/MUC1 glycosylation limits aggressive progression of pancreatic cancer. World J Gastroenterol, 2014. 20(34): 12171-81.
120. Kalra, A.V., R.B. Campbell. Mucin impedes cytotoxic effect of 5-FU against growth of human pancreatic cancer cells: overcoming cellular barriers for therapeutic gain. Br J Cancer, 2007. 97(7): 910-8.
121. Jeong, S.J., J.H. Kim, B.J. Lim, I. Yoon, J.A. Song, H.S. Moon, et al. Inhibition

of MUC1 biosynthesis via threonyl-tRNA synthetase suppresses pancreatic cancer cell migration. Exp Mol Med, 2018. 50(1): e424.
122. Hisatsune, A., H. Nakayama, M. Kawasaki, I. Horie, T. Miyata, Y. Isohama, et al. Anti-MUC1 antibody inhibits EGF receptor signaling in cancer cells. Biochem Biophys Res Commun, 2011. 405(3): 377-81.
123. Moniaux, N., G.C. Varshney, S.C. Chauhan, M.C. Copin, M. Jain, U.A. Wittel, et al. Generation and characterization of anti-MUC4 monoclonal antibodies reactive with normal and cancer cells in humans. J Histochem Cytochem, 2004. 52(2): 253-61.
124. Wu, J., J. Wei, K. Meng, J. Chen, W. Gao, J. Zhang, et al. Identification of an HLA-A*0201-restrictive CTL epitope from MUC4 for applicable vaccine therapy. Immunopharmacol Immunotoxicol, 2009. 31(3): 468-76.
125. Amedei, A., E. Niccolai, D. Prisco. Pancreatic cancer: role of the immune system in cancer progression and vaccine-based immunotherapy. Hum Vaccin Immunother, 2014. 10(11): 3354-68.
126. Hossain, M.K., K.A. Wall. Immunological evaluation of recent MUC1 glycopeptide cancer vaccines. Vaccines (Basel), 2016. 4(3).
127. Guo, M., C. You, J. Dou. Role of transmembrane glycoprotein mucin 1 (MUC1) in various types of colorectal cancer and therapies: current research status and updates. Biomed Pharmacother, 2018. 107: 1318-25.
128. Bhatia, R., S.K. Gautam, A. Cannon, C. Thompson, B.R. Hall, A. Aithal, et al. Cancer-associated mucins: role in immune modulation and metastasis. Cancer Metastasis Rev, 2019. 38: 223-36.
129. Teramoto, K., Y. Ozaki, J. Hanaoka, S. Sawai, N. Tezuka, S. Fujino, et al. Predictive biomarkers and effectiveness of MUC1-targeted dendritic-cell-based vaccine in patients with refractory non-small cell lung cancer. Ther Adv Med Oncol, 2017. 9(3): 147-57.
130. Stergiou, N., N. Gaidzik, A.S. Heimes, S. Dietzen, P. Besenius, J. Jakel, et al. Reduced breast tumor growth after immunization with a tumor-restricted MUC1 glycopeptide conjugated to tetanus toxoid. Cancer Immunol Res, 2019. 7(1): 113-22.
131. Saltos, A., F. Khalil, M. Smith, J. Li, M. Schell, S.J. Antonia, et al. Clinical associations of mucin 1 in human lung cancer and precancerous lesions. Oncotarget, 2018. 9(86): 35666-75.
132. Stergiou, N., M. Glaffig, H. Jonuleit, E. Schmitt, H. Kunz. Immunization with a synthetic human MUC1 glycopeptide vaccine against tumor-associated MUC1 breaks tolerance in human MUC1 transgenic mice. ChemMedChem, 2017. 12(17): 1424-8.
133. Lakshminarayanan, V., N.T. Supekar, J. Wei, D.B. McCurry, A.C. Dueck, H.E. Kosiorek, et al. MUC1 vaccines, comprised of glycosylated or non-glycosylated peptides or tumor-derived MUC1, can circumvent immunoediting to control tumor growth in MUC1 transgenic mice. PLoS One, 2016. 11(1): e0145920.
134. Mehla, K., J. Tremayne, J.A. Grunkemeyer, K.A. O'Connell, M.M. Steele, T.C. Caffrey, et al. Combination of mAb-AR20.5, anti-PD-L1 and PolyICLC inhibits tumor progression and prolongs survival of MUC1.Tg mice challenged with pancreatic tumors. Cancer Immunol Immunother, 2018. 67(3): 445-57.
135. Liu, L., Y. Wang, L. Miao, Q. Liu, S. Musetti, J. Li, et al. Combination immunotherapy of MUC1 mRNA nano-vaccine and CTLA-4 blockade effectively inhibits growth of triple negative breast cancer. Mol Ther, 2018. 26(1): 45-55.

136. Liu, C., Z. Lu, Y. Xie, Q. Guo, F. Geng, B. Sun, et al. Soluble PD-1-based vaccine targeting MUC1 VNTR and survivin improves anti-tumor effect. Immunol Lett, 2018. 200: 33-42.
137. Hillman, G.G., L.A. Reich, S.E. Rothstein, L.M. Abernathy, M.D. Fountain, K. Hankerd, et al. Radiotherapy and MVA-MUC1-IL-2 vaccine act synergistically for inducing specific immunity to MUC-1 tumor antigen. J Immunother Cancer, 2017. 5: 4.
138. Scheid, E., P. Major, A. Bergeron, O.J. Finn, R.D. Salter, R. Eady, et al. Tn-MUC1 DC vaccination of Rhesus Macaques and a Phase I/II trial in patients with nonmetastatic castrate-resistant prostate cancer. Cancer Immunol Res, 2016. 4(10): 881-92.
139. Rong, Y., X. Qin, D. Jin, W. Lou, L. Wu, D. Wang, et al. A phase I pilot trial of MUC1-peptide-pulsed dendritic cells in the treatment of advanced pancreatic cancer. Clin Exp Med, 2012. 12(3): 173-80.
140. Ge, C., R. Li, H. Song, T. Geng, J. Yang, Q. Tan, et al. Phase I clinical trial of a novel autologous modified-DC vaccine in patients with resected NSCLC. BMC Cancer, 2017. 17(1): 884.
141. Antonilli, M., H. Rahimi, V. Visconti, C. Napoletano, I. Ruscito, I.G. Zizzari, et al. Triple peptide vaccination as consolidation treatment in women affected by ovarian and breast cancer: clinical and immunological data of a phase I/II clinical trial. Int J Oncol, 2016. 48(4): 1369-78.
142. Vergati, M., C. Intrivici, N.-Y. Huen, J. Schlom, K.Y. Tsang. Strategies for cancer vaccine development %. J Biomed Biotechnol, 2010. 2010.
143. Chen, J., X.Z. Guo, H.Y. Li, X. Liu, L.N. Ren, D. Wang, et al. Generation of CTL responses against pancreatic cancer in vitro using dendritic cells co-transfected with MUC4 and survivin RNA. Vaccine, 2013. 31(41): 4585-90.
144. Wei, J., W. Gao, J. Wu, K. Meng, J. Zhang, J. Chen, et al. Dendritic cells expressing a combined PADRE/MUC4-derived polyepitope DNA vaccine induce multiple cytotoxic T-cell responses. Cancer Biother Radiopharm, 2008. 23(1): 121-8.
145. Cai, H., B. Palitzsch, S. Hartmann, N. Stergiou, H. Kunz, E. Schmitt, et al. Antibody induction directed against the tumor-associated MUC4 glycoprotein. Chembiochem, 2015. 16(6): 959-67.
146. Torres, M.P., S. Chakraborty, J. Souchek, S.K. Batra. Mucin-based targeted pancreatic cancer therapy. Curr Pharm Des, 2012. 18(17): 2472-81.
147. Macha, M.A., S. Rachagani, S. Gupta, P. Pai, M.P. Ponnusamy, S.K. Batra, et al. Guggulsterone decreases proliferation and metastatic behavior of pancreatic cancer cells by modulating JAK/STAT and Src/FAK signaling. Cancer Lett, 2013. 341(2): 166-77.
148. Torres, M.P., M.P. Ponnusamy, S. Chakraborty, L.M. Smith, S. Das, H.A. Arafat, et al. Effects of thymoquinone in the expression of mucin 4 in pancreatic cancer cells: implications for the development of novel cancer therapies. Mol Cancer Ther, 2010. 9(5): 1419-31.
149. Torres, M.P., S. Rachagani, V. Purohit, P. Pandey, S. Joshi, E.D. Moore, et al. Graviola: a novel promising natural-derived drug that inhibits tumorigenicity and metastasis of pancreatic cancer cells in vitro and in vivo through altering cell metabolism. Cancer Lett, 2012. 323(1): 29-40.
150. Goh, S.K., G. Gold, C. Christophi, V. Muralidharan. Serum carbohydrate antigen 19-9 in pancreatic adenocarcinoma: a mini review for surgeons. ANZ J Surg, 2017. 87(12): 987-92.

151. Madsen, C.B., P. Jess, M. Harndahl, H.H. Wandall, A.E. Pedersen. MUC4-specific CTLs. Immunopharmacol Immunotoxicol, 2013. 35(1): 202-3.
152. Wittel, U.A., A. Goel, G.C. Varshney, S.K. Batra. Mucin antibodies – new tools in diagnosis and therapy of cancer. Front Biosci, 2001. 6: D1296-310.
153. Jain, M., G. Venkatraman, N. Moniaux, S. Kaur, S. Kumar, S. Chakraborty, et al. Monoclonal antibodies recognizing the non-tandem repeat regions of the human mucin MUC4 in pancreatic cancer. PLoS One, 2011. 6(8): e23344.
154. Aithal, A., W.M. Junker, P. Kshirsagar, S. Das, S. Kaur, C. Orzechowski, et al. Development and characterization of carboxy-terminus specific monoclonal antibodies for understanding MUC16 cleavage in human ovarian cancer. PLoS One, 2018. 13(4): e0193907.

CHAPTER 13

Targeted Therapy for Cancer Stem Cells

Rama Krishna Nimmakayala[1], Saswati Karmakar[1], Garima Kaushik[1], Sanchita Rauth[1], Srikanth Barkeer[1], Saravanakumar Marimuthu[1] and Moorthy P. Ponnusamy[1,2]*

[1] Department of Biochemistry and Molecular Biology, University of Nebraska Medical Center, Omaha, NE 68198-5870, USA
[2] Eppley Institute for Research in Cancer and Allied Diseases, Fred & Pamela Buffett Cancer Center, University of Nebraska Medical Center, Omaha, NE 68198-5870, USA

Introduction

Cancer develops when the regular cellular growth control is disrupted; cells lose contact inhibition and develop resistance to cell death, leading to uncontrolled cellular proliferation and metastasis. The structure and composition of the tumor are incredibly complex with heterogeneous cell populations, which display a large number of mutations and consequent dysregulated gene expressions [1]. Cancer stem cells (CSCs) [2], by residing at the top of the cell hierarchy, can self-renew and undergo multi-lineage differentiation to produce heterogeneous cell population within the tumor [1].

Historical Milestones in the Emergence of Cancer Stem Cell Theory

In the mid-1800, Joseph Claude and Robert Remak observed that cancer cells resemble embryonic cells in their phenotype, and this marks the foundation for the emergence of the current notion of CSCs. Later in the 1900s, a de-differentiation theory of CSCs has emerged, in which it was proposed that changes in differentiated somatic cells transform these cells into tumor cells [3]. Also, it was proposed that exposure to chemicals, toxic substances, and infectious organisms induce carcinogenesis through de-differentiation [3, 4]. In 1994, Lapidot and co-workers showed for the first time the role of stem cells in tumor formation. They isolated and characterized the tumor-

---

*Corresponding author: mpalanim@unmc.edu

initiating properties of CD34+CD38- cells in acute myeloid leukemia [5]. Recent studies also focused on another new theory of cancer origin. The core concept of this theory is the transformation of normal stem cell or progenitor cell into more aggressive tumor-initiating cells (TICs). However, this theory has not yet gained many pieces of evidence and require further research. The existence of CSCs in the many solid tumors of the organs such as the brain, head, neck, lung, liver, mammary glands, pancreas, ovary and prostate was also established in subsequent studies.

Current studies based on functional characterization shows that the CSCs are a small subset of the cell population within the tumor that is chemo- and radio-resistant; therefore, these cells can escape from conventional therapies and are involved in tumor recurrence [1]. There are several mechanisms which protect these CSCs from chemotherapy and radiotherapy mediated cellular death. These include drug efflux through ABC family of drug transporters such as ABCB1, ABCC1 and ABCG2, and cellular detoxification systems such as increased aldehyde dehydrogenase (ALDH) activity and elevated DNA repair mechanism. Majority of the studies show that conventional cancer therapies such as chemotherapy, radiotherapy, and even targeted anti-cancer drugs are not able to ameliorate the cancerous effects [2].

Sources of Therapy Resistance in CSCs

Tumor microenvironment (TME) mediated therapy resistance in CSCs

CSC interactions with surrounding cancer cells, extracellular matrix (ECM), immune cells, cytokines, hormones, and various growth factors show an impact on their sensitivity to therapy. Other micro-environmental conditions such as hypoxia and acidity can also affect the sensitivity of these CSCs to apoptosis.

Therapy resistance caused by hypoxia and acidity

Hypoxia, or low oxygen condition, is one of the critical factors which contributes to the therapy resistance in CSCs by increasing cell survival and proliferation. Hypoxia increases the expression of hypoxia-inducible factor (HIF) mediated signaling pathway, which activates cellular responses such as increased survival of CSCs. HIF1α has been shown to induce drug resistance by activating multidrug resistance-associated protein 1 (MRP1) [6]. Besides, the pH of the microenvironment surrounding CSCs can show a dramatic effect on drug uptake. Since the extracellular pH of cancer cells is acidic and intracellular pH is neutral to alkaline, drugs such as mitoxantrone and doxorubicin, which are weakly basic can be protonated in the extracellular environment leading to their decreased drug uptake [7, 8].

Immunoresistance in CSCs

CSCs or TICs divide and produce differentiated cancer cells during the progression of the tumor. The immune system detects and interacts with these cells and tries to eliminate them; however, CSCs or TICs escape this elimination and survive. This immune escape may be due to many mechanisms, and one such mechanism is the release of immunosuppressive molecules by immune-resistant CSCs, which inhibit the immune system and recruit cells that are involved in immunosuppression. Besides, major stemness signaling such as Hedgehog (HH), Wnt/β-catenin, EGFR and Notch signaling pathways also mediate the immune escape of CSCs [9, 10].

Therapy Resistance Conferred by ABC Transporters

CSCs within the tumor mediate drug resistance and metastasis of cancer cells, resulting in a more aggressive disease. Several studies propose that various drug-resistant mechanisms confer resistance to CSCs; among these, ABC transporters mediated drug resistance is notable.

ABC transporters comprise 49 members, which are classified into seven gene subfamilies (ABCA-G). Most of these act as active transporters (ATP dependent) and mediate the transport of substances through the plasma membrane and intracellular membrane. Several transporters are involved in the efflux of certain chemotherapeutic drugs leading to multidrug resistance (MDR) of CSC population. The most well studied MDR family members include ABCB1, ABCC1, and ABCG2. Those cells, which efflux Hoechst 33342 dye through ABCB1 and ABCG2 transporters, were identified as drug-resistant side population (SP) or CSCs [11, 12]. Primarily, four pathways regulate drug resistance genes in CSCs: HH, Wnt/β-catenin, EGFR, and Notch signaling pathways [13, 14].

The canonical Wnt/β-catenin signaling pathway is crucial for the regulation of normal development and stem cell maintenance, and studies show that mutations in these pathway molecules could make this pathway constitutively active leading to the drug resistance and uncontrolled proliferation of CSCs in various cancers [15]. Another drug resistance signaling mechanism, Hh signaling is required for normal embryonic development and tissue homeostasis. However, the pathway also regulates drug resistance genes in various cancers. For example, the Hh signaling transcription factor, Gli2 increases the expression of ABCC1 drug-resistant transporter leading to increased therapy resistance in hepatoma cells [16]. Moreover, aberrant activation of Notch signaling, which is required for normal development and tissue homeostasis, increases the expression levels of major CSC genes such as Myc, Nanog, OCT-4 and Sox2 [17]. EGFR mediated signaling is also known to regulate drug resistance in CSCs. Studies report that EGFR signaling is constitutively active in CSCs, and the inhibition of the signaling sensitizes glioma stem cells to radiation mediated cytotoxicity [18]. Also, OCT-4, a pluripotent transcription factor, regulates the genes of ABC transporters, especially ABCG2 drug resistant protein [19]. Myc driven signaling also induces ABC transporter genes such as ABCC1 [20].

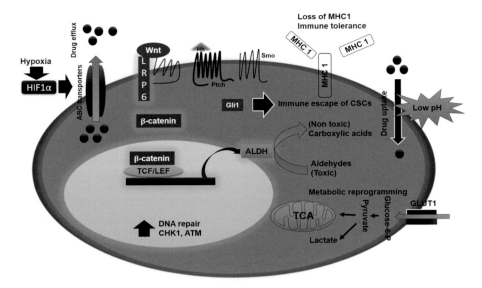

Fig. 1: Drug resistance in cancer stem cells (CSCs). Hypoxia induces HIF1α, leading to increased levels of ABC transporters (drug-resistant transporters). Wnt/β-catenin signaling increases the gene expression of ALDH1, which reduces aldehyde toxicity resulting in increased drug resistance in CSCs. Other signaling mechanisms such as hedgehog (HH) signaling induces cancer stemness and helps CSCs in the immune escape. Extracellular lower pH levels also reduce drug uptake capacity leading to increased survival and maintenance in CSCs. CSCs display increased DNA repair capability and immune-tolerance. Metabolic reprogramming (glycolysis to oxidative or vice versa) induces CSCs.

Therapy Resistance Caused by Cellular Detoxification Systems

There are 19 ALDH proteins in humans, and many of these have a function related to cancers. Major functions of these proteins are to catalyze aldehyde oxidation, cellular detoxification, and protection from reactive oxygen species [9, 21]. It was shown that the β-catenin/TCF transcriptional complex directly binds to its target region on the promoter of ALDH1A1 gene suggesting that Wnt/β-catenin signaling regulates ALDH1A1 gene [22]. Also, MUC1-C (the cytosolic side of MUC1) activates ERK-C/EBPβ-ALDH1A1 signaling axis leading to the induction of ALDH activity in breast cancer [23]. CSCs also hold an efficient inactivation system to eradicate ROS. Studies show that ALDH protects drug-tolerant cell population from increased levels of ROS, conferring an additional resistance to CSCs against therapeutic drugs [24].

CSCs also possess another important defensive mechanism that is enhanced DNA damage repair system. Studies show a strong link between enhanced DNA repair mechanism and therapy resistance in CSCs. For example, CD133+ glioblastoma stem cells showed increased activation of DNA damage response elements such as Chk1 and Ataxia telangiectasia mutated (ATM) than CD133- cells. The study also confirmed that the CD133+ cells show increased radio-resistance than CD133- population [25].

Metabolism Mediated Therapy Resistance in CSCs

Cancer cells undergo an extensive metabolic adaptation, which is one of their significant features. Glycolysis, which converts glucose into pyruvate through a series of steps, produces two ATP molecules per one glucose molecule. In the presence of oxygen, cells use oxidative phosphorylation, which produces 36 molecules of ATP per molecule of glucose. Majority of the studies suggest that cancer cells undergo Warburg effect, where cancer cells depend on glycolysis even in the presence of oxygen and convert pyruvate to lactate. If sufficient levels of glucose are available, glycolysis can produce ATPs more rapidly as compared to oxidative phosphorylation.

Metabolism of CSCs is one of the less studied areas so far, and the exact metabolic nature of CSCs has not been completely delineated. However, few studies report that CSCs follow fermentative glycolysis. For example, Ciavardelli et al. have observed a shift from oxidative phosphorylation into fermentative glycolysis in breast CSCs suggesting that inhibition of glycolysis in these breast CSCs could be a novel target to eradicate breast CSCs [26]. In contrast, few other studies proposed that CSCs are mainly oxyolytic. Ovarian CSCs isolated from patients which showed increased expression of oxidative phosphorylation associated genes suggesting that CSCs are oxyolytic [27]. Vlashi et al. also showed that breast CSCs rely on oxidative phosphorylation, whereas their differentiated progeny cells display glycolytic phenotype [28].

Studies also provided evidence for the existence of distinct CSC population with differential metabolic profiles. Breast tumors consist of two CSC subpopulations, EMT CSCs (ALDH ^{High}) and MET CSCs (CD44$^+$CD24$^-$); the EMT CSCs are more glycolytic, whereas MET CSCs are oxyolytic. Overall, distinct CSC subpopulation shows different metabolic profiles and targeting the CSC subtype specific metabolic pathway could be an efficient therapeutic anti-cancer strategy.

Taken together, drug resistance in CSCs is caused by multiple factors. Therapeutic strategies targeting CSCs need to be developed in order to destroy the tumor bulk and reduce recurrence.

Therapeutic Strategies Targeting CSCs

Targeting CSCs TME

CSCs microenvironment is a specialized cellular niche that provides an environment conducive for maintaining the stemness of CSCs by secreting various paracrine factors or by direct cell-cell contact that influences the self-renewal and differentiation pathways. The TME is composed of stromal cells (cancer-associated fibroblasts (CAFs), adipocytes, and perivascular cells), immune cells, secreted cytokines and growth factors, hypoxic regions, pH and the ECM. These environmental factors collectively play an important role in the maintenance of CSC self-renewal and impact their sensitivity to therapy by altering Wnt/Catenin, Notch, and HH (self-renewal pathways) signaling, or by regulating transcriptional master regulators such as OCT-4, NANOG, and SOX-2 that sustain embryonic stem cell (ESC) self-renewal [29, 30]. The crosstalk between CSCs and the niche cells is dynamic, with an exchange of cytokines and growth factors. For instance, CSCs can secrete vascular endothelial growth factor (VEGF) and stromal-derived factor1 (SDF-1) to recruit perivascular cells, or generate granulocyte colony-stimulating factor (G-CSF) to recruit myeloid-derived cells (MDSCs) into the CSC microenvironment [25, 31, 32].

Immune Therapy to Target CSCs

Immunotherapy, when activated and employed for the whole tumor, results in inducing immune response towards antigens on the tumor, which is mostly comprised of differentiated cells, consequently leading to an immune response against differentiated cancer cells. Hence, to target CSCs and enhance the efficacy of immunotherapy in aggressive cancers, we need to activate immune response against CSCs specifically.

Innate Immune Response

Potential of natural killer (NK cells) for targeting CSCs has been a subject of debate. Some studies have found that stem cells lack expression of Major histocompatibility (MHC) class 1 and MHC class 2 molecules making them resistant to NK cell-mediated killing. Likewise, CD133 + brain tumor cells were found lacking in the expression of MHC class 1 and MHC class 2 molecules and low expression of MHC class 1 related chain A and B and other NK cell activating ligands, making these cells resistant to immunotherapy [33, 34]. Conversely, glioma stem cells have also been reported to express several NK cell activating ligands and hence showed that CD24+/CD44+, CD133+ and ALDH+CSCs are preferentially killed by NK cells [35, 36].

Another kind of unconventional T cells, Gamma delta T cells, are innate effector immune cells. These cells have been shown to express a higher amount of CCR5 and CXCR3 receptors and enhance CSC specific immune response. These cells have also been reported to proliferate and secrete TNF-α and IFN-γ, and produce apoptotic and cytotoxic granzymes, and target CSCs when administered with zoledronate in colon cancer [37, 38].

CSCs Primed T Cells

CSCs specific CD8T cells can be generated to target specific CSC specific markers, to generate cytotoxic T cell response and generate a memory T cells response. Studies have reported the development of CD8 T cells against an antigenic peptide from ALDH1A1, resulting in T cells that specifically target CSCs [38].

Targeted Therapy for Cancer Stem Cells

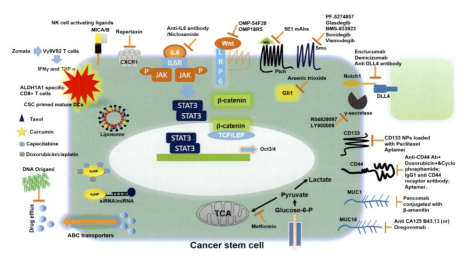

Fig. 2: Therapeutic strategies and drugs that target CSCs. Zometa-induced Vγ9Vδ2 T cells release IFNγ and TNF α that kill CSCs. Also, ALDH1A1 specific CD8(+) T cells and CSC primed mature dendritic cells (DCs) (DC vaccine) show cytotoxic effects on CSCs. CSCs express increased levels of NK cell activating ligands, MICA/B leading to NK cell-mediated CSC death. An IL8 receptor, CXCR1 is increased in its expression in CSCs, and Rapertaxin has been shown to inhibit CXCR1 receptor leading to decreased cancer stemness. An anti-IL6 antibody or Niclosamide has been shown to inhibit IL6/JAK/STAT3/Oct3/4 signaling axis leading to diminished stem cell features. Anti-cancer stemness agents, OMP-54F28 and OMP18R5, inhibit Wnt signaling resulting in the inhibition of stemness. 5E1 monoclonal antibody interferes with Hedgehog (Hh) ligand binding with patched receptor leading to decreased stemness. Moreover, inhibitors such as PF-5274857, Glasdegib, BMS-833923, Sonidegib, and Vismodegib specifically target Smoothened (Smo), leading to decreased signaling and stemness signature. Notch signaling, one of the essential stemness pathways, can be disrupted using anti DLL4 antibody, demicizumab, and Enctucumab. Cell surface oncogenic proteins, CD133 and CD44, can also be inhibited using various nanomedical based strategies (CD133 nanoparticles loaded with paclitaxel, and aptamers) and antibodies (anti CD44 antibodies along with doxorubicin and cyclophosphamide). Pancomab conjugated with β-amanitin, and Oregovomab can target mucins (MUC1 and MUC16), which are involved in the enrichment of CSCs. Liposomes loaded with taxol or curcumin can also target CSCs. Also, gold nanoparticles carrying drugs (Capecitabine, doxorubicin or cisplatin) or siRNAs or micro RNAs efficiently target CSCs. DNA origami inhibits drug resistance mechanisms. Increased oxidative phosphorylation in CSCs cab is inhibited by metformin leading to the reduction in the stemness in CSCs.

CSCs Based Vaccines

Dendritic cells (DCs), which are antigen-presenting cells, can be used to initiate anti-tumor T cell response, and hence can be utilized for the production of cancer vaccines. Human immature DCs can be co-cultured with irradiated cancer cells or CSCs to generate CSCs primed mature DCs that express high levels of CD80, CD86, and CD40 and stimulate high Th1 response. Present evidence suggests that CSCs are antigenic and can be used to generate vaccines against CSCs [39].

Inhibition of Cytokines and Myeloid-derived Suppressor Cells (MDSCs) for CSCs Elimination

MDSCs have been known to be pro-angiogenic, pro-metastatic and pro-tumorous in several cancers. It has also come to light that specific cytokines like IL6, IL1, and IL8 have roles in maintaining CSC niche and hence maintain self-renewal and maintenance of CSCs. Researchers are working to inhibit the cytokines and their receptors from disturbing the CSC niche. Rapertaxin was used to inhibit CXCR1, a receptor for IL8, which selectively depleted CSCs in human breast cancer cell lines [40]. Similarly, inhibition of IL6 using antibody inhibited STAT3 and OCT-4 gene expressions, thereby inhibiting CSCs [41].

Immune Checkpoint (Programmed Cell Death-1 (PD1) and Programmed Death-ligand 1 (PDL1))

It is now widely known that PD1/PDL1 and CTLA4/B7 axis are inhibitory signaling pathways that inhibit immune response generation and help cancer cells in immune evasion [42, 43]. Some of the recent studies have found a higher expression of PDL-1 on CSCs and hence suggest CSCs may be downregulating T cell responses via such immune evasion co-inhibitory signals. Further research and proof would be needed in this area to take this to a therapeutic scale, but this suggests a therapeutic window for immune-based targeting of CSCs [44].

Targeting Drug Resistance Signaling in CSCs

The signaling pathways in normal stem cell homeostasis are highly regulated and are highly intricate, with many intrinsic and extrinsic molecular signals and regulatory elements. Dysregulation or abnormality of the developmental and homeostatic stemness signaling pathways is responsible for abnormal self-renewal and differentiation of CSCs, which leads to cancer development [45-47]. Also, multiple CSC models have been proposed for tumor heterogeneity including the classical CSCs unidirectional differentiation model and the plastic CSCs bidirectional differentiation model [48, 49]. According to the first model, the CSCs that are differentiated to non-CSC tumor cells are unable to convert into and acquire CSC-like activity; however, in the plastic CSCs bidirectional differentiation model, non-CSC tumor cells can undergo a de-differentiation process and acquire CSC-like properties [50-53]. The most common deregulated signaling pathways in CSCs are the Hedgehog [10], Notch, Wnt, Janus-activated kinase/signal transducer and activator of transcription (JAK/STAT), Nuclear factor Kappa B (NF-κB) and phosphatidylinositol 3-Kinase/phosphate and tensin homolog (PI3K/PTEN) signaling systems [54].

As discussed previously, CSCs are capable of resisting the conventional therapies and thus results in tumor formation, drug resistance and recurrence, therefore targeting the CSCs in tumors may present an effective anti-cancer therapeutic strategy to suppress tumor growth and recurrence. To achieve this, various significant therapeutic agents have been developed that target critical steps in HH, Notch, Wnt, JAK/STAT, and NF-κB pathways. .

Therapeutic Drugs Targeting HH Signaling

HH signaling is an important pathway during embryogenesis and development that communicates the information to embryonic cells required for proper cell differentiation [55, 56]. The major molecules of HH signaling that regulate the activation and repression of this pathway are Sonic, Desert and Indian, their receptor Patched, transmembrane protein Smoothened (SMO), and three Gli transcription factors (Gli 1-3) [57]. Experimental studies on samples from mouse, patient, and cancer cell lines have shown the activation of abnormal HH signaling in CSCs of various cancers, including multiple myeloma, glioblastoma, colon cancer, chronic myeloid leukemia (CML), etc.[49]. HH signaling plays an essential role in epithelial to mesenchymal transformation (EMT); for example, knockdown of Gli resulted in decreased CSCs' viability, motility, clonogenicity, and self-renewal in claudin-low breast CSCs [58]. As HH signaling plays a significant role in CSCs self-renewal and regulation, inhibiting HH signaling can thus interrupt CSCs stemness and induce their differentiation, which is desirable for cancer treatment. Researchers have developed numerous therapeutic agents for targeting the HH pathway. Vismodegib is the first FDA approved HH inhibitor developed by Genentech, which targets SMO and is used for the metastatic basal cell carcinoma treatment [59, 60]. At present, Vismodegib in clinical trials is used as monotherapy or in combination with other therapeutic agents to treat several cancers, such as metastatic pancreatic cancer, metastatic prostate cancer, recurrent glioblastoma, acute myeloid leukemia, small cell lung cancer and medulloblastoma [61]. Sonidegib is another FDA approved HH inhibitor in 2015, which also targets the SMO and is used for the treatment of adult patients with locally advanced BCC [62]. However, the drug has grade ½ adverse effects, comprising nausea, anorexia, vomiting, muscle spasms, fatigue and grade ¾ adverse effects, such as weight loss, hyperbilirubinemia, myalgia and dizziness [62]. Several phase I/II trials of Sonidegib to treat other solid tumors and hematological malignancies are still underway [61]. Other inhibitors targeting the HH signaling that were tested in clinical trials include SMO inhibitors (BMS-833923, Glasdegib, and PF-5274857) and Gli inhibitor (arsenic trioxide). Along with these, a monoclonal antibody 5E1 which inhibits the binding of all three HH mammalian ligands to PTCH is also developed [63]. However, this antibody has not yet been entered in clinical trials.

Therapeutic Agents Targeting Wnt Signaling

Wnt has become a substantial new target for drug development to treat cancer because of its signaling cascade that plays a central role in regulating significant functions of malignant epithelial cells. Wnt ligands and signals drive the Wnt signaling pathway through canonical (β-catenin dependent) or non-canonical (β-catenin independent) paths. The Wnt ligand binds to various transmembrane receptors, such as Frizzled (FZD), Receptor tyrosine kinases (RTKs) and Receptor tyrosine kinase-like orphan receptor (ROR) 1 or ROR2. The pathway is activated with the binding of Wnt ligand to its receptor, followed by the activation of β-catenin. In the absence of Wnt ligand, β-catenin undergoes phosphorylation by a destruction complex containing glycogen synthase kinase 3b (GSK3b), adenomatous polyposis coli (APC) and axin, followed by degradation of β-catenin.

Differential expression of Wnt signaling molecules has been implicated in a plethora of CSC's phenotype, such as self-renewal, tumorigenesis, and dedifferentiation. This leads to the development of several targeted therapies (like monoclonal antibodies and small molecule agents) which have been tested preclinically over the year. However, no drugs have been approved for clinical usage [64]. Small molecules targeting the Wnt/β-catenin signaling can be classified into four groups: β-catenin/TCF antagonists, transcriptional coactivator modulators, Dvl binders, and other mechanism-based inhibitors [65]. Among all the small molecules, Porcupine inhibitors are the leading class, as represented by LGK974 (Novartis) [66]. Porcupine is a membrane-bound O-acetyltransferase that catalyzes the post-translational acylation of Wnt molecules and downregulation of this enzyme leads to decreased or abolished Wnt secretion [67]. In addition to small molecules, various biological agents such as OMP18R5 (also known as Vanticumab) have entered the clinical trials. OMP18R5 targets the FZD receptor and showed open label phase 1 dose escalation for solid tumors [68]. OMP-54F28 is another biologic agent and a fusion protein, which binds the Wnt ligands, thus blocking the binding of Wnt ligands with FZD receptors [68].

Therapeutic Agents Targeting Notch Signaling

Notch signaling played a role in cell-cell communication and was found to be essential for cell proliferation and apoptosis during embryonic development. Several pathways such as nuclear factor-κB (NF-κB), HH, mammalian target of rapamycin (mTOR), Wnt, and epidermal growth factor have been shown to cross-talk with notch pathway, and thus, play significant functions in CSCs and tumor growth [69-71]. Therefore, Notch can be an attractive therapeutic target for CSCs treatment. Therapeutics aiming Notch signaling mainly consist of anti-Delta-like ligand 4 (DLL4) antibodies and g-secretase inhibitors. The main function of monoclonal antibodies against DLL4 is to block the binding of a ligand. Encticumab (REN421) is a DLL4 targeting antibody and has been used to treat advanced ovarian solid tumors overexpressing DLL4 [72]. Demcizumab is another anti-DLL4 antibody developed by OncoMed Pharmaceuticals and Celgene, and has completed the phase I clinical trial. In addition to inhibitors against DLL4, various g-secretase inhibitors such as RO4929097 and LY900009 have been undergoing phase I trial. RO4929097 suppresses Notch signaling genes, such as Hey1, Hey L and Hes1, thus blocking activation of self-renewal genes [10].

Cell Surface Markers as Targets

CSCs present specific cell surface markers that distinguish them from the other tumor cell and are used for identification and isolation of CSCs, and are also important targets for therapy. Antibodies against CSCs surface markers are often used as an adjunct to chemotherapy, surgery and radiation therapy, and thus enhance the therapeutic strategies. Few of the most important markers associated with CSCs are CD133, CD44, IL-3R, and the immunoglobulin mucin TIM-3.

CD133, also called Prominin-1, is a transmembrane glycoprotein of 120 kDa, which is overexpressed in solid tumors such as glioma [73], liver [74], lung [75] and colorectal cancer [76]. Overexpression of CD133 is associated with resistance in treatment and poor prognosis; therefore, various therapeutic strategies such as targeted immunotoxins have been developed to selectively target this molecule. CD133NPs, a polymeric nanoparticle loaded with paclitaxel, were tested *in vitro* on colorectal adenocarcinoma Caco-2 cells and have shown efficient anti-cancer activity. Also, *in vivo* study on xenograft model has demonstrated better anti-tumor activity of CD133NPs when compared to free paclitaxel treatment [77]. Anti-CD133$^+$ cell therapy was also tested on xenografted ovarian cancer mice model, sarcoma CSCs, pancreatic CSCs and hepatic CSCs [37, 78-80].

CD44 like CD133 is a transmembrane glycoprotein that facilitates cell-cell and cell to ECM interactions and is found to be overexpressed in several cancers. Study on patient-derived AML blasts has revealed that treatment with anti-human CD44 monoclonal antibody to stimulate myeloid differentiation alters stem cell fate, and impedes homing to the microenvironment niche [81]. Doxorubicin and cyclophosphamide, in combination with anti-human CD44 antibody, have shown synergistic anti-cancerous activity in breast cancer and prevent relapse of aggressive breast cancer [82]. Another study by Tang et al. [83] used a mouse IgG1 anti-human CD44 receptor antibody on MiaPaCa-2 cells and was shown to inhibit CSCs tumorsphere formation, also decreasing tumor growth, metastasis and recurrence in pancreatic cancer xenografted nude mice.

Apart from CD133 and CD44, one of the studies has examined the levels of mucin-like MUC1 on human ESC H9 and H14 and showed the expression of full-length MUC1 on newly differentiated human embryonic stem cells [84]. Further, expression of MUC1 has also been studied on CSCs of breast cancer cell line MCF7 [84] and pancreatic cancer (PC) cells [85]. MUC16 is another mucin which is found to be overexpressed in CSCs of various cancers such as ovarian and PC. Based on their role in different cancers, various therapeutic drugs have been generated to target these mucins. PankoMab is a novel MUC1 antibody, which was conjugated with a toxin β-amanitin and has been shown to induce specific cytotoxicity of tumor cells in *in vitro* experiments [86]. Further, another study investigating 137 surgical specimens of different types of carcinomas by immunohistochemistry reported that humanized form of PankoMab reacted strongly towards a tumor-related MUC1 epitope (TA-MUC1) [87]. One of the promising drug, targeting the MUC16, and currently in use is monoclonal anti-CA125 B43.13, also known as oregovomab [88]. Other targeted antibodies against cell surface markers that are approved by FDA are rituximab (anti-CDC20), cetuximab (anti-EGFR), trastuzumab (anti-HER2), bevacizumab (anti-VEGEF-A), and pembrolizumab (anti-PD-1) which are currently in use [89]. Although great advancement has been made in aiming the CSC surface molecule, many issues remain to be addressed. We should understand that CSC markers are not perfect and not all CSCs express the markers.

Nanomedical Strategies of CSC Targeting

CSCs possess enhanced diverse cellular processes that allow them to exhibit resistance towards most standard therapies [90]. Most conventional therapies like radio- and chemotherapy fail to have an impact on CSCs due to their properties like entering the quiescent stage, expression of ABC transporters allowing efflux of chemotherapeutic drugs, enhanced DNA damage repair response and increased expression of anti-apoptotic and detoxifying enzymes [25, 91-93]. This has made them very attractive and viable therapeutic targets. Nanomedicine has recently come into the spotlight due to its exceptional properties that allow precise control in delivery, targets and quantity of the drugs administered. Nanomaterials possess unique features and have a high surface to volume ratio that allows them to have outstanding surface characteristics that can benefit anti-cancer therapies to generate molecularly directed therapies [94]. Nanoparticles have the ability to sequester the chemotherapeutic drugs to higher concentrations and then maintain controlled drug release once they reach the target sites or CSCs [94]. This provides researchers a means to bypass a few of the resistance mechanisms utilized by CSCs.

Nanomaterials: Carbon-based

Carbon-based nanomaterials have found their place in nanomedicine in the last decade including its allotropes like diamond, graphene, and carbon nanotubes due to their relatively inert and nontoxic nature. Graphene, when oxidized, forms graphene oxide (GO) and exhibits high chemical versatility and hence can be modified to link a variety of biochemical like drugs, nucleic acids or ligands to enable targeted therapy. GO has been utilized in several studies for targeted cancer therapies; however, limited studies have employed them to target CSCs [95]. A study by Fiorillo et al. showed the efficacy of GO

Targeted Therapy for Cancer Stem Cells

in inhibiting CSCs in six different independent cancer cell lines including glioblastoma (brain), lung, breast, pancreatic, ovarian, and prostate cancers. They performed tumorsphere assay to examine self-renewal and tumorigenic potential of CSCs and showed GO to inhibit tumorsphere formation of CSCs selectively over non-stem or differentiated cells [95]. Their work suggested that GO has the potential to induce differentiation and inhibit proliferation of CSCs, providing evidence towards the potential of GO-based therapies.

Graphene is also utilized in the form of cylindrical nanostructures called nanotubes. These nanotubes can carry high drug load, and penetrate the cell membrane, selectively be retained in the tumor, and are water soluble and non-toxic [96, 97]. They also possess photothermal, photoacoustic, and Raman properties, making them valuable tools for nanotechnological clinical research. A recent study by Burke et al. showed that carbon nanotubes mediated thermal therapy could be utilized to overcome the breast CSCs resistance towards traditional hypothermia therapy [96]. In this study, stem cells lost their long-term self-renewal potential *in vitro* and induced complete tumor regression and long-term survival of mice bearing cancer stem cell-driven breast tumors. Nanotube thermal therapy promotes rapid membrane permeabilization and necrosis of cells and simultaneously inhibits both CSCs and differentiated cancer cells [96]. These nanotube systems can also be modified to enhance their specificity for CSCs by coating with chitosan and loading with salinomycin activated with hyaluronic acid or functionalized antibiotic-loaded nanotubes with anti-CD44 antibodies [98]. These nanotubes can also carry chemotherapeutic drugs like paclitaxel or a combination of antibiotics and chemotherapeutic drugs. Nanodiamonds are another form of carbon nanoparticles, which can be modified, like nanotubes to develop targeted therapy with a high concentration of drug load [99, 100].

DNA Origami

DNA origami is another nanoscale delivery system that utilizes DNA self-assembly and controlled pattern formation using computational algorithms and software for origami design and analysis [101]. DNA origami is a technique that uses long single-stranded DNA molecules and folds them into arbitrary two-dimensional shapes [101, 102]. The desired shape is achieved by raster-filling the shape with a 7-kilobase single-stranded scaffold and by choosing over 200 short oligonucleotides 'staple strands' to hold the scaffold in place. Once synthesized, the staple strands are mixed with scaffold resulting in their self-assembly into approximate desired shapes like squares, discs etc. roughly 100 nm in diameter. Finally, individual DNA structures can be further programmed to form larger assemblies [101]. These structures provide high drug loading capacity and biocompatibility. Studies have demonstrated these structures to have the capacity to evade resistance mechanism in stem cells like ATP dependent drug efflux in breast cancer and leukemia [103, 104].

Gold Nanoparticles (AuNPs)

Nanoparticles can also be formulated with gold due to its properties of localized surface plasmon resonance, facile synthesis and functionalization, and excellent biocompatibility [105]. It has been well established that AuNPs are non-toxic and non-immunogenic [106]. These days, AuNPs can be easily synthesized in high yields like nano-spheres, nano-rods, and nano-cages. Gold has radiative properties like absorption, scattering and localized surface plasmon resonance making it very suitable for photothermal therapy and easy surface modifications make them attractive options for drug delivery and cancer therapies [105]. People have successfully coated and loaded these particles with drugs like capecitabine, doxorubicin, and cisplatin [107, 108]. AuNPs coated with thio-polyethylene glycol, and thio-glucose have been used to target CSCs specifically due to their high glucose uptake. AuNPs can also be modified with microRNA duplexes specific for CSCs to allow precise target delivery [109].

Gold nano-rods are better for photothermal therapy to target CSCs since they exhibit localized surface plasmon resonance at near-infrared wavelength. Near-infrared wavelength is relatively less damaging and possesses high penetrability. Gold nanorods can be conjugated with aptamers to specifically target CSCs [110]. Studies also report a higher uptake of gold nanoparticles by CSCs once they are polyelectrolyte conjugated [111, 112]. Researchers have employed this for the targeted killing of CSCs by modifying polyelectrolyte conjugated gold particles with salinomycin. Researchers have integrated these tools further to develop multi-level therapeutic approaches by combining AuNPs loaded with chemotherapeutic drugs and siRNA (for example, against KRAS) in hydrogels or porous silicon. Such strategies can be upgraded further by incorporating porous silicon and modified AuNPs, and DNA origami loaded with different targeted and chemotherapeutic drugs and generating an emulsion system [104, 113, 114]

Liposomes

Liposomes are highly biocompatible and can enhance the bioavailability and stability of drugs to reach the tumor within the body [115]. Liposomes can be used to administer targeted therapy against CSCs by functionalizing with glucose and loading them with CSC specific inhibitors [116, 117]. In an alternative approach, liposome-mediated delivery also enhances the therapeutic efficacy of curcumin and chemotherapeutic drugs like taxol [118]. These liposomes could also be functionalized with anti-CD44 antibody or similar CSCs specific modifications to allow targeted therapy [116].

Polymeric Nanoparticles

Polyethylene glycol (PEG) is a common polymeric platform utilized for developing nanomedical approaches. PEG nanoparticles loaded with HH inhibitor or thioridazine have explicitly been used to target CSCs [119]. It is also possible to load these nanoparticles with salinomycin and attach them

Targeted Therapy for Cancer Stem Cells 305

to CD133 aptamers to generate a CSCs specific therapeutic response [120]. Similar methodologies can be used to load these particles with CSCs specific siRNAs (U87MG, U251, SiGLUT3) or miRNAs or chemotherapeutic drugs and functionalize them with CD44 or CD133 specific monoclonal antibodies or aptamers [121].

Target Molecules

Various target molecules are being presently used to device CSCs specific nanomedical approaches. Carbon-based nanoparticles have been modified to target hyaluronic acid receptors, whereas gold nanorods have been used to target TGF-β and ALDH1 [122]. Organic liposomes have been functionalized to target NF$_\kappa$B pathway, CCL-23R, UM-SCC-1R, and CD44, whereas polymeric nanoparticles have been modified towards CD133 and CD44 [120, 121].

Oligonucleotide Aptamers

Aptamers are small nucleic acid ligands, which can comprise RNA or single-stranded DNA, and possess high affinity and selectivity for their target molecules [123]. These strands of DNA or RNA fold into concrete 3-dimensional structures and recognize their targets based on this structure with a dissociation constant in pico to nanomolar range. Their low molecular weight (8-25 kDa) allows better, faster, and more efficient tissue penetration [123]. They are non-immunogenic since they are oligonucleotides and are thermally stable. They are also easily synthesized with low production cost and batch variation. They can also be modified with various functional moieties, making them ideal candidates for selective therapeutic approaches [123].

Aptamers-specific for Cell Surface Markers

Aptamers that are developed for a specific target molecule are generated with SELEX technology using repetitive enrichment and selection process. This technology works by selecting aptamers specific for a target molecule from a random aptamer library. Selected aptamers are selected by affinity binding towards their targets and then again dissociated and amplified for the next round of selection cycle. After several rounds of positive and negative selection to eliminate non-target binding antigens and select specifically binding antigens, an apatmer has to go through 4-20 rounds of selection amplifications and enrichment cycles [124].

Cell-SELEC technology uses whole living cells to select cell-specific aptamers to incorporate post-translational modification specific aptamers [125, 126]. Such technology would be an excellent means to generate aptamers against biomarkers like Erb superfamily, EpCAM, and CD133 to target CSCs specifically. Such specific aptamers can be conjugated with drugs like doxorubicin, gemcitabine, photosensitizers, or nanoparticles that have been described above. Such modifications enhance the bioavailability and stability of aptamers [124].

Aptamers can also be used to target stem cells using gene therapy. SiRNAs and miRNAs are powerful gene silencing tools and can be conjugated with aptamers for an enhanced therapeutic specific response for CSCs [123]. Although siRNA and miRNA alone can also be used for generating therapies, combining them with aptamers or nanoparticles enhances their efficacy by improving bioavailability, enhancing specificity and stability, and maintaining a controlled release. Oncogenic miRNAs vital for CSCs can be silenced like miR-302 (melanomas) and tumor suppressing miRNAs can be upregulated like Let7 and Mir34a [123]. Such microRNAs can be effectively delivered in liposomes or polymeric nanoparticles as delivery vehicles like miR145 to CD133+ CSCs in glioblastoma [127]. Virus-like particles have also been developed from MS2 bacteriophage capsids for RNA delivery [127]. Blocking oncogenic miRNAs requires delivery of anti-miRNA nucleotides, which may include several modifications to enhance stability like the incorporation of locked nucleic acids or 2'-O-methyl oligonucleotides, miRNA sponges and miRNA masks.

Targeting CSCs Based on Metabolism

Cancer cells depend on glycolysis pathway for their energy needs, converting glucose to lactate, even in the presence of oxygen, a phenomenon known as 'Warburg effect' [128]. This is considered as an adaptive requirement to cope with their high proliferation rate. The higher glycolytic flux of Cancer cell/CSCs in comparison to normal cells has been historically exploited for detection of tumors, monitoring of disease stage, and therapy response of patients using 2-[^{18}F] fluoro-2-deoxyglucose positron emission tomography (FDG-PET) scans [129]. Moreover, accumulating evidence suggests that CSCs exhibit a distinct metabolic phenotype as opposed to the glycolytic differentiated tumor cells. Depending on the type of cancer, the metabolic phenotype of CSCs can either be primarily glycolytic or oxidative phosphorylation (OXPHOS) dependent. Regardless of the metabolic phenotype, CSCs seem to be dependent on intact mitochondrial function, and this represents an Achilles heel that can be used for metabolic targeting of CSCs for therapeutic intervention.

Cellular Metabolism

In normal or non-transformed cells, the main store of energy production is the mitochondria, where ATP is produced through the tricarboxylic acid (TCA) cycle coupled to oxidative phosphorylation. As carbon fuels such as pyruvate, glutamine, and fatty acids pass through the TCA cycle, reducing equivalents including nicotinamide adenine dinucleotide phosphate (NADH) and flavin adenine dinucleotide (FADH$_2$) are generated that are subsequently used as electron donors for the electron transport chain (ETC). Proton motive force is generated via coupling of movement of electrons across the different complexes of the electron transport chain and is used by ATP synthase to generate ATPs.

Rapidly proliferating cancer cells need to reprogram their metabolism to fulfill three critical tasks: (1) rapid ATP generation to maintain energy status, (2) elevated rate of macromolecules' biosynthesis and (3) regulation of the cellular redox status. As carbon sources are limited and distributed in a spatially and temporally heterogeneous manner, tumor cells adapt their metabolic machinery to meet their energy needs during tumor growth and metastasis. Therefore, tumor cells shift to ATP generation via glycolysis from oxidative phosphorylation, despite the presence of sufficient oxygen. The advantage is that energy generation via glycolysis is much more rapid than via oxidative phosphorylation, even though it is less efficient in terms of number of ATPs generated per molecule of glucose consumed. Glucose is also metabolized via pentose phosphate pathways and other alternative pathways to generate the basic building blocks required for sustaining high cell division rates [128].

Interconnection of Stemness with Metabolism: Metabostemness

Metabolism of cancer and CSCs is an emerging area of cancer research. Emerging evidence alludes to the metabolic infrastructural changes being a key player in dictating cellular differentiation states instead of just being a passive by-product of oncogene-driven transformation. In fact, metabolic reprogramming appears to be fundamental in deciding cell fate, as the transformation of cellular metabolism precedes changes in stemness [130, 131]. Lessons learned from induced pluripotent stem cells (iPSCs) research reinforce the notion that metabolic reprogramming is essential for transcription factor induced stemness [132]. Indeed, tight regulation of the metabolic master switches contribute to the metabolic changes that occur in the transition between a differentiated cell and a stem cell, and to the exit from the pluripotent state to become primed for differentiation, and to the maintenance of stemness in a stem cell niche [132, 133]. Since there are many parallels mechanistically in the process of dedifferentiation of somatic cells to iPSCs as well as *de novo* generation of CSCs cellular states from non-CSCs, it is an emerging belief that metabolic reprogramming phenomena that include epigenetic remodeling (that is the activation of genes related to maintenance of stemness, and/or inhibiting the determinants related to cell differentiation) might be co-opted in the absence of functional tumor suppressing mechanisms. In other words, the acquisition of and departure from stemness in pre-malignant and cancer tissues is not exclusively regulated by genetic and epigenetic controllers, but also by cellular metabotype that functions as a molecular constraint controlling the kinetics of stemness reprogramming during cancer genesis and progression. In this context, specific metabolic shifts might occur very early in the course of malignant transformation of a non-CSC to a CSC state that might render a differentiated cell more susceptible to the transcriptional and epigenetic rewriting necessary for the acquisition of stemness. Definitive state of cancer stemness can subsequently be acquired through variable orders, combinations or intensities of subsequent hits. The term 'metabostemness' has been coined to refer to the metabolic parameters causally controlling or functionally substituting the epigenetic orchestration of the genetic program that directs normal and non-CSCs tumor cells towards a less-differentiated CSCs cellular states [134]. A new phenotypic cancer hallmark refers to the metabolic parameters at the cell-intrinsic, tissue-micro environmental, and systemic levels that enable the functional properties of CSCs.

Also, small molecule-components or enantiomers of normal metabolism can be oncogenic themselves, known as oncometabolite. The accumulation of these oncometabolites establishes an environment that initiates and drives carcinogenesis by epigenetically inducing the expression of stem cell maintenance genes while blocking induction of differentiation markers. However, research in the field of metabostemness is still in its infancy. Several key questions remain unanswered: How does the metabolome of cell changes as it undergoes a transformation from a normal stem/progenitor cell to a CSC state and then subsequently as it acquires metastatic capabilities? How can metabolites exert influence over the transcription factors, the chromatin structure, and the epigenetic circuits that establish and maintain self-renewal and differentiation capacities of CSCs?

Metabolic Phenotype of CSCs: Conclusions and Controversies

The metabolism of CSCs is an emerging area of research and has been a subject of intense investigation in recent years. Of interest, depending on the type of cancer, CSCs have been found to rely primarily either on glycolysis or on oxidative phosphorylation. However, contradictory findings have been reported for the same tumor type. These contradictory findings can be reconciled with the fact that different reports used different isolation techniques for CSCs and, therefore, did their metabolic analyses on distinct subsets of CSCs. Also, some of the early studies were conducted using cell lines that lack proper TME. Indeed, metabolic phenotypes were found to be different between *in vivo* and *in vitro* conditions in case of lung cancer [135], emphasizing the relevance of microenvironment for the metabolic phenotype.

Metabolism and stemness are intrinsically linked, and the relation is evident in case of iPSCs, wherein the switch from oxidative phosphorylation to glycolysis has been deemed essential for effective acquisition of a pluripotent state [133]. CSCs in breast cancer, hepatocellular, and nasopharyngeal carcinomas have been reported to rely on glycolysis [110, 136]. Interestingly, the main driver for stemness in these cancers is an elevated expression of oncogenic Myc, suggesting that Myc-driven glycolytic program and stemness are interlinked in these tumors [137]. CSCs from lung cancer, pancreatic ductal adenocarcinoma, glioblastoma, and leukemia have been convincingly shown to use oxidative phosphorylation as the preferred energy production process [138-141]. Besides carbohydrate metabolism, mitochondrial fatty acid oxidation is also utilized by CSCs for ATP and NADPH generation. Of note, the self-renewal process of leukemia-initiating cells is also dependent on fatty acid oxidation, suggesting that inhibition of fatty acid oxidation can be used as a strategy to target the mitochondrial metabolism of CSCs.

Mitochondrial Function in CSCs

Evidence suggests that stemness is regulated by mitochondria, regardless of the underlying metabolic phenotype in cells. Therefore, increased mitochondrial biogenesis is vital for CSC functionality in glycolytic and oxidative phosphorylation-dependent CSCs. Cells with enhanced self-renewal capacity and chemoresistance exhibit increased mitochondrial mass, which is indicative of elevated mitochondrial biogenesis [138, 140].

Targeting Metabolism in CSCs

Targeting mitochondrial metabolism could be an effective pharmacological strategy for eradication of CSCs, given the dependence of CSCs on oxidative phosphorylation and regulation of stemness by mitochondria. Pre-clinical and clinical trials are being conducted with pharmacological agents targeting oxidative phosphorylation. Few elegant studies have attempted to demonstrate the efficacy of combinatorial targeting of metabolic nodes and genetic drivers for effective targeting of pancreatic CSCs. Sancho et al. have demonstrated that while the differentiated non-CSCs are highly glycolytic, pancreatic CSCs are dependent on oxidative phosphorylation [140].

Furthermore, targeting oxidative phosphorylation in CSCs via metformin stalled tumor progression or even induced regression in patient-derived xenografts, owing to the reduced metabolic plasticity of CSCs, which rendered them unable to switch to glycolysis upon mitochondrial inhibition. This is due to reduced levels of c-Myc expression in CSCs compared to non-CSCs. However, eventually, a resistant CSC population with an intermediate metabolic phenotype and increased c-Myc expression emerges that mediates tumor relapse. This resistant population of CSCs could be targeted via c-Myc inhibition, indicating that a multimodal strategy is most useful for metabolic targeting of CSCs. In another study, comprehensive proteomic analysis and metabolite profiling indicated that fatty acid synthesis and mevalonate pathway are essential for the survival of pancreatic CSCs grown as tumorsphere cultures of PC cell line Panc1 [142].

We now know that glycolysis and oxidative phosphorylation are not the only metabolic pathways utilized by CSCs for their energy requirements. Amino acid metabolism, especially glutaminolysis and fatty acid metabolism, also play an essential role in cancer development and growth. Currently, there is a lack of knowledge on the compensatory effect between metabolic pathways, the effect of blocking one part of the pathway on the overall signaling, and most importantly, the metabolic phenotype of patients. Therefore, a comprehensive analysis of the metabolic landscape of CSCs from early to late stages of PC progression is needed. Given that only a few permitted metabolic types will be compatible with the functional properties of CSCs, metabostemness trait is very relevant for tackling the problem of tumor heterogeneity and, ultimately, is capable of discovering drugs aimed at targeting CSCs themselves.

Summary and Conclusion

Tumor metastasis, aggressiveness, therapy resistance, and recurrence are mainly associated with CSCs population of cancer. CSCs display resistance to conventional therapies like chemo- and radio-therapy due to many factors such as the increased expression of ABC transporters, higher ALDH activity, enhanced DNA repair mechanism, altered TME, increased HIF1alpha, altered cell pH, expression of immune suppressive molecules, activation of stemness signaling and altered cellular metabolism. CSCs activation depends on stemness signaling pathways like hh, Wnt, Notch and EGFR signaling for their self-renewal and growth, and also aberrant activation of this stemness signaling is associated with drug resistance in many carcinomas. Hence, drug resistance in cancer can be overcome by developing therapeutic strategies that specifically target CSCs. Therefore, therapies have been developed to eradicate CSCs by inhibiting the processes mentioned above. Many therapeutic agents targeting critical molecules in HH, Notch, Wnt, JAK/STAT, and NF-κB pathways have been developed to target CSCs, in turn inhibiting the drug resistance. Specific antibodies targeting CSC cell surface markers like CD133, CD44, and MUC1 have been generated and have shown promising results in inhibiting the CSCs population. CSCs are also targeted by specific immune therapeutics such as activation of NK cells, Gamma delta T cells, and CD8T cells, inhibition of cytokines and MDSCs. Inhibition of PD1/PDL1 and CTLA4 axis and generation of CSCs specific mature DCs for vaccine preparation is also a good therapy for targeting CSCs, which needs to be explored.

In recent years, nanomedicine is gaining greater importance and is being identified as a potent strategy for targeted therapeutics. Nanoparticles have shown promising results because of their high surface to volume ratio, biocompatibility, higher drug loading capacity, and precise and controlled drug release to target CSCs, including cancer cells. Different nanomaterials have been developed, such as carbon-based nanomaterials, DNA origami, AuNPs, PEG, and oligonucleotide aptamers for targeted therapeutics. Carbon-based nanoparticles like GO and nanotubes are non-toxic in nature, load variety of drugs and are identified as a potential inhibitor of CSCs. DNA origami-based nanoparticles have been shown to inhibit ABC transporters and evade drug resistance mechanism of CSCs. One of the well-studied nanoparticle for targeted therapeutics is AuNPs because of their surface plasmon resonance and biocompatibility. AuNPs loaded with various drugs, and thio-glucose have been used to target cancer and specifically CSCs, and have shown fruitful results. CSCs have also been targeted by

PEG nanoparticle loaded with Hh specific inhibitor. Nanoparticles can also be modified for specific targeting of CSCs such as liposomes functionalized with glucose and CSCs markers, and loading with CSCs specific inhibitor or chemo drugs. PEG and aptamers can also be functionalized with antibodies against CSC markers and loading with CSCs specific Si- or miRNA.

Cellular metabolism is an emerging area of research, and scientists are working to understand the mechanism of CSCs metabolism to target them. CSCs from various cancers depend on distinct metabolic profile like glycolytic and OXPHOS, and also on intact mitochondria for cellular functions. Cellular metabolism also controls the process of stemness, and metabolic change and synthesis of oncometabolites mark the early event in the transformation of non-CSCs to CSCs, leading to the activation of stemness and cancer development. CSCs depend on either glycolysis or OXPHOS or fatty acid as the energy source in different cancer types. Increased mitochondria biogenesis is also shown to regulate stemness and drug resistance in CSCs. Hence, the important molecules of the metabolism pathways can be targeted for the eradication of CSCs. Recently, pre-clinical and clinical trials have been conducted targeting OXPHOS to eradicate CSCs and showed promising results in tumor regression. C-Myc overexpressing resistant CSCs population has been targeted by using c-Myc specific inhibitor. As some of the CSCs population also relies on the fatty acid synthesis, glutaminolysis, and mevalonate pathways as an energy source, it is good to develop combinatorial therapy for targeting multimodal pathways to inhibit different subpopulation of CSCs and in turn inhibit cancer.

References

1. Prasetyanti, P.R., J.P. Medema. Intra-tumor heterogeneity from a cancer stem cell perspective. Molecular Cancer, 2017. 16: 41.
2. Abdullah, L.N., E.K.-H. Chow. Mechanisms of chemoresistance in cancer stem cells. Clin Transl Med, 2013. 2: 3-3.
3. Bakhshinyan, D., A.A. Adile, M.A. Qazi, M. Singh, M.M. Kameda-Smith, N. Yelle, et al. Introduction to cancer stem cells: past, present, and future. pp. 1-16. *In*: Papaccio, G., V. Desiderio (eds.). Cancer Stem Cells: Methods and Protocols. Springer New York, New York, NY. 2018.
4. Hermann, P.C., P. Sancho, M. Canamero, P. Martinelli, F. Madriles, P. Michl, et al. Nicotine promotes initiation and progression of KRAS-induced pancreatic cancer via Gata6-dependent dedifferentiation of acinar cells in mice. Gastroenterology, 2014. 147: 1119-33. e1114.
5. Lapidot, T., C. Sirard, J. Vormoor, B. Murdoch, T. Hoang, J. Caceres-Cortes, et al. A cell initiating human acute myeloid leukaemia after transplantation into SCID mice. Nature, 1994. 367: 645-8.
6. Lv, Y., S. Zhao, J. Han, L. Zheng, Z. Yang, L. Zhao. Hypoxia-inducible factor-1α induces multidrug resistance protein in colon cancer. Onco Targets Ther, 2015. 8: 1941-8.

7. Gerweck, L.E., S. Vijayappa, S. Kozin. Tumor pH controls the in vivo efficacy of weak acid and base chemotherapeutics. Mol Cancer Ther, 2006. 5: 1275-9.
8. Tannock, I.F., D. Rotin. Acid pH in tumors and its potential for therapeutic exploitation. Cancer Res, 1989. 49: 4373-84.
9. Codony-Servat, J., R. Rosell. Cancer stem cells and immunoresistance: clinical implications and solutions. Transl Lung Cancer Res, 2015. 4: 689-703.
10. Debeb, B.G., E.N. Cohen, K. Boley, E.M. Freiter, L. Li, F.M. Robertson, et al. Preclinical studies of Notch signaling inhibitor RO4929097 in inflammatory breast cancer cells. Breast Cancer Res Treat, 2012. 134: 495-510.
11. Scharenberg, C.W., M.A. Harkey, B. Torok-Storb. The ABCG2 transporter is an efficient Hoechst 33342 efflux pump and is preferentially expressed by immature human hematopoietic progenitors. Blood, 2002. 99: 507-12.
12. Shapiro, A.B., A.B. Corder, V. Ling. P-glycoprotein-mediated Hoechst 33342 transport out of the lipid bilayer. Eur J Biochem, 1997. 250: 115-21.
13. Begicevic, R.-R., M. Falasca. ABC Transporters in cancer stem cells: beyond chemoresistance. Int. J. Mol. Sci, 2017. 18: 2362.
14. Patrawala, L., T. Calhoun, R. Schneider-Broussard, J. Zhou, K. Claypool, D.G. Tang. Side population is enriched in tumorigenic, stem-like cancer cells, whereas ABCG2+ and ABCG2− cancer cells are similarly tumorigenic. Cancer Res, 2005. 65: 6207-19.
15. de Sousa e Melo, F., L. Vermeulen. Wnt signaling in cancer stem cell biology. Cancers, 2016. 8: 60.
16. Ding, J., X.T. Zhou, H.Y. Zou, J. Wu. Hedgehog signaling pathway affects the sensitivity of hepatoma cells to drug therapy through the ABCC1 transporter. Lab Invest, 2017. 97: 819-32.
17. Miele, L. Notch signaling. Clin Cancer Res, 2006. 12: 1074-9.
18. Pang, L.Y., L. Saunders, D.J. Argyle. Epidermal growth factor receptor activity is elevated in glioma cancer stem cells and is required to maintain chemotherapy and radiation resistance. Oncotarget, 2017. 8: 72494-512.
19. Oliveira, B.R., M.A. Figueiredo, G.S. Trindade, L.F. Marins. OCT4 mutations in human erythroleukemic cells: implications for multiple drug resistance (MDR) phenotype. Mol Cell Biochem, 2015. 400: 41-50.
20. Porro, A., M. Haber, D. Diolaiti, N. Iraci, M. Henderson, S. Gherardi, et al. Direct and coordinate regulation of ATP-binding cassette transporter genes by Myc factors generates specific transcription signatures that significantly affect the chemoresistance phenotype of cancer cells. J. Biol. Chem, 2010. 285: 19532-43.
21. Clark, D.W., K. Palle. Aldehyde dehydrogenases in cancer stem cells: potential as therapeutic targets. Ann Transl Med, 2016. 4: 518.
22. King, T.D., M.J. Suto, Y. Li. The Wnt/beta-catenin signaling pathway: a potential therapeutic target in the treatment of triple negative breast cancer. J Cell Biochem, 2012. 113: 13-8.
23. Alam, M., R. Ahmad, H. Rajabi, A. Kharbanda, D. Kufe. MUC1-C oncoprotein activates ERK-->C/EBPbeta signaling and induction of aldehyde dehydrogenase 1A1 in breast cancer cells. J Biol Chem, 2013. 288: 30892-903.
24. Raha, D., T.R. Wilson, J. Peng, D. Peterson, P. Yue, M. Evangelista, et al. The cancer stem cell marker aldehyde dehydrogenase is required to maintain a drug-tolerant tumor cell subpopulation. Cancer Res, 2014. 74: 3579-90.
25. Bao, S., Q. Wu, R.E. McLendon, Y. Hao, Q. Shi, A.B. Hjelmeland, et al. Glioma stem cells promote radioresistance by preferential activation of the DNA damage response. Nature, 2006. 444: 756-60.

26. Ciavardelli, D., C. Rossi, D. Barcaroli, S. Volpe, A. Consalvo, M. Zucchelli, et al. Breast cancer stem cells rely on fermentative glycolysis and are sensitive to 2-deoxyglucose treatment. Cell Death Dis, 5: 2014. e1336.
27. Pasto, A., C. Bellio, G. Pilotto, V. Ciminale, M. Silic-Benussi, G. Guzzo, et al. Cancer stem cells from epithelial ovarian cancer patients privilege oxidative phosphorylation, and resist glucose deprivation. Oncotarget, 2014. 5: 4305-19.
28. Vlashi, E., C. Lagadec, L. Vergnes, K. Reue, P. Frohnen, M. Chan, et al. Metabolic differences in breast cancer stem cells and differentiated progeny. Breast Cancer Res Treat, 2014. 146: 525-34.
29. Borah, A., S. Raveendran, A. Rochani, T. Maekawa, D.S. Kumar. Targeting self-renewal pathways in cancer stem cells: clinical implications for cancer therapy. Oncogenesis, 4: 2015. e177.
30. Kim, J., S.H. Orkin. Embryonic stem cell-specific signatures in cancer: insights into genomic regulatory networks and implications for medicine. Genome Medicine, 2011. 3: 75.
31. Monzani, E., F. Facchetti, E. Galmozzi, E. Corsini, A. Benetti, C. Cavazzin, et al. Melanoma contains CD133 and ABCG2 positive cells with enhanced tumourigenic potential. Eur. J. Cancer (Oxford, England: 1990), 2007. 43: 935-46.
32. Welte, T., I.S. Kim, L. Tian, X. Gao, H. Wang, J. Li, et al. Oncogenic mTOR signalling recruits myeloid-derived suppressor cells to promote tumour initiation. Nat. Cell Biol, 2016. 18: 632-44.
33. Wang, Y., X. Chen, W. Cao, Y. Shi. Plasticity of mesenchymal stem cells in immunomodulation: pathological and therapeutic implications. Nat Immunol, 2014. 15: 1009-16.
34. Wu, A., S. Wiesner, J. Xiao, K. Ericson, W. Chen, W.A. Hall, et al. Expression of MHC I and NK ligands on human CD133+ glioma cells: possible targets of immunotherapy. J Neurooncol, 2007. 83: 121-31.
35. Ames, E., R.J. Canter, S.K. Grossenbacher, S. Mac, R.C. Smith, A.M. Monjazeb, et al. Enhanced targeting of stem-like solid tumor cells with radiation and natural killer cells. Oncoimmunology, 2015. 4: e1036212.
36. Castriconi, R., A. Daga, A. Dondero, G. Zona, P.L. Poliani, A. Melotti, et al. NK cells recognize and kill human glioblastoma cells with stem cell-like properties. J Immunol, 2009. 182: 3530-9.
37. Pan, Q., Q. Li, S. Liu, N. Ning, X. Zhang, Y. Xu, et al. Concise review: targeting cancer stem cells using immunologic approaches. Stem Cells, 2015. 33: 2085-92.
38. Visus, C., Y. Wang, A. Lozano-Leon, R.L. Ferris, S. Silver, M.J. Szczepanski, et al. Targeting ALDH (bright) human carcinoma-initiating cells with ALDH1A1-specific CD8(+) T cells. Clin Cancer Res, 2011. 17: 6174-84.
39. Xu, Q., G. Liu, X. Yuan, M. Xu, H. Wang, J. Ji, et al. Antigen-specific T-cell response from dendritic cell vaccination using cancer stem-like cell-associated antigens. Stem Cells, 2009. 27: 1734-40.
40. Ginestier, C., S. Liu, M.E. Diebel, H. Korkaya, M. Luo, M. Brown, et al. CXCR1 blockade selectively targets human breast cancer stem cells in vitro and in xenografts. J Clin Invest, 2010. 120: 485-97.
41. Kim, S.Y., J.W. Kang, X. Song, B.K. Kim, Y.D. Yoo, Y.T. Kwon, et al. Role of the IL-6-JAK1-STAT3-Oct-4 pathway in the conversion of non-stem cancer cells into cancer stem-like cells. Cell Signal, 2013. 25: 961-9.
42. Lyford-Pike, S., S. Peng, G.D. Young, J.M. Taube, W.H. Westra, B. Akpeng, et al. Evidence for a role of the PD-1:PD-L1 pathway in immune resistance of

HPV-associated head and neck squamous cell carcinoma. Cancer Res, 73: 2013. 1733-41.
43. Pardoll, D.M. The blockade of immune checkpoints in cancer immunotherapy. Nat Rev Cancer, 2012. 12: 252-64.
44. Schatton, T., U. Schutte, N.Y. Frank, Q. Zhan, A. Hoerning, S.C. Robles, et al. Modulation of T-cell activation by malignant melanoma initiating cells. Cancer Res, 2010. 70: 697-708.
45. Fang, D., T.K. Nguyen, K. Leishear, R. Finko, A.N. Kulp, S. Hotz, et al. A tumorigenic subpopulation with stem cell properties in melanomas. Cancer Res, 2005. 65: 9328-37.
46. Kroon, P., P.A. Berry, M.J. Stower, G. Rodrigues, V.M. Mann, M. Simms, et al. JAK-STAT blockade inhibits tumor initiation and clonogenic recovery of prostate cancer stem-like cells. Cancer Res, 2013. 73: 5288-98.
47. Singh, S.K., C. Hawkins, I.D. Clarke, J.A. Squire, J. Bayani, T. Hide, et al. Identification of human brain tumour initiating cells. Nature, 2004. 432: 396-401.
48. Lin, L., A. Liu, Z. Peng, H.J. Lin, P.K. Li, C. Li, et al. STAT3 is necessary for proliferation and survival in colon cancer-initiating cells. Cancer Res, 2011. 71: 7226-37.
49. Merchant, A.A., W. Matsui. Targeting hedgehog – a cancer stem cell pathway. Clin Cancer Res, 2010. 16: 3130-40.
50. Bachoo, R.M., E.A. Maher, K.L. Ligon, N.E. Sharpless, S.S. Chan, M.J. You, et al. Epidermal growth factor receptor and Ink4a/Arf: convergent mechanisms governing terminal differentiation and transformation along the neural stem cell to astrocyte axis. Cancer Cell, 2002. 1: 269-77.
51. Friedmann-Morvinski, D., E.A. Bushong, E. Ke, Y. Soda, T. Marumoto, O. Singer, et al. Dedifferentiation of neurons and astrocytes by oncogenes can induce gliomas in mice. Science (New York, N.Y.), 2012. 338: 1080-4.
52. Marjanovic, N.D., R.A. Weinberg, C.L. Chaffer. Cell plasticity and heterogeneity in cancer. Clinical Chemistry, 2013. 59: 168-79.
53. Schwitalla, S. Tumor cell plasticity: the challenge to catch a moving target. J. Gastroenterol, 2014. 49: 618-27.
54. Schwitalla, S., A.A. Fingerle, P. Cammareri, T. Nebelsiek, S.I. Goktuna, P.K. Ziegler, et al. Intestinal tumorigenesis initiated by dedifferentiation and acquisition of stem-cell-like properties. Cell, 2013. 152: 25-38.
55. Chen, K., Y.H. Huang, J.L. Chen. Understanding and targeting cancer stem cells: therapeutic implications and challenges. Acta Pharmacologica Sinica, 2013. 34: 732-40.
56. Jin, Z., T. Schwend, J. Fu, Z. Bao, J. Liang, H. Zhao, et al. Members of the Rusc protein family interact with Sufu and inhibit vertebrate hedgehog signaling. Development (Cambridge, England), 2016. 143: 3944-55.
57. Varjosalo, M., J. Taipale. Hedgehog: functions and mechanisms. Genes Dev, 2008. 22: 2454-72.
58. Colavito, S.A., M.R. Zou, Q. Yan, D.X. Nguyen, D.F. Stern. Significance of glioma-associated oncogene homolog 1 (GLI1) expression in claudin-low breast cancer and crosstalk with the nuclear factor kappa-light-chain-enhancer of activated B cells (NFkappaB) pathway. Breast Cancer Res., 2014. 16: 444.
59. Sekulic, A., M.R. Migden, A.E. Oro, L. Dirix, K.D. Lewis, J.D. Hainsworth, et al. Efficacy and safety of vismodegib in advanced basal-cell carcinoma. N Engl J Med, 2012. 366: 2171-9.

60. Von Hoff, D.D., P.M. LoRusso, C.M. Rudin, J.C. Reddy, R.L. Yauch, R. Tibes, et al. Inhibition of the hedgehog pathway in advanced basal-cell carcinoma. N Engl J Med, 2009. 361: 1164-72.
61. Rimkus, T.K., R.L. Carpenter, S. Qasem, M. Chan, H.W. Lo. Targeting the sonic hedgehog signaling pathway: review of smoothened and GLI inhibitors. Cancers (Basel), 2016. 8: 22.
62. Doan, H.Q., S. Silapunt, M.R. Migden. Sonidegib, a novel smoothened inhibitor for the treatment of advanced basal cell carcinoma. Onco Targets Ther, 2016. 9: 5671-8.
63. Bosanac, I., H.R. Maun, S.J. Scales, X. Wen, A. Lingel, J.F. Bazan, et al. The structure of SHH in complex with HHIP reveals a recognition role for the Shh pseudo active site in signaling. Nat. Struct. Mol. Biol, 2009. 16: 691-7.
64. Zhang, X., J. Hao. Development of anticancer agents targeting the Wnt/beta-catenin signaling. Am J Cancer Res, 2015. 5: 2344-60.
65. Takahashi-Yanaga, F., M. Kahn. Targeting Wnt signaling: can we safely eradicate cancer stem cells? Am J Cancer Res, 2010. 16: 3153-62.
66. Liu, J., S. Pan, M.H. Hsieh, N. Ng, F. Sun, T. Wang, et al. Targeting Wnt-driven cancer through the inhibition of Porcupine by LGK974. Proc Natl Acad Sci USA, 2013. 110: 20224-29.
67. Takada, R., Y. Satomi, T. Kurata, N. Ueno, S. Norioka, H. Kondoh, et al. Monounsaturated fatty acid modification of Wnt protein: its role in Wnt secretion. Developmental Cell, 2006. 11: 791-801.
68. Smith, D.C., L.S. Rosen, R. Chugh, J.W. Goldman, L. Xu, A. Kapoun, et al. First-in-human evaluation of the human monoclonal antibody vantictumab (OMP-18R5; anti-Frizzled) targeting the WNT pathway in a phase I study for patients with advanced solid tumors. J Clin Oncol, 2013. 31: 2540.
69. Mungamuri, S.K., X. Yang, A.D. Thor, K. Somasundaram. Survival signaling by Notch1: mammalian target of rapamycin (mTOR)-dependent inhibition of p53. Cancer Res, 2006. 66: 4715-24.
70. Nakamura, T., K. Tsuchiya, M. Watanabe. Crosstalk between Wnt and Notch signaling in intestinal epithelial cell fate decision. J. Gastroenterol, 2007. 42: 705-10.
71. Nickoloff, B.J., J.Z. Qin, V. Chaturvedi, M.F. Denning, B. Bonish, L. Miele. Jagged-1 mediated activation of notch signaling induces complete maturation of human keratinocytes through NF-kappaB and PPARgamma. Cell Death Differ, 2002. 9: 842-55.
72. Huang, J., W. Hu, L. Hu, R.A. Previs, H.J. Dalton, X.Y. Yang, et al. Dll4 inhibition plus aflibercept markedly reduces ovarian tumor growth. Mol. Cancer Ther, 2016. 15: 1344-52.
73. Singh, S.K., I.D. Clarke, M. Terasaki, V.E. Bonn, C. Hawkins, J. Squire, et al. Identification of a cancer stem cell in human brain tumors. Cancer Res, 2003. 63: 5821-8.
74. Yamashita, T., X.W. Wang. Cancer stem cells in the development of liver cancer. J Clin Invest, 2013. 123: 1911-8.
75. Alamgeer, M., C.D. Peacock, W. Matsui, V. Ganju, D.N. Watkins. Cancer stem cells in lung cancer: evidence and controversies. Respirology (Carlton, Vic.), 2013. 18: 757-64.
76. Vaiopoulos, A.G., I.D. Kostakis, M. Koutsilieris, A.G. Papavassiliou. Colorectal cancer stem cells. Stem Cells, 2012. 30: 363-71.
77. Swaminathan, S.K., E. Roger, U. Toti, L. Niu, J.R. Ohlfest, J. Panyam. CD133-targeted paclitaxel delivery inhibits local tumor recurrence in a mouse model of breast cancer. J Control Release, 2013. 171: 280-7.

78. Huang, J., C. Li, Y. Wang, H. Lv, Y. Guo, H. Dai, et al. Cytokine-induced killer (CIK) cells bound with anti-CD3/anti-CD133 bispecific antibodies target CD133(high) cancer stem cells in vitro and in vivo. Clinical Immunology (Orlando, Fla.), 2013. 149: 156-68.
79. Skubitz, A.P., E.P. Taras, K.L. Boylan, N.N. Waldron, S. Oh, A. Panoskaltsis-Mortari, et al. Targeting CD133 in an in vivo ovarian cancer model reduces ovarian cancer progression. Gynecologic Oncology, 2013. 130: 579-87.
80. Stratford, E.W., M. Bostad, R. Castro, E. Skarpen, K. Berg, A. Hogset, et al. Photochemical internalization of CD133-targeting immunotoxins efficiently depletes sarcoma cells with stem-like properties and reduces tumorigenicity. Biochimica et Biophysica Acta, 2013. 1830: 4235-43.
81. Jin, L., K.J. Hope, Q. Zhai, F. Smadja-Joffe, J.E. Dick. Targeting of CD44 eradicates human acute myeloid leukemic stem cells. Nature Medicine, 2006. 12: 1167-74.
82. Marangoni, E., N. Lecomte, L. Durand, G. de Pinieux, D. Decaudin, C. Chomienne, et al. CD44 targeting reduces tumour growth and prevents post-chemotherapy relapse of human breast cancers xenografts. Br. J. Cancer, 2009. 100: 918-22.
83. Tang, W., X. Hao, F. He, L. Li, L. Xu. Abstract 565: anti-CD44 antibody treatment inhibits pancreatic cancer metastasis and post-radiotherapy recurrence. Cancer Res, 2011. 71: Abstract nr 565.
84. Engelmann, K., H. Shen, O.J. Finn. MCF7 side population cells with characteristics of cancer stem/progenitor cells express the tumor antigen MUC1. Cancer Res, 2008. 68: 2419-26.
85. Curry, J.M., K.J. Thompson, S.G. Rao, D.M. Besmer, A.M. Murphy, V.Z. Grdzelishvili, et al. The use of a novel MUC1 antibody to identify cancer stem cells and circulating MUC1 in mice and patients with pancreatic cancer. J. Surg. Oncol, 2013. 107. 10.1002/jso.23316.
86. Danielczyk, A., R. Stahn, D. Faulstich, A. Löffler, A. Märten, U. Karsten, et al. PankoMab: a potent new generation anti-tumour MUC1 antibody. Cancer Immunol Immunother, 2006. 55: 1337-47.
87. Fan, X.N., U. Karsten, S. Goletz, Y. Cao. Reactivity of a humanized antibody (hPankoMab) towards a tumor-related MUC1 epitope (TA-MUC1) with various human carcinomas. Pathol Res Pract, 2010. 206: 585-89.
88. Noujaim, A.A., B.C. Schultes, R.P. Baum, R. Madiyalakan. Induction of CA125-specific B and T cell responses in patients injected with MAb-B43.13—evidence for antibody-mediated antigen-processing and presentation of CA125 in vivo. Cancer Biother Radiopharm, 2001. 16: 187-203.
89. Kwiatkowska-Borowczyk, E.P., A. Gabka-Buszek, J. Jankowski, A. Mackiewicz. Immunotargeting of cancer stem cells. Contemporary Oncology (Poznan, Poland), 2015. 19: A52-A59.
90. Dean, M., T. Fojo, S. Bates. Tumour stem cells and drug resistance. Nat Rev Cancer, 2005. 5: 275-84.
91. Eyler, C.E., J.N. Rich. Survival of the fittest: cancer stem cells in therapeutic resistance and angiogenesis. J Clin Oncol, 2008. 26: 2839-45.
92. Singh, V.K., A. Saini, R. Chandra. The implications and future perspectives of nanomedicine for cancer stem cell targeted therapies. Front Mol Biosci, 2017. 4: 52.
93. Vinogradov, S., X. Wei. Cancer stem cells and drug resistance: the potential of nanomedicine. Nanomedicine (Lond), 2012. 7: 597-615.
94. Qin, W., G. Huang, Z. Chen, Y. Zhang. Nanomaterials in targeting cancer stem cells for cancer therapy. Front Pharmacol, 2017. 8: 1.

95. Fiorillo, M., A.F. Verre, M. Iliut, M. Peiris-Pages, B. Ozsvari, R. Gandara, et al. Graphene oxide selectively targets cancer stem cells, across multiple tumor types: implications for non-toxic cancer treatment, via "differentiation-based nano-therapy". Oncotarget, 2015. 6: 3553-62.
96. Burke, A.R., R.N. Singh, D.L. Carroll, J.C. Wood, R.B. D'Agostino Jr., P.M. Ajayan, et al. The resistance of breast cancer stem cells to conventional hyperthermia and their sensitivity to nanoparticle-mediated photothermal therapy. Biomaterials, 2012. 33: 2961-70.
97. Wu, H., H. Shi, H. Zhang, X. Wang, Y. Yang, C. Yu, et al. Prostate stem cell antigen antibody-conjugated multiwalled carbon nanotubes for targeted ultrasound imaging and drug delivery. Biomaterials, 2014. 35: 5369-80.
98. Al Faraj, A., A.S. Shaik, B. Al Sayed, R. Halwani, I. Al Jammaz. Specific targeting and noninvasive imaging of breast cancer stem cells using single-walled carbon nanotubes as novel multimodality nanoprobes. Nanomedicine (Lond), 2016. 11: 31-46.
99. Wang, X., X.C. Low, W. Hou, L.N. Abdullah, T.B. Toh, M. Mohd Abdul Rashid, et al. Epirubicin-adsorbed nanodiamonds kill chemoresistant hepatic cancer stem cells. ACS Nano, 2014. 8: 12151-66.
100. Zhang, Y., Z. Cui, H. Kong, K. Xia, L. Pan, J. Li, et al. One-shot immunomodulatory nanodiamond agents for cancer immunotherapy. Adv Mater, 2016. 28: 2699-2708.
101. Rothemund, P.W. Folding DNA to create nanoscale shapes and patterns. Nature, 2006. 440: 297-302.
102. Zhang, Q., Q. Jiang, N. Li, L. Dai, Q. Liu, L. Song, et al. DNA origami as an in vivo drug delivery vehicle for cancer therapy. ACS Nano, 2014. 8: 6633-43.
103. Halley, P.D., C.R. Lucas, E.M. McWilliams, M.J. Webber, R.A. Patton, C. Kural, et al. Daunorubicin-loaded DNA origami nanostructures circumvent drug-resistance mechanisms in a leukemia model. Small, 2016. 12: 308-20.
104. Jiang, Q., C. Song, J. Nangreave, X. Liu, L. Lin, D. Qiu, et al. DNA origami as a carrier for circumvention of drug resistance. J Am Chem Soc, 2012. 134: 13396-13403.
105. X.Huang, M.A. El-Sayed. Gold nanoparticles: optical properties and implementations in cancer diagnosis and photothermal therapy. J Adv Res, 2010. 1:13-28
106. P. Ghosh, G. Han, M. De, C.K. Kim, V.M. Rotello. Gold nanoparticles in delivery applications. Adv Drug Deliv Rev, 2008. 60: 1307-1315.
107. Elbialy, N.S., M.M. Fathy, W.M. Khalil. Doxorubicin loaded magnetic gold nanoparticles for in vivo targeted drug delivery. Int J Pharm, 2015. 490: 190-9.
108. Tomuleasa, C., O. Soritau, A. Orza, M. Dudea, B. Petrushev, O. Mosteanu, et al. Gold nanoparticles conjugated with cisplatin/doxorubicin/capecitabine lower the chemoresistance of hepatocellular carcinoma-derived cancer cells. J Gastrointestin Liver Dis, 2012. 21: 187-96.
109. Kouri, F.M., L.A. Hurley, W.L. Daniel, E.S. Day, Y. Hua, L. Hao, et al. miR-182 integrates apoptosis, growth, and differentiation programs in glioblastoma. Genes Dev, 2015. 29: 732-45.
110. Dong, C., T. Yuan, Y. Wu, Y. Wang, T.W. Fan, S. Miriyala, et al. Loss of FBP1 by snail-mediated repression provides metabolic advantages in basal-like breast cancer. Cancer Cell, 2013. 23: 316-31.
111. Conde, J., N. Oliva, N. Artzi. Implantable hydrogel embedded dark-gold nanoswitch as a theranostic probe to sense and overcome cancer multidrug resistance. Proc Natl Acad Sci USA, 2015. 112: E1278-87.

112. Xu, Y., J. Wang, X. Li, Y. Liu, L. Dai, X. Wu, et al. Selective inhibition of breast cancer stem cells by gold nanorods mediated plasmonic hyperthermia. Biomaterials, 2014. 35: 4667-77.
113. Kong, F., H. Zhang, X. Qu, X. Zhang, D. Chen, R. Ding, et al. Gold nanorods, DNA origami, and porous silicon nanoparticle-functionalized biocompatible double emulsion for versatile targeted therapeutics and antibody combination therapy. Adv Mater, 2016. 28: 10195-203.
114. Conde, J., N. Oliva, Y. Zhang, N. Artzi. Local triple-combination therapy results in tumour regression and prevents recurrence in a colon cancer model. Nat Mater, 2016. 15: 1128-38.
115. Colson, Y.L., M.W. Grinstaff. Biologically responsive polymeric nanoparticles for drug delivery. Adv Mater, 2012. 24: 3878-86.
116. Arabi, L., A. Badiee, F. Mosaffa, M.R. Jaafari. Targeting CD44 expressing cancer cells with anti-CD44 monoclonal antibody improves cellular uptake and antitumor efficacy of liposomal doxorubicin. J Control Release, 2015. 220: 275-86.
117. Liu, P., Z. Wang, S. Brown, V. Kannappan, P.E. Tawari, W. Jiang, et al. Liposome encapsulated Disulfiram inhibits NFkappaB pathway and targets breast cancer stem cells in vitro and in vivo. Oncotarget, 2014. 5: 7471-85.
118. Shen, Y.A., W.H. Li, P.H. Chen, C.L. He, Y.H. Chang, C.M. Chuang. Intraperitoneal delivery of a novel liposome-encapsulated paclitaxel redirects metabolic reprogramming and effectively inhibits cancer stem cells in Taxol®-resistant ovarian cancer. Am J Transl Res, 2015. 7: 841-55.
119. Ke, X.Y., V.W. Lin Ng, S.J. Gao, Y.W. Tong, J.L. Hedrick, Y.Y. Yang. Co-delivery of thioridazine and doxorubicin using polymeric micelles for targeting both cancer cells and cancer stem cells. Biomaterials, 2014. 35: 1096-1108.
120. Ni, M., M. Xiong, X. Zhang, G. Cai, H. Chen, Q. Zeng, et al. Poly(lactic-co-glycolic acid) nanoparticles conjugated with CD133 aptamers for targeted salinomycin delivery to CD133+ osteosarcoma cancer stem cells. Int J Nanomedicine, 2015. 10: 2537-54.
121. Xu, C.F., Y. Liu, S. Shen, Y.H. Zhu, J. Wang. Targeting glucose uptake with siRNA-based nanomedicine for cancer therapy. Biomaterials, 2015. 51: 1-11.
122. He, L., J. Gu, L.Y. Lim, Z.X. Yuan, J. Mo. Nanomedicine-mediated therapies to target breast cancer stem cells. Front Pharmacol, 2016. 7: 313.
123. Zhou, G., O. Latchoumanin, M. Bagdesar, L. Hebbard, W. Duan, C. Liddle, et al. Aptamer-based therapeutic approaches to target cancer stem cells. Theranostics, 2017. 7: 3948-61.
124. Chen, C., S. Zhou, Y. Cai, F. Tang. Nucleic acid aptamer application in diagnosis and therapy of colorectal cancer based on cell-SELEX technology. NPJ Precis Oncol, 2017. 1: 37.
125. Kim, J.W., E.Y. Kim, S.Y. Kim, S.K. Byun, D. Lee, K.J. Oh, et al. Identification of DNA aptamers toward epithelial cell adhesion molecule via cell-SELEX. Mol Cells, 2014. 37: 742-6.
126. Shigdar, S., J. Lin, Y. Yu, M. Pastuovic, M. Wei, W. Duan. RNA aptamer against a cancer stem cell marker epithelial cell adhesion molecule. Cancer Sci, 2011. 102: 991-8.
127. Pan, Y., T. Jia, Y. Zhang, K. Zhang, R. Zhang, J. Li, et al. MS2 VLP-based delivery of microRNA-146a inhibits autoantibody production in lupus-prone mice. Int J Nanomedicine, 2012. 7: 5957-67.

128. Vander Heiden, M.G., L.C. Cantley, C.B. Thompson. Understanding the Warburg effect: the metabolic requirements of cell proliferation. Science (New York, N.Y.), 2009. 324: 1029-33.
129. Kelloff, G.J., J.M. Hoffman, B. Johnson, H.I. Scher, B.A. Siegel, E.Y. Cheng, et al. Progress and promise of FDG-PET imaging for cancer patient management and oncologic drug development. Clin Cancer Res, 2005. 11: 2785-2808.
130. Folmes, C.D., T.J. Nelson, A. Terzic. Energy metabolism in nuclear reprogramming. Biomark Med, 2011. 5: 715-29.
131. Clifford, D.L. Folmes, Timothy J. Nelson, A. Martinez-Fernandez, D.K. Arrell, Jelena Z. Lindor, Petras P. Dzeja, et al. Somatic oxidative bioenergetics transitions into pluripotency-dependent glycolysis to facilitate nuclear reprogramming. Cell Metabolism, 2011. 14: 264-71.
132. Prigione, A., B. Fauler, R. Lurz, H. Lehrach, J. Adjaye. The senescence-related mitochondrial/oxidative stress pathway is repressed in human induced pluripotent stem cells. Stem Cells, 2010. 28: 721-33.
133. Clifford, D.L. Folmes, Petras P. Dzeja, Timothy J. Nelson, A. Terzic. Metabolic plasticity in stem cell homeostasis and differentiation. Cell Stem Cell, 2012. 11: 596-606.
134. Menendez, J.A., T. Alarcón. Nuclear reprogramming of cancer stem cells: corrupting the epigenetic code of cell identity with oncometabolites. Mol Cell Oncol, 2016. 3: e1160854.
135. Davidson, S.M., T. Papagiannakopoulos, B.A. Olenchock, J.E. Heyman, M.A. Keibler, A. Luengo, et al. Environment impacts the metabolic dependencies of ras-driven non-small cell lung cancer. Cell Metab, 2016. 23: 517-28.
136. Shen, L., J.M. O'Shea, M.R. Kaadige, S. Cunha, B.R. Wilde, A.L. Cohen, et al. Metabolic reprogramming in triple-negative breast cancer through MYC suppression of TXNIP. PNAS, 2015. 112: 5425-30.
137. Folmes, C.D., A. Martinez-Fernandez, R.S. Faustino, S. Yamada, C. Perez-Terzic, T.J. Nelson, et al. Nuclear reprogramming with c-MYC potentiates glycolytic capacity of derived induced pluripotent stem cells. J. Cardiovasc. Transl. Res, 2013. 6: 10-21.
138. Janiszewska, M., M.L. Suva, N. Riggi, R.H. Houtkooper, J. Auwerx, V. Clement-Schatlo, et al. Imp2 controls oxidative phosphorylation and is crucial for preserving glioblastoma cancer stem cells. Genes Dev, 2012. 26: 1926-44.
139. Lagadinou, E.D., A. Sach, K. Callahan, R.M. Rossi, S.J. Neering, M. Minhajuddin, et al. BCL-2 inhibition targets oxidative phosphorylation and selectively eradicates quiescent human leukemia stem cells. Cell Stem Cell, 2013. 12: 329-41.
140. Sancho, P., E. Burgos-Ramos, A. Tavera, T. Bou Kheir, P. Jagust, M. Schoenhals, et al. MYC/PGC-1α balance determines the metabolic phenotype and plasticity of pancreatic cancer stem cells. Cell Metabolism, 2015. 22: 590-605.
141. Ye, X.Q., Q. Li, G.H. Wang, F.F. Sun, G.J. Huang, X.W. Bian, et al. Mitochondrial and energy metabolism-related properties as novel indicators of lung cancer stem cells. Int. J. Cancer, 2011. 129: 820-31.
142. Brandi, J., I. Dando, E.D. Pozza, G. Biondani, R. Jenkins, V. Elliott, et al. Proteomic analysis of pancreatic cancer stem cells: functional role of fatty acid synthesis and mevalonate pathways. J. Proteom, 2017. 150: 310-22.

Index

A

Active surveillance, 236
Androgen deprivation therapy, 234, 237, 241, 246, 250
Androgen receptor, 208, 209, 211, 213, 216
Apoptosis, 25, 27, 29, 32, 33, 34-36

B

Biomarker, 50, 52, 56, 59, 62, 63, 66
Bisphosphonates, 236, 251, 252
Bone metastasis, 245, 250, 251-253

C

Cabazitaxel, 238, 240, 247, 248
Cancer stem cells, 167, 169, 172, 291, 294
Cancer, 23-28, 32-38, 50-66, 73-86, 88, 90-96
Castration-resistant prostate cancer, 247, 248
Checkpoint blockade, 2, 4, 14-17
Chemoprevention, 34
Copy number variation, 209
ctDNA, 109-115
Cytokines, 23, 25, 26, 31-33, 35, 38

D

Docetaxel, 238, 240, 245-248
Drug-resistance, 292, 293, 294, 295, 297, 298, 299, 310, 311

E

EGFR, 185, 187, 188, 190, 191, 193, 194, 195, 197-199
ErbB, 164-167, 169-173
ERG fusion, 210
Exosomes, 109, 119, 122-126

G

GnRH agonists, 236, 237, 239, 240
Group 3 medulloblastoma, 135-138
Group 4 medulloblastoma therapeutics, 138

H

Head and neck squamous cell carcinoma, 163
HER2, 185, 198, 199
Human papilloma virus, 163

I

Immunotherapy, 2-5, 8, 13-17, 236, 240, 242, 250, 264, 265
Inflammation, 23-26, 31, 32, 34, 37, 50, 55, 58, 63, 66
Intraductal papillary mucinous neoplasia (IPMN), 153

K

Kras, 186, 187, 191, 193, 194, 197 198

L

Liquid biopsy, 108, 109, 115, 119
Lung adenocarcinoma, 185, 186, 188, 190, 193, 194, 195, 197, 198, 199

M

Medulloblastoma, 132-142
MET pathway, 194, 195

Metabolomics, 109, 115, 116, 117, 118, 119, 120, 123
Metastasis, 51-54, 56-62, 66, 75-77, 81, 88, 89, 90, 92, 93, 95
Methylation specific electrophoresis (MSE), 154, 161
Methylation, 151-159
MUC1, 151-153, 156, 158
MUC2, 152, 153, 156, 157
MUC4, 151-153, 156-158
Mucin regulation, 265, 270, 271

N

NCOA1, 74, 75, 77-80, 85, 95, 97
NCOA3, 74, 75, 77, 78, 83-90, 94-97
NF-κB, 23, 24, 27-38
Nuclear receptor co-activator, 74

P

Pancreatic cancer, 264-269, 271, 273-276, 278
Pancreatic ductal carcinoma (PDAC), 151
Pancreatic intraepithelial neoplasia (PanIN), 153
Pancreatic juice, 154, 156, 157

PDAC, 1-8, 10-17
Pediatric brain tumors, 144
Prostate cancer, 207-217, 234-236, 237, 239, 241, 242, 244, 245, 247, 248, 250

R

Radiation therapy, 243, 251
Radical prostatectomy, 237
RAGE, 52, 54, 55, 58, 59, 61-64

S

S100A7, 50-60, 66
S100A8, 50, 53, 59-66
S100A9, 50, 53, 59-66
SHH medulloblastoma, 136, 137, 139
Somatic mutation, 213
Stroma, 50, 61, 65

T

Targeted therapy, 302, 303, 304
Targeting mucins, 274, 278, 280

W

WNT medulloblastoma, 132, 134, 136, 138, 139, 142